Thank You Fossil Fuels and Good Night

GREGORY MEEHAN

THANK YOU FOSSIL FUELS AND GOOD NIGHT

The Twenty-First Century's Energy Transition

THE UNIVERSITY OF UTAH PRESS
Salt Lake City

The Defiance House Man colophon is a registered trademark
of The University of Utah Press. It is based on a four-foot-tall
Ancient Puebloan pictograph (late PIII) near Glen Canyon, Utah.

21 20 19 18 17   1 2 3 4 5

LIBRARY OF CONGRESS CATALOGING-IN-PUBLICATION DATA
Names: Meehan, Gregory, 1954- author.
Title: Thank you fossil fuels and good night : the twenty-first
   century's energy transition / Gregory Meehan.
Description: Salt Lake City : The University of Utah Press, [2017]
   | Includes bibliographical references and index.
Identifiers: LCCN 2016040355| ISBN 9781607815396 (pbk. : alk.
   paper) | ISBN 9781607815402 (ebook)
Subjects: LCSH: Fossil fuels. | Renewable energy sources.
Classification: LCC TP318 .M44 2017 | DDC 621.042--dc23 LC
   record available at https://lccn.loc.gov/2016040355

Printed and bound by Edwards Brothers Malloy,
Ann Arbor, Michigan.

Table of Contents

Preface   vii
Acknowledgments   xi
Definition of Energy and Power   xiii

Chapter 1.   The Seven Signals Heralding an Energy Transition   1

Part I.   The Sources of Energy We Draw Upon: Alternative Energies to the Front

Chapter 2.   Fossil Fuels: Stored Solar Energy   23

Chapter 3.   Nuclear Energy and the Mass Defect   33

Chapter 4.   The Power behind Renewables   42

Chapter 5.   The Potential and Toll of Alternative Energy Sources   60

Part II.   Age-Old Technologies Find Application: Seeds of Innovation, Dormant for Centuries, Are Now Germinating

Chapter 6.   Technology Introduction   73

Chapter 7.   Solar   78

Chapter 8.   Wind   89

Chapter 9.   Nuclear   96

Chapter 10.   Advanced Energy Storage Solutions   107

Chapter 11.   Smart Grids   122

Chapter 12.   Transportation   130

Chapter 13.   Heat Generation   146

Chapter 14.   Technological Readiness   157

Part III.   A Scan of Country Energy Plans: The Globe We Share

Chapter 15.   Fossil Fuel Wealth and Energy Consumption Framework   167

Chapter 16.   The United States   183

Chapter 17.   Canada   199

Chapter 18.   Brazil   205

Chapter 19.   Germany, Norway, Spain, and Italy   209

Chapter 20. Russia   218

Chapter 21. Saudi Arabia   223

Chapter 22. Japan   227

Chapter 23. China   233

Chapter 24. India   240

Chapter 25. South Korea   245

Chapter 26. Country Roundup   250

Part IV. History, Inertia, Interventions, and Adaptation: Energy Policy Conundrums

Chapter 27. Our Response to Limited Resources   263

Chapter 28. Global Energy Price and Cost Points   277

Chapter 29. Energy Policy Tool Kit   286

Chapter 30. Four Repressive Forces on the Use of Fossil Fuels   308

Chapter 31. The Only Path Forward Departs from the Use of Fossil Fuels   314

Appendix A. Primary Energy Consumption Factor and Fossil Fuel Wealth Factor   325

Appendix B. Energy Platforms of Presidential Candidates 1992–2016   335

Selected References   343

Index   365

Preface

The future is but the present a little farther on.
—Jules Verne

Products, technology, and even basic ways of doing things inescapably adhere to a life cycle, no differently than nature. As to the topic of this book, it seems that our massive dependence on fossil fuels must then necessarily run its course, not because stocks of petroleum will be entirely depleted but because more attractive alternatives will emerge, disrupting and ultimately replacing them. Decline comes to all things, so when will our world's reliance on fossil fuels naturally change?

From an outsider's perspective unbridled by any strong bias other than an innate interest, this book shares signals in today's market that spell the end of the fossil fuel era and estimates when and how that change will occur. Already, powerful forces are changing the energy field irreversibly. This book describes these forces collectively and reveals that an exciting wholesale change in energy sourcing is afoot. Climate change, just one of the many forces, has seized the public discourse, obscured our grasp of these other forces, and clouded our view of the dangers and opportunities that lie ahead.

One of the principal goals of this book is to broaden the energy conversation to illuminate these other forces. This book looks forward through the twenty-first century to forecast the inevitable arrival of new energy solutions and examines the challenges to bring them on line. It also salutes the role fossil fuels have played in advancing the human condition over the last several hundred years.

Even in the face of an obvious life cycle decline, often the market incumbent wastefully plows money and effort into a weakening product and fails to participate in the market's revitalization. The petroleum industry is tightly holding onto the present as if it can walk it into the future; you can't embrace something new until you let go of the old. We are disproportionately investing in the end stage of fossil fuels to such an extent that we are suppressing superior alternatives, to our great detriment. The supplies and price of fossil fuels will sequentially become volatile and less affordable with time. As the future comes to meet us, our combined reluctance to adapt will risk a more disruptive transition.

The journey of this book begins by investigating the time line of when accessible and affordable fossil fuels will be depleted, with *depletion* defined as referring to economic attractiveness and not physical supplies. The transition naturally happens as alternatives acquire superior benefits, and the world doesn't all follow the same clock. Seven signals clue the world that the energy field is reinventing itself.

We will explore the potency of alternative sources and the toll for their access. On a global scale, the research strongly indicates that alternatives indeed have the necessary potential to replace fossil fuels. But are there commercially viable technologies? It is essential that the critical technologies are ready and launched while we still have sufficient fossil fuels to make a smooth transition. We have the pieces, but it's the continued economic refinement of interrelated technologies that will cultivate the widespread deployment of alternatives.

We are disproportionately investing in the end stage of fossil fuels to such an extent that we are suppressing superior alternatives, to our great detriment.

As with any endeavor with global implications, it is necessary to know what other countries are doing. So I sketch the energy plans of 13 representative countries, and the differences prove enlightening. We do live on the same planet, air and ocean currents circulate around the globe, and the differences in behaviors, priorities, and approaches will touch us all. What matters most are the collective implications for the global security, prosperity, health, and environment of individual country-level energy plans.

Behind the hard technologies for tapping alternative energy sources are important soft technologies that guide behavior, and we need both. Throughout our past and to the present we have shown character flaws whereby we hold onto discernibly harmful patterns of living. Making smooth transitions and living in a sustainable manner is not necessarily in our DNA. Globalization, which has brought worldwide access to reserves and technologies, has postponed our conscious realization that our lifestyles and business models will need to adapt along with advancing technology for our long-term well-being.

A truism in this book's story is that the sun bestows the vast majority of the energy we exploit. Although we have recently taken advantage of solar radiation stored chemically in fossil fuels, our longer history will be about directly harnessing the sun's energy. With the exception of geothermal energy and the tidal effects of the moon, all renewables can be traced to the sun. The creativity and vibrancy directed toward the invention of alternative ways to capture energy is inspiring. Parts of the world have already entered the first stages of an exciting and challenging refashioning of energy sourcing. This book aspires to move us beyond the paralyzing extremism that often dominates the discussion of energy to invite and embrace the positive and inevitable transformation that is beginning to occur.

It is essential that the critical technologies are ready and launched while we still have sufficient fossil fuels to make a smooth transition.

Acknowledgments

I am first inclined to admit a hopefulness that the pages that follow take the reader on a worthwhile journey. Without the encouragement and support of my wife, Sarah, researching and writing this book would've been just an idea. My daughter Sally has taken a keen interest in environmental sustainability, and with persistent and renewable energy she reminds that this is more than an ideal—it is the principle necessary for our continued flourishing. For me this undertaking has opened my eyes to the big picture and has brought into sharp focus what I hope and expect will be recorded as humankind's amazing twenty-first-century pivot in energy sourcing. We don't need to set fossil fuels afire to thrive! My daughter Rebecca, the anthropologist, correctly informs me that technology alone will not allow us to dodge necessary changes in our human behavior. A constant companion during this project, my grandson Kingston has been more influential than he will ever know. There is no doubt that switching between researching hydrogen fuel cell technology to stacking blocks and reading about Piglet and Pooh Bear kept the writing tone inquisitive and innocent.

Family and dear friends have suffered early versions of the material, and for this I thank them and apologize. Special thanks go to Mike Meehan, Clay Parr, Robert Keiter, Les Meredith, Lea O'Gorman, and especially Freckles. The University of Utah's 2014 Wallace Stegner Center symposium on the U.S. national parks propelled me to a volunteer stint in southeastern Utah, where coincidentally so many energy sources are conspicuously on display. I saw firsthand the intensity of high-altitude sunlight, remnants of uranium mining along with active petroleum development, and finally the famous Colorado River, which is harnessed at the Hoover and Glen Canyon dams to produce hydroelectric power. Moab, Utah, where all this converges, inspired the ponderings that became the book that follows. Enjoy.

Definition of Energy and Power

Throughout the book there are many references to both energy and power. The terms are perhaps vague to many readers, so a brief description prior to getting underway will help avoid unnecessary confusion.

Energy

Energy is the capacity of a system to perform work. It takes energy to drive your car and light your home. Energy can be transferred between objects or converted into different forms, but it is always conserved. Think of all the different types of energy. A water reservoir used by a hydroelectric power plant has potential energy, a flowing river has kinetic energy, a piece of carbon charcoal holds chemical energy, while the battery for your computer stores energy electrochemically. The types continue, with atomic energy, thermal energy, our solar system's electromagnetic energy, and the electrical energy delivered by our power industry. Even sound is a form of energy; you can literally hear energy.

Each of these forms has a preferred unit of measure. Our energy sectors touch most of these forms. Our fossil-fueled, large hydro, and nuclear power plants convert chemical, potential, and atomic energy into electricity. Our solar panels capture electromagnetic energy and convert this to either thermal or electrical energy. Those tall wind turbines we see are capturing the kinetic energy of air and converting this to electricity. Most of the energy we use in our daily lives is derived from fuels with their own intrinsic chemical or atomic forms.

Our bodies are power plants; we require a minimal amount of energy every day to function, referred to as the basal metabolic rate, and obviously this varies for each of us. For instance, a 55-year-old woman weighing 130 lbs and measuring 5 ft, 6 in, would have a basal metabolic rate of about 1,200 kcal/day. The kcal is commonly referred to as the food calorie.

When we fill an automobile with 15 gal of gasoline, we don't identify with the 1.8 million British thermal units (BTUs) of chemical energy but, rather, with the mileage the fuel delivers, the work the fuel is able to perform. Accepting that this depends on driving conditions and the type of vehicle, a fuel-efficient car would convert that amount of chemical energy into 400+ mi. Today, fuels provide most of the world's energy, and BTUs per kg is used to convey their heat content. When we discuss nuclear, the energy released from a single reaction is measured in mega-electron volts (MeV). Electricity is perhaps the most common source of energy in our daily lives, so we will see many references to the kilowatt-hour (kWh) unit of measure. In 2000 the worldwide average kWh per person per year was 2,200, and this increased to 2,700 by 2010, well below the average per-person consumption in the United States, which is nearly fivefold higher. For further reference, an energy-efficient refrigerator/freezer consumes about 500 kWh/year.

A power plant's annual electricity generation is generally rated in megawatt-hours (MWh), where *mega* is a million. The amount of electrical energy consumed in a residential environment is expressed in kilowatt-hours, where *kilo* is a thousand. Chemical energy and atomic energy are often expressed in BTUs or MeV, respectively.

Since energy is always conserved and the energy sector is fundamentally in the conversion business, the following relationships will be useful (though not mandatory) throughout the book. A single kWh is equivalent to 3,412 BTU, 2.247×10^{19} MeV, and 0.001 MWh.

Power

Power, by definition, is the rate at which work is done or energy is converted. The amount of energy depends on what power levels are applied over what period of time. When you go to the gym, you may choose an intense (high-power) exercise program and burn much more energy than the following day, when you choose a less strenuous (low-power) program over the same amount of time. The intensity or power level of your exercise would be measured in kcal per hour, and the total energy is determined by the length of your workout and measured in kcal. Again depending on the individual, 600 kcal per hour is a typical exercise power level. Facilities that generate electricity are called power plants because they are built with a certain capacity to convert a source of energy into electricity. A coal power plant converts chemical energy into thermal energy as an intermediate. The thermal energy is subsequently used to drive turbines that convert the thermal energy into mechanical and finally electrical energy. A power plant's capacity is rated in megawatts (MW). A coal power plant with a 250 MW rating operating at 75% capacity for a day (250 MW × 75% × 24 hours) would produce 4,500 MWh of electricity. How much fuel the coal plant consumes to achieve its power rating depends on the heat content of the coal (BTUs/kg) and the efficiency of converting chemical energy into electricity (typically one-third). If a 250 MW coal plant operated at full capacity for a year, it would produce 250 megawatt-years of electricity. A 2 MW wind turbine is a fluctuating energy source, only able to take what the winds offer. A capacity factor of 33% for this turbine and location would produce 16 MWh of electricity daily (8 hours × 2 MW). Solar power, like wind, also fluctuates at the earth's surface.

Watts are units of power that convey the capacity of a plant to produce electricity. A typical nuclear power plant has a power rating on the scale of 1,000 MW. It is implied that the power is used to generate electricity, and literature sometimes introduces a suffix—"MWe"—to distinguish it from other forms of output energy. High-temperature industrial processes employ plants or equipment to provide thermal energy. A thermal plant may have a BTU per hour or "MWth" rating. The internal combustion engines in our cars convert chemical energy into mechanical energy. We use the unit horsepower rather than watt to express an automobile's engine power capacity.

Scale

I've defined the notions of power and energy that will be used throughout this book and their units of measure. However, the specific prefix for the units will depend on the scale within the discussion. At a residential scale, power is on the watt scale. A power plant is measured on an MW scale. Country power capacity is generally in the gigawatt range, while worldwide power is on a scale of terawatts (see Table 0.1). The same is true of energy.

Table 0.1. Scale.

VALUE	DENOMINATION	PREFIX	SCIENTIFIC NOTATION
1,000	Thousand	*Kilo-*	1.00E + 03
1,000,000	Million	*Mega-*	1.00E + 06
1,000,000,000	Billion	*Giga-*	1.00E + 09
1,000,000,000,000	Trillion	*Tera-*	1.00E + 12
1,000,000,000,000,000	Quadrillion	*Peta-*	1.00E + 15

1

The Seven Signals Heralding an Energy Transition

"There is no such thing as peak oil," said the petrochemical executive even though he wasn't asked.

How long can fossil fuels be depended upon to satiate our worldwide appetite? Will the fuels that took nature millions of years to create be used up in just a few generations? The questions themselves betray the world's utter reliance on fossil fuels and the mistaken sense that we lack alternatives beyond these prehistoric soups.

The fact is, fossil fuels are in their decline phase, and the world will transition to alternative sources over the course of the next 50 to 100 years. The answer to the question doesn't simply stem from understanding the quantities of fossil fuels in the ground but, rather, when forces will converge to render those supplies unattractive as a source of energy. These forces include the economics of oil and gas production, ineluctable evidence related to health and environmental consequences, the emergence of viable alternative energies, and the pursuit of energy self-sufficiency by nations around the world.

Many genuinely smart people believe that or at least act as if fossil fuels have centuries of supply and securing those supplies is simply a matter of unlocking resources through advanced technology, less regulation, and greater access to restricted areas. That position underestimates the strength of the opposing forces. Two of the three fossil fuels, coal and petroleum, are already losing appeal as energy sources. Natural gas is replacing coal in the power sector, and zero-carbon transportation vehicles are replacing the petroleum-based internal combustion engine. Although this seems like a good thing, it will place new and ultimately unmanageable demands on the world's supply of natural gas. Noncarbon energy technologies are not just environmentally better choices; they are becoming economically preferable. Some countries and municipalities are already executing energy plans that entirely displace the use of fossil fuels in the power sector.

A fuel's attractiveness includes the summation of its energetic qualities and other points of value, including its impact on health and environment and whether it is domestically accessible.

Whether a nation is rich in fossil fuel resources, poor, or self-sufficient, there are consequences to ignoring the signals of change in the world's energy markets. It will take decades to adapt our infrastructures, and when we cling to our reliance on fossil fuel, we squander the time needed to smoothly alter our sources of energy. Replacing fossil fuels will mean whole new markets for products that harvest, store, and deliver energy and new transportation vehicles, smart electronics, and more. Forward-looking communities will stake commercial positions in some of these markets, while nations determinedly bound to fossil fuels will realize the opportunities too late. On both the supply and demand sides, the flow of fossil fuels will become erratic, and the price will be subject to high and low swings. The world's petroleum exporters face the greatest challenges ahead—adapting their economies to a post–fossil fuel era without painful social and economic upheaval.

Our sun and earth are both about 4.5 billion years of age, and the sun at least will continue for another five billion years. Modern humankind, on the other hand, has existed for a mere 200,000 years. For 199,850 of humans' 200,000 years of existence, we harnessed human or animal muscle as a primary source of energy. We used biomass for heating and whale oil, animal fat, and beeswax for illumination. Of the fossil fuels, surface deposits of coal have long been used as an energy source for heating. Things changed dramatically 150 years ago, when the world collectively shifted to all three fossil fuels—coal, oil, and gas. The first oil extraction well was dug by Edwin Drake in 1859 in western Pennsylvania, marking the beginning of the age of petroleum. Since then, around the world, access to energy has been the key to modernization, to lifting people's quality of life.

Even as pictures from space show night lights blazing brightly across the continents, 2.9 billion people still lack access to clean forms of cooking, and one billion people around the world lack access to electricity.[1] A startling statistic is that all of Africa has less capacity in the power sector than Germany, and this is an inequity that renewables can help correct.

As noted earlier, the term *depletion* used throughout the book does not refer to the physical exhaustion of fossil fuels. There are far more stocks of fossil fuels in the ground than can ever be recovered. A fair portion of technically recoverable fossil fuels will remain in the ground due to the high cost of recovery, access limitations, and resistance to certain extraction methods—but most importantly, because better options will exist. Depletion of fossil fuels refers to the accessibility and economic limits of supplying fossil fuels.

Estimating the world's limit of economically recoverable fossil fuels is a daunting task, but to paraphrase Mark Twain, we should at least attempt to "first get the facts. Then you can manipulate them as you please." Complicating the estimates are supply-side assumptions that include extraction technology, undiscovered and recoverable resources, access for development, environmental and health-related resistance, and future price points. Demand-side assumptions include world population, civil nuclear power, renewable technology, and consumption patterns. Take any combination of assumptions and the model could tell a different tale. An excerpt from BP's *Outlook 2030* published in January 2013 is illustrative. According to this international petrochemical giant, "The world has ample proved reserves of oil and natural gas to meet expected future demand growth." Whew! "At the end of 2011, global proved reserves of oil were sufficient to meet 54 years of current (2011) production; for natural gas that figure is 64 years."[2] Wait, what? Fifty-four years? BP later goes on to say that reserves are a poor indicator for production estimates.

Evolving extraction technologies, newly opened access, and new discoveries have held forward supplies of oil and gas in the 50+-year range since 1980. How much longer can the future supplies of oil and gas hold

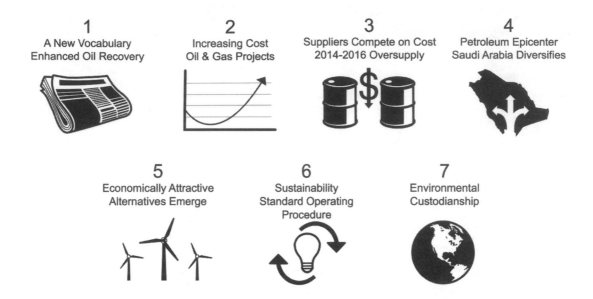

Figure 1.1. Seven signals.

this pattern before they inevitably shrink? Trends in reserve levels ultimately reflect the relative economic attractiveness of fossil fuels compared with alternatives, and those trends are changing. Attractiveness is the summation not only of energetic qualities but of other points of value, including health and environmental impact and whether or not the energy source is domestically accessible. The economic equation is altering the incumbent's forward outlook; the market share of fossil fuels has only one direction to go, and that is to accede position to alternatives.

For those of us who like to see the numbers, a quantitative examination based on authoritative sources for petroleum resources and reserves along with several reputable projections of global energy demand will follow. Beyond the hard numbers, however, seven clear signals (Figure 1.1) provide clues that a major change in energy sourcing is under way and reveal the forces that will cause the global decline in the use of fossil fuels. The time line for the decline of fossil fuels depends in part on the price we are willing to pay, the safety of advanced extraction methods, the emergence of economically viable alternatives, and our response time to the signals. We will find that for some the "when" comes much earlier than for others.

Signal 1: A New Vocabulary Has Arrived—Enhanced Recovery for Higher Yields

A century ago we were discovering resources around the world from Iran to Venezuela. But by the mid-twentieth century discoveries had peaked. We are still finding new fields, but we have quietly entered the era of tertiary enhanced oil recovery (EOR) and hydraulic fracturing for both oil and natural gas. Primary recovery is typically only able to recover 10% of the original oil, while tertiary EOR is able to recover 30% to 60%. So, even after tertiary recovery, a large portion of the oil remains hard to reach and/or extract. There is a point of diminishing economic returns, and the signal in this new era highlights the limits to which we can realistically push fossil fuel production. We are reaching the bottom of the proverbial barrel. Literally, we can practically squeeze only so much oil from a rock and frack only so much oil and gas.

The new lexicon summons a very different image from that of a conventional well pumping liquid crude oil from an immense underground reservoir. Terms such as *shale gas* and *tight gas*, *oil sands*, *light oil* and *tight oil*, and *coal-bed methane* have entered our vocabulary, and they describe different types of reserves. The same goes for extraction technologies; we now recognize terms such as *hydraulic fracturing, steam injection,* and *horizontal drilling*. We are advancing technologies to recover offshore fields, and we are applying a large arsenal of EOR and hydraulic fracturing methods onshore. An ExxonMobil report, *Outlook for Energy: A View to 2040*, reveals that the world's future oil and gas supplies will increasingly flow from tight oil, oil sands, and deepwater resources, and, consequently, satiating our global appetite will increasingly depend on the use of advanced extraction methods.[3]

Disquietingly, these new oil and gas extraction technologies are increasingly invasive. For example, three techniques are now common to EOR today: thermal, gas flooding, and chemical flooding. Thermal, the most prevalent, involves the injection of steam in conventional fields and in situ combustion and electric heating in unconventional fields. Carbon dioxide injection, also quite prevalent, is now popular in gas flooding because it reduces the viscosity of the oil, allowing it to flow. The least common method, chemical flooding, uses a polymer to increase the viscosity of the displacement fluid (typically water) and an alkaline surfactant polymer to mobilize the oil. Hydraulic fracturing, commonly referred to as fracking, applies advanced recovery techniques to release natural gas and oil trapped in shale, sandstone, and coal seams. Deep wells are drilled into underground rock formations, and fluids are pumped inside under extremely high pressure, causing the rock to shatter, thereby releasing the trapped oil and gas.

A few decades ago, vast untapped fields supplied our swelling appetites. Now the language is about applying technological prowess to wring more from known deposits and unlocking previously inaccessible resources.

Signal 2: The Marginal Costs of Oil and Gas Projects Are Increasing

The implication of the first signal is that the marginal cost of extraction projects will grow as these advanced technologies are necessarily brought to bear. Fossil fuel production and development are entering a new phase marked by wide ranges in the costs of projects. For clarification, the costs include the direct traceable costs of extraction.

Downward shifts in the price of both crude oil and gas can occur as lower-cost suppliers seek to regain market share from competitors reliant on a basket of projects with higher costs. As long as lower prices are sustained, producers applying expensive extraction technologies and/or working in less accessible sites will be sidelined. This market dynamic is hard evidence that the project costs are rising and supplies in the long-term will come at higher prices.

A concerning aspect of developing some technically challenging reserves is the trend toward faster drops in production. Horizontal drilling, for instance, accelerates production but comes with faster declines. Average oil field production deterioration from first-year levels rose from a 6% decline pre-1970 to a more than 14% decline in the 2000–2007 period.[4] Some shale fields utilizing hydraulic fracturing have reported even higher first-year declines—as much as 70%. Petroleum producers are investing in projects with smaller potentials as the larger fields are tapped. That we now have three articulated phases of recovery and multiple tertiary methods is another signal that we are reaching a limit. And with each new phase the methods become more intrusive and costly.

Signal 3: The 2014–2016 Global Oversupply of Petroleum— Suppliers Competing on Cost

A perplexing phenomenon has been occurring with the 2014–2016 global oversupply of fossil fuels. If we are entering the end stage of the fossil fuel era, why are there such plentiful and cheap supplies? Let's take an exporting country's perspective and briefly consider two scenarios. We can assume that the exporter will always seek to maximize the economic benefits of its in-ground assets and avoid sitting on top of lost wealth, perhaps the biggest fear of petroleum-rich countries.

In the first scenario, the exporters take a stable and positive forward outlook for both the supply and the price of fossil fuels; they patiently produce and export supplies in the hopes of enjoying a long and stable period of prosperity. Supply would be metered to meet demand at the highest possible price. Alternatively, in the second scenario, sensing a future with disruptive alternatives emerging and taking hold, an exporter would accelerate production to translate as much in-ground asset as possible into revenue to avoid the unthinkable—unrealized affluence.

Products entering their decline phase lose appeal, and this is reflected in less market price leverage. Successful decline-phase suppliers compete by managing costs to remain profitable at lower prices. It would be natural but wrong to read the signal to mean that the oversupply represents the existence of plentiful low-cost sources and consequently remain comfortably wedded to fossil fuels. Instead, the oversupply we are now enjoying reflects a different market dynamic. As viable alternatives are emerging and taking hold, low-cost exporters everywhere are pumping while they can. It is a grave mistake to let this oversupply allow us to lose urgency. The wise interpretation is to utilize this period to execute a smooth transition.

Signal 4: The Epicenter of Petroleum—Saudi Arabia Is Diversifying

Saudi Arabia, home to 16% of the world's crude oil reserves, is beginning to employ the use of EOR in its superfields, a sure sign that we are progressing through viable resources. When the world's top producer begins to apply EOR techniques, the signal should prompt a response. Further evidence is that Saudi Arabia's energy plan calls for major investments in solar, wind, and nuclear to be on line by 2040. When the epicenter of petroleum exports makes plans to diversify its domestic energy supplies, it is foolish to ignore the larger message that decline comes to all things. The Kingdom is going one step further and investigating transmitting the excess electricity generated from these alternative sources through a multinational linkage network. Unlike ExxonMobil, which defines itself as a petrochemical company, Saudi Arabia is exploring the inclusion of alternatives as part of its overall energy export economy—a smart play in refashioning its economy.

Signal 5: Emergence of Cost-Competitive Alternative Sources

Large investments in both research and deployment of alternative technologies are a further indication that fossil fuel production will be limited, that the world will acquire new and affordable choices. According to a 2015 REN21 report, investment in renewable power capacity surpassed that of fossil fuel five years running. Many countries have instituted renewable portfolio standards that call for a certain amount of renewably sourced electricity in a utilities generation mix. Renewable installations in Western Europe are already economically competitive with conventionally sourced supplies of electricity. And nuclear alongside renewable investment will deliver new choices that will limit the extent to which we choose to recover fossil fuels.

Signal 6: Sustainability—Placing Higher Value on Renewable Energy Sources

The unbridled use of resources isolated from systems to reuse, along with a disregard for natural regenerative processes, is a big problem and also wrong. A century ago perhaps we could be forgiven as we were celebrating new discoveries of petroleum and powering a new world. Today, for the most part, we know the limits of supply, and we are changing demand-side behavior to be more prudent in the use of carbon fuels while developing the ultimate long-term solution, renewables. Many countries, particularly in Western Europe, see these signals and have accelerated efforts to conserve energy and build renewable capacity.

As part of this growing awareness comes the recognition that most of our business models are incongruent with our long-term societal health. The extreme consumerism evident in our throwaway culture is poor behavior; our parents and grandparents would confirm this assessment. Behavior today is guided by economics and supported with the notion that technology can solve all our problems. Now, though, a new consciousness is emerging that believes that our human and business value systems will need to adapt as well. If we only replace fossil fuels with new age renewables and do not advance the human side, we will solve one problem and allow others to fester.

The current disproportionate use of fossil fuels that are depleted upon use is becoming the negative symbol for this new awareness. Lifestyles and business practices are embracing a new responsibility, to seek a sustainable and respectful integration within the natural world. Renewable energy more readily aligns with this awakening, while the depletion characteristic of petroleum will prove completely incompatible.

Signal 7: Environmental Custodianship Is Diminishing the Appeal of Fossil Fuels

Energy policies from countries around the world have commonly included priorities to build independence, diversify supply chains to mitigate perceived risk, and create an attractive economic environment that supports industry and affordable access. Now we are seeing a new class of energy priorities allied to the environment and human health. Environmentally focused policies are creating forces of change toward noncarbon energy sources along with efficiency and conservation measures. For instance, the 2015 United Nations Framework Convention on Climate Change held in Paris convened with 195 nations in agreement on the threat and necessity of holding global temperature to 2°C above preindustrial levels, and this will mean pressures to change the world's energy mix.

According to many sources, including the U.S. Environmental Protection Agency, global fossil fuel consumption is responsible for 65% of global greenhouse gas emissions. Policies that target reductions in greenhouse gas emissions are forcing fossil fuel mix changes (i.e., coal to gas), higher efficiency (i.e., combined cycle power), and shifts to non-fossil energy sources. The policies are executed through regulation and/or carbon charges that increase costs. The effect of these new energy policies will be lower shares for fossil fuels.

As further evidence, climate change policy is already placing substantial pressure on the continued use of coal in the power sector. For perspective, a power plant that converts from coal to natural gas can reduce flue greenhouse emissions by 55%. Crude oil refinement for transportation fuels is similarly under pressure within climate and health policies around the globe. Alternative-energy vehicles with zero carbon emissions are already entering the market, which will, decade by decade, cause a decline in the use of liquid petroleum products. The last fossil fuel standing will be natural gas, the cleanest of the three.

However, with cleaner-burning natural gas comes the controversial production and development practice of fracking. Around the world local- and country-level initiatives to ban the use of "fracking" are formidable. In late 2014, a citizen-led initiative in the city of Denton, Texas, just north of Dallas in the heart of the U.S. oil and gas industry, banned the use of hydraulic fracturing within city limits. The ban was quickly overturned by the state, but a certainty remains: as long as groundwater mechanisms have potential health and environmental risks, stiff resistance will persist. We can live in the dark but not without clean water. As advocacy groups are poised and ready to challenge new fields and new recovery methods, legal challenges will provide constant opposition and add costs to oil and gas development.

Taken together, these signals tell us that the age when fossil fuel supplies dominate our energy use is coming to an end. So many years ago the earth began to incubate massive quantities of what became fossil fuels. Unfortunately, once we consume them, they are depleted, so it is natural and obvious that economical and acceptable supplies are limited. What is vital now is that we heed these signals and fully develop alternative sources in a timely manner. We need to step away from the immobilizing politics of the climate change debate, act on the fact that fossil fuel supplies and pricing will become volatile this century, and acknowledge that it would be unfortunate at best if we do not commit to building large capacities of alternative energy sources now. Unfortunately, human history is filled with examples of our willingness to ignore signs of trouble until they are associated with an intolerable level of pain.

When Fossil Fuels Lose Their Appeal

For some, signals taken from fossil fuel resource and reserve data will be more persuasive. However, readers less interested in this type of analysis can skim the following or skip ahead to the section titled "The Quantitative Signal," and I will take no offense.

In the process of developing fossil fuel depletion estimates, it is helpful to begin by drawing a distinction between the resource and the proven reserves. Estimates of the resource reflect the amount of the fuel underground. Reserves, on the other hand, overlay technology and economics to estimate that portion of the resource that can be recovered. The portion of the resource considered recoverable increases in three ways: as technology improves and is accepted, as new areas such as the Arctic and offshore sites are opened for development, and as energy price increases make an investment profitable. So, resources are estimates of physical supplies, while reserves are estimates of economically recoverable supplies. A crucial point in the following exercise is to distinguish technically recoverable oil and gas from that portion that is both economically and acceptably accessible. Historically, it has been technology and access, not economics, that has driven the growth in reserves. Now, the emergence of viable alternatives will promote economics as a key influence in the growth (and shrinkage) of reserves. Where technology and access have been positive factors, economics will be a negative factor in reserve levels.

Petrochemical companies tell us that there is a lot more in the ground, and this is true, but recovering those resources will be increasingly expensive and require further access. Resources that are locked in shale or buried under ocean floors can only be recovered using advanced technologies that generally increase the costs of extraction. As alternative energy technologies are developed and deployed, economics will determine the amount of fossil fuels we actually recover.

"Clean" Natural Gas—The Preferred Fossil Fuel, but Long-Term Supply Is Untenable

Despite the rapid growth in the consumption of fossil fuels, the petroleum industry has thus far been able to scour the world to find resources, develop technologies to convert more resources to reserves, and manage projects that have held the world's future supply at half a century. But examining reserve data reveals how incautious it would be to assume that the future supply of reserves holds to the half-century pattern.

The data also reveal how the world's supply chains depend on just a few countries, and many of these countries have associated risks. Worldwide net recoverable petroleum reserves increased by 298 billion barrels between 2010 and 2015, but this increase unfortunately came from just a few countries. Two-thirds of the increase came from Venezuela, and another 23% came from Iraq, Iran, and Russia. This scenario of growth dependent on just a few countries doesn't exude stability as the rest of the world tries to secure supply channels to meet rising demand. In Europe, petroleum reserves are shrinking, while other regions with measurable reserves of petroleum are holding level. Specifically, 30 countries reported increases in their petroleum reserves since 2010, while another 20 countries reported declines. So, while it is comforting that worldwide reserves are growing, declining reserves are reminders of the inevitable limits of fossil fuels. According to the numbers, we haven't seen peak oil, but reliance on just a few countries for increases is concerning even in the near-term.

Natural gas reserves tell a similar tale but are more precarious going forward given the amplitude of global resistance to projects dependent on fracking, the sensitivity of the reserves to market price points, and increasing worldwide demand. Between 2010 and 2015, worldwide net recoverable reserves of natural gas increased 334 trillion ft^3, but nearly 50% of the gain came from Iran, followed by Mozambique (major new discovery), the United States, and China. Europe's reserves are shrinking, while Eurasia is holding steady with modest gains from natural gas powerhouse Russia. There were as many countries reporting declines as increases in natural gas reserves since 2010. On first blush this may feel balanced, but in actuality it is concerning that so many countries are reporting declines.

Where hydraulic fracturing is applied in the recovery of natural gas, resistance is passionate and effective. Of the countries reporting declines, a fair number have instituted bans or moratoriums on the use of fracking.[5] Germany, registering a decline, has limited fracking in sensitive areas and at depths above 3,000 ft. Queensland and Western Australia are rich in natural gas resources trapped within shale rock, coal seams, and tight rock (i.e., sandstone), and extracting this resource with hydraulic fracturing is facing stiff resistance. Natural gas reserves in Australia posted an 80 trillion ft^3 decline in 2012. Forward supplies of natural gas, the preferred fossil fuel for electricity generation, are particularly vulnerable because demand is steadily increasing while reserves are becoming erratic. The warning sounds for the long-term supply of natural gas are quite loud.

Here's another example validating Mark Twain's prognostication. The U.S. Geological Survey (USGS) conducted an extensive study to estimate undiscovered and technically recoverable quantities for oil and gas. The 2012 report, "An Estimate of Undiscovered Conventional Oil and Gas Resources of the World," shared recoverable estimates for oil, natural gas, and natural gas liquids.[6] The quantities of undiscovered oil and natural gas represented a 55% and 110% increase, respectively, above 2000 reserve levels. Furthermore, these increases of conventional oil and gas resources far exceeded worldwide consumption over the same time period. This sounds great until we closely examine the marginal rates of growth in the reserve data. While oil reserves are increasing faster than historical or projected consumption rates, the same data reveal that recent natural gas consumption rates and projections for the future are higher than reserve growth rates. For instance, natural gas reserves for the period 2000 to 2014 grew at an annual rate of 2.15%, while the rate between 2005 and 2014 fell to 1.6%, followed by an even lower rate of 1.25% between 2010 and 2014. On the other hand, annual consumption growth rates were consistently near 2.5% within each of these time periods.

Despite growth projections, coal will likely be the first fossil fuel to see global demand decrease.

The USGS and U.S. Energy Information Administration (EIA) data seem to tell different stories. Annual rates as reported in the granular EIA reserve data disclose cracks within natural gas that are masked in the USGS information. There are menacing signs for the world's ability to sustain forward supplies of natural gas, the least offensive fossil fuel environmentally.

On the demand side of the fossil fuel reserve equation, the primary drivers are population growth and the rapid rise in access and consumption occurring in developing countries. Conservation and efficiency efforts in developed countries are helping to slow worldwide consumption growth, but it is still steadily expanding. "Outlook 2040" reports from the EIA and ExxonMobil share projections for the consumption growth rates of petroleum and natural gas for the period between 2010 and 2040,[7] averaging roughly 0.95% and 1.75%, respectively. But as shown in Figure 1.2, the two reports depart significantly on their relative projections for coal and renewables. It would be nice to realize ExxonMobil's lower consumption estimates for coal, but indications point to higher rates in the near term. As for renewables, ExxonMobil wouldn't be the first market leader to underestimate the relevance of disruptive technologies.

As noted earlier, natural gas projections, displayed in Figure 1.3, are more concerning since it is the most efficient of the fossil fuels in energy conversion, has the lowest CO_2 emission rates, and is the preferred energy source within the power sector. Applying natural gas demand and reserve growth rates of 1.75% and 1.25%, respectively, going forward, the

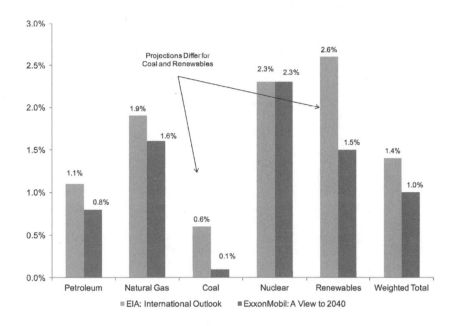

depletion of accessible supplies would occur somewhere around 2080, which aligns almost exactly with BP's *Outlook* date of 2075. It may be the first of the fossil fuels to be fully depleted of economical supplies. Satiating natural gas demand beyond this point would require overcoming worldwide resistance to production techniques, and undoubtedly it would be supplied at higher prices, reflecting the increased costs of recovery. The clock is ticking in a measurable way for natural gas.

Coal is the cheapest, dirtiest, and most prevalent fossil fuel. It also has the lowest percentage of reserves to resources. There is more than a century's supply of coal if we choose to harvest the rock. Coal, though, is quickly losing global appeal due to its relatively high emissions of CO_2 and other air pollutants and its toxic waste products (ash). Despite the growth projections for coal, it will likely be the first of the fossil fuels to see its global demand shrink.

Figure 1.2. Projected growth rates in global primary energy consumption by source between 2010 and 2040. EIA = U.S. Energy Information Administration.

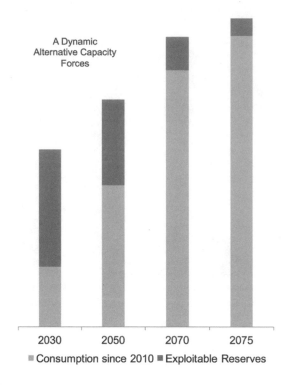

A Dynamic
Alternative Capacity
Forces

| | 2030 | 2050 | 2070 | 2075 |

■ Consumption since 2010 ■ Exploitable Reserves

Figure 1.3. Pace of depletion of exploitable natural gas supplies.

The Quantitative Signal

To summarize the quantitative assessment, data point to the second half of the twenty-first century as the answer to the question of when economical and acceptable fossil fuels run out. If you are age 50 years or older, this provides sufficient time for you to wrap up your residency here on earth. However, if you are less than 50 years of age, this has the potential to create significant pain points in your lifetime, absent huge leaps forward in alternative energy capacity. To underestimate the real likelihood of the affordable depletion of oil and gas by the end of the century on the hopes of additional exploitable reserves seems reckless at best.

Signals Lost in the Noise

If the practical depletion of fossil fuels is 50 to 100 years away, why doesn't energy policy and planning explicitly acknowledge this deadline and effectively account for the time it will take to develop and deploy alternatives and the fact that we will likely need a transition period in which fossil fuels back up alternatives? What is the holdup? Is it the tautological climate change debate that is diverting our attention from planning the transition? It's also possible that the urgency around the depletion of economical fossil fuels is lessened by the hope that we will continue to develop more exploitable resources. Take, for example, Shell Company's "Energy Strategy" identifying growth priorities as integrated gas and deepwater extraction operations, with future opportunities beyond 2020 in underdeveloped and high-risk areas such as the Arctic and Iraq.[8] Targeting "future opportunities" in Iraq, where instability has been unleashed with no end in sight, and the Arctic, where environmental resistance is already substantial, highlights the uncertainty of some petrochemical business strategies. Imagine that board meeting discussion of one or both of those choices. Someone must be asking the business strategy group, "Anybody have another idea?"

Now is the time to acknowledge that easy-to-extract fossil fuels are undermining our sense of urgency.

With virtually no competition from alternative energies other than hydropower and nuclear, petrochemical companies have become fine-tuned and sophisticated hunters of fossil fuels. Opening up more lands and applying new technology has allowed petrochemical companies to sustain forward supplies. What is changing now is that alternative energy will place a ceiling on the price companies can attract, and that lid will thereby limit future expansion of reserves. Investors will choose between increasing costs of oil and gas projects and declining renewable costs. Already in some places, and in time elsewhere, the equation will tip to alternative sources. Further conversion of resources to reserves will slow and eventually contract. At this point the page will turn on fossil fuels as the world's primary energy source.

Whether or not fossil fuels can sustain our appetite beyond this century, they will be recorded as a transitional energy source in hopefully the long history of humankind. Will our grandchildren have a manageable task, or by failing to act with urgency now, are we leaving future generations vulnerable? Think of our time horizon as an hourglass rather than a clock. The moment we began to pump fossil fuels, the hourglass was turned.

Our Energy Appetite Is Still Growing

Worldwide population in 2000 was just over six billion, and per capita energy consumption was 65 million BTUs. By 2012 the world's population had climbed to seven billion, and per capita energy consumption had increased to 74 million BTUs. Continued population growth and increasing per capita energy consumption will make it even more difficult to transition to alternatives and therefore should raise the level of urgency. Not only do we need to replace fossil fuels; we need to do it at a speed that matches the rising worldwide consumption rate.

Unfortunately, not much has changed in the world's energy mix within the power sector. Back in 1980 roughly 69.7% of electricity was generated from fossil fuels, and 32 years later, in 2012, that percentage was 67.3%. Encouragingly, nonhydro renewables in 1980 generated a meager 31 TWh of electricity, and by 2012 they were producing slightly more than 1,000 TWh. Still, renewables have a long way to go to replace fossil fuels in the power sector, let alone across all energy sectors. After decades of investment, nonhydro renewables supplied only 18 days of the world's annual electricity in 2012.

Does this trend look likely to close the fossil fuel gap this century? Nonhydro renewables electricity generation doubled between 2000 and 2007 and again between 2007 and 2012. If renewable generation follows Moore's law and doubles every five years, we would be fossil fuel–independent within the depletion window.[9] Hopefully, from humble beginnings come great things.

Securing new sources for energy will be difficult but doable, and now is the time to acknowledge that easy-to-extract fossil fuels are undermining our sense of urgency. It reminds me of the annual invasion of ants at my house. I place an ant trap and notice a long parade of ants collecting and delivering the easy food source to the colony. But every day the line of ants is shorter, and soon all signs are gone. We are feeding on fossil fuels in much the same way it seems. Readily available fuels can be just as lethal as the ant bait in a different way because they create a false sense of security. They are so convenient, and building alternative energy sources is much harder than going back to the oil well. At the minimum, we need to make a good effort as we pass the baton to the next generation and avoid the bait. If we delay in building new energy sources, we may not have time to make the jump without severe and enduring economic, environmental, and social upheavals.

Notes

1. REN21 Renewable Energy Policy Network for the 21st Century, *Renewables 2015 Global Status Report* (Paris: REN21 Secretariat, 2015), http://www.ren21.net /wp-content/uploads/2015/07/REN12-GSR2015__Onlinebook__low1.pdf.

2. British Petroleum, *BP Energy Outlook 2030*, January 2013, http://www.bp.com /content/dam/bp/pdf/energy-economics/energy-outlook-2015/bp-energy-outlook-booklet__2013.pdf, 71.

3. ExxonMobil, *The Outlook for Energy: A View to 2040*, 2016, http://cdn.exxonmobil .com/~/media/global/files/outlook-for-energy/2016/2016-outlook-for-energy.pdf.

4. International Energy Agency, *Resources to Reserves 2013: Oil, Gas and Coal Technologies for the Energy Markets of the Future* (Paris: Organisation for Economic Co-operation Development/International Energy Agency, 2013), https://www.iea.org /publications/freepublications/publication/Resources2013.pdf.

5. Keep Tap Water Safe, "List of Bans Worldwide," June 2, 2015, accessed August 6, 2015, http://keeptapwatersafe.org/global-bans-on-fracking/.

6. U.S. Geological Survey, "An Estimate of Undiscovered Conventional Oil and Gas Resources of the World, 2012," Fact Sheet 2012-3042 (U.S. Department of the Interior, U.S. Geological Survey, 2012), http://pubs.usgs.gov/fs/2012/3042 /fs2012-3042.pdf.

7. U.S. Energy Information Administration, *International Energy Outlook 2016: With Projections to 2040*, report no. DOE/EIA-0484(2016) (Washington, D.C.: U.S. Energy Information Administration, Office of Energy Analysis, U.S. Department of Energy, May 2016), http://www.eia.gov/forecasts/ieo/pdf/0484(2016) .pdf, 165, Table A2; ExxonMobil, *Outlook for Energy*.

8. "Shell's Energy Strategy," January 1, 2013, accessed March 1, 2015, http://www .shell.com/content/dam/shell-new/local/corporate/corporate/downloads/pdf /investor/strategy/shells-energy-strategy-2013.pdf.

9. Encyclopaedia Britannica, "Moore's Law," September 22, 2013, accessed April 17, 2015, https://www.britannica.com/topic/Moores-law.

PART I

The Sources of Energy
We Draw Upon

Alternative Energies to the Front

2

Fossil Fuels: Stored Solar Energy

So long as you have food in your mouth, you have solved
all questions for the time being.
—Franz Kafka

For most of us, energy is something we plug into, switch on, or pump into our cars but rarely take the time to think too deeply about. But let's ask a few questions. Where do these fossil fuels come from? How long did it take to create them, and how much do we use? Fossil fuels have been extremely useful, so it's beneficial to consider their source, properties, and consumption history in the context of the earth's life cycle.

Petroleum, natural gas, and coal store energy chemically. As we discuss renewables, it is interesting to reflect that fossil fuels are formed from processes dependent on the conversion of solar energy—photosynthesis in the case of plants and animals feeding upon those plants. Because the energy content is stored chemically, they can be transported and used as needed—handy qualities. What is not so handy, in addition to their combustion emissions and extraction damage, is the fact that they are depleted upon use. Oppositely, the term *renewable* is used to describe energy sources that are not depleted upon use. Although we harvest solar and wind energy, the sun still shines, and new winds are stirred.

Coal

The earth has gone through many profound variations in climate and topography over geologic time.[1] Much of the earth has been covered from time to time in swamps. Typically when a plant dies it decays and releases its CO_2 directly into the atmosphere. The beginnings of coal occurred when many plants were buried in swamps before they could decay. Years upon years of deposition fostered higher pressures and temperatures, and eventually coal was formed.

The combustion of coal releases energy manifested as heat. This heat can be used directly or employed to generate electricity in the case of steam power plants and mechanical energy in other cases. Electricity is ubiquitous in our personal lives, but state-of-the-art coal power plants are only able to convert roughly one-third of the heat content into electrical energy. Convenient, yes, but tapping coal's chemical energy to generate electricity involves a lot of waste.

So, where does the heat come from? If you were to burn a block of coal, you would witness a chemical reaction. During combustion the carbon in coal reacts with atmospheric oxygen to produce carbon dioxide and a release of energy. The actual combustion of coal is more complicated; it also includes the release of oxides and the volatile material contained in the coal, and this varies a lot. In the principal reaction, breaking the carbon and oxygen bonds takes energy, while the formation of the CO_2 molecule results in the net release of energy. That's why you have to strike a match to coal to get the reaction going! The amount of thermal energy released varies by the grade of coal; the average in the United States is 20 million BTUs per short ton. A coal power plant is able to convert about one-third of the released energy to electricity, so a short ton of coal produces about seven million BTUs of electricity, which translates to 2,000 kWh. The heat values for different fuel types shown in Figure 2.1 reveal an astonishing range, from 15,000 BTUs/kg for firewood to 26 trillion BTUs/kg for a nuclear fast reactor using natural uranium, punctuating the point that not all kilograms of fuel are the same energetically. Take notice of hydrogen, which has a value three times that of natural gas. At a standard temperature and pressure, hydrogen is the lightest of all gases but packs quite a punch by weight, and we will hear much more in the chapters ahead.

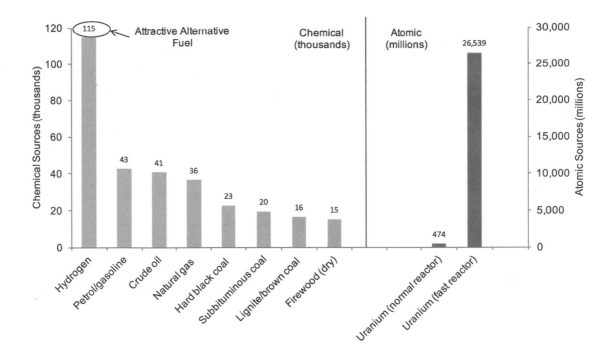

The quality of coal varies, based on the time, pressure, and temperature of its underground incubation. The types of coal are characterized by carbon content and moisture. Carbon content decreases while moisture increases, going from anthracite to bituminous to subbituminous and finally to lignite. Peat is plant mass not yet incubated to the coal stage. Heat content is directly related to the carbon content, and higher is better. Bituminous coal has anywhere from 65% to 85% carbon, while anthracite has greater than 85% carbon.

For hundreds if not thousands of years, coal has been used like wood for heating. Coal was a big advantage in that stands of trees were spared the ax and its heat content was a third more than that of the best wood (i.e., hickory). My mother, born in 1919, remembers shoveling bituminous coal into her home's heating unit. She still revels in a memory that most of us can't relate to—the day the family converted from dirty bituminous coal, which left soot and dirt everywhere, to a cleaner grade of coal, anthracite.

Figure 2.1. Heat values by fuel type (BTUs/kg). *Source Data:* World Nuclear Association. "Heat Values for Various Fuels." March 2010. Accessed April 9, 2016. http://www.world-nuclear.org/information-library/facts-and-figures/heat-values-of-various-fuels.aspx

Coal burners were invented in the early 1800s. They produced steam from burning the coal and converted this to mechanical energy powering trains and steamboats. Today most coal is used to generate electricity, while it also continues to be used for industrial processes that depend on heat generation. How much do we use? Worldwide, 8.2 billion short tons (2,000 lbs) of coal were consumed in 2012.[2] It would take 75 million railcars to haul this quantity of coal. If you were waiting for a train traveling at 100 mi an hour to pass with the world's annual supply of coal, you would be a year older for the experience. Some forecasts have the global consumption of coal increasing annually by as much as 0.6% through 2040. By then, the train would have to add 13 million railcars and increase its speed to 118 mi per hour to deliver the coal. Although coal compares unfavorably with petroleum and natural gas in terms of heat content, CO_2 emissions, and pollutants, countries burn what they can access and afford, which is why coal consumption, for now at least, is projected to increase over the next several decades.

Like the other fossil fuels, coal is disproportionately deposited around the globe. Sharing the wealth in roughly equal measure are North America, Asia and Oceana, and Eurasia, at 27%, 32%, and 26%, respectively. Europe possesses 9% of reserves, with Central and South America, the Middle East, and Africa holding the balance of less than 6%. The United States has 26% of the world's coal reserves and obviously the vast majority of North America's deposits. China, Australia, and India hold most of Asia and Oceana's world share of coal. Similarly, Russia, Ukraine, and Kazakhstan hold the majority of Eurasia's coal reserves. It is fair to say that the more concentrated a valuable resource, the greater the potential for politically induced supply disruptions, and unfortunately we find that all three fossil fuels are fixed in a select few countries.

The CO_2 emission factor (lbs of CO_2 per million BTUs) allows comparison between different types of fuel. Figure 2.2 displays the emission factors for common carbon-based sources. Unfortunately, coal emits more atmospheric CO_2 per heat unit of energy than the other fossil fuels. In the United States bituminous and subbituminous coal make up most of the coal deposits. Bituminous coal has a CO_2 factor of 206, compared

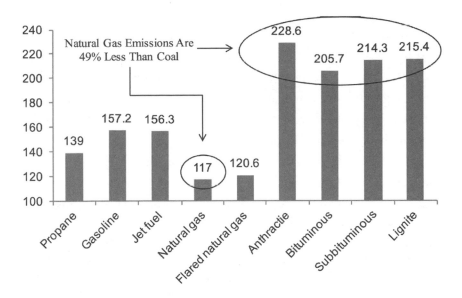

Figure 2.2. CO$_2$ emissions (lbs of CO$_2$ per million BTUs). *Source Data:* U.S. Energy Information Administration, "Carbon Dioxide Emissions Coefficients," accessed March 14, 2016, https://www.eia.gov/environment/emissions/co2_vol_mass.cfm.

with 117 for natural gas. Replacing a coal power plant with a natural gas plant can reduce CO$_2$ emissions given both its lower emission factor (49%) and its higher conversion efficiency (6%). Later, however, we learn that, upon closer examination, both the upstream flaring and direct emissions of methane from oil and gas production may nullify a large portion of this advantage.

Petroleum

Petra in Latin translates to "rock," while *oleum* translates to "oil." Over geologic time, there have been episodes when an abundance of single-celled planktonic plants and animals lived in aquatic environments.[3] The rapid burial of remains at the bottom of the oceans created ever deeper and deeper sediment layers containing the raw organic material, called protopetroleum. Like coal, petroleum was brewed under high pressure and temperature over millions of years. Protopetroleum went through various stages of maturation until, voilà, it became petroleum.

Mais non, pas si vite! An alternate theory suggests that primordial hydrocarbons were part of the earth from the very beginning, not formed by past deposits of biological matter undergoing geologic processes, and that hydrocarbons are far more abundant than the biogenic theories estimate.[4] According to this theory, hydrocarbons are continuously migrating upward through the earth's crust. Deep in the earth, a biosphere exists where life feeds on this hydrocarbon soup, creating the range of hydrocarbons found in fossil fuels. However, if the pace by which this process occurs were faster than our consumption rate, then the USGS fossil fuel resource estimates would be higher. The theory that primordial hydrocarbons are far more abundant is intriguing but unfortunately does not deliver new sources at a rate to secure sufficient energy to meet our needs here on the earth's surface. So we are back to the drawing board.

The most common use for petroleum (crude oil) is fueling our transportation vehicles. Hydrocarbons that are liquid at normal temperatures with boiling points between 100°F and 400°F are refined from crude oils as gasoline. When gasoline is used in an internal combustion engine, chemical energy is converted into mechanical energy. As gasoline is burned the hot gases that are formed have more thermal energy than the original fuel and air mixture. The gases are at extremely high temperatures and pressures. It is this high pressure that drives the pistons, which completes the process from chemical to mechanical energy. Since this book is about energy it is useful to get acquainted with scale. Petroleum possesses about 1,700 kWh of chemical energy per barrel, and for perspective the monthly consumption of electricity in the United States is just over 1,000 kWh per person.

Crude oil, or alternately liquid petroleum, is chemically complex. That is why the world has built high-throughput oil refineries to separate the molecules into functional groups. The crude oil feedstock delivered to refineries is made up of hundreds of different hydrocarbon molecules. By weight, greater than 90% of crude oil is composed of carbon and hydrogen, followed by varying amounts of nitrogen, sulfur, and oxygen and finally small amounts of metals. Refineries produce fuels, lubricants, asphalt and tar, and feedstocks that are subsequently used to produce plastics, detergents, solvents, and even fibers such as nylon and polyester. The fuels include gasoline, diesel, heating oil, kerosene, and liquid petroleum gas (i.e., propane and butane). Refining also creates an opportunity to remove undesirable elements prior to combustion in the case of gasoline, most notably sulfur and lead. In the case of sulfur, this is associated with the generation of sulfur oxides during combustion (acid rain). Regulatory bodies around the world have placed limits on the sulfur (and also lead) content of fuels, thereby reducing downstream emissions to help address air quality concerns.

Modern man has had access to surface supplies of petroleum for thousands of years. Some of the earliest uses of petroleum involved medicine and weaponry, an unfortunate but complementary pair of applications. Petroleum was used to treat wounds and consumed as a laxative. In 430 B.C., Persian warriors wrapped fibers saturated with oil around their arrows and lit them before loosing them on their enemies. Distillation of petroleum to produce illuminants was introduced in the twelfth century, and by the twentieth century petroleum had become a principal source of energy.

Worldwide, every day 91 million barrels of petroleum were consumed in 2013. What is a barrel? A barrel of crude oil is the equivalent of 42 gal, about the volume used when we draw a bath. Furthermore, a barrel of crude oil weighs about 300 lbs, while a barrel of water, which is denser, would weigh 350 lbs. In the United States, almost three-fourths of petroleum supplies are used in the transportation sector. Since 2000, worldwide consumption of petroleum has increased steadily at 1.3%, and projections are estimating roughly 1% through 2040.

The Middle East drew a royal flush, sitting atop nearly half the world's proven reserves, with the largest portions in Saudi Arabia, while Iran, Iraq, Kuwait, and United Arab Emirates each have double-digit shares. The next region of the world is Central and South America, holding a 20% world share, primarily in Venezuela. North America follows with 13% of the world's reserves, mainly in Canada. Africa, led by Libya and Nigeria, holds an 8% world share. Finally, Asia and Oceana have only 2% of the world's reserves, while Europe has the short end of the dipstick, with 1%.

Back to the topic of a resource's geographic concentration. Oil fields are classified by their size. The largest class is supergiant, with five billion or more barrels, of which there are 40 around the world. Two-thirds of the supergiant fields are in the Arabian/Iranian fields, and these have some of the lowest costs of production.

Natural Gas

Natural gas, as the name suggests, is the gaseous form of petroleum. Odorless, tasteless, colorless, and lighter than air, it is recovered from dedicated gas wells and alongside crude oil. Since oil and gas were formed at the same time, it is natural that they are found within the same fields. This is pertinent because methane, the main constituent of natural gas, is a potent greenhouse gas and susceptible to direct atmospheric emissions during production and development.

Strike a match to natural gas and you will get quite an intense response. As with the other fossil fuels, a chemical reaction occurs during combustion. When burned, methane (CH_4) reacts with atmospheric oxygen to produce carbon dioxide and water and the release of energy. It takes about 10 ft^3 of natural gas to produce 1 kWh of electricity. A person in the United States who utilizes 1,000 kWh of electricity per month would consume more than 10,000 ft^3 of natural gas.

The concentration of fossil fuel wealth in just a few countries negatively affects global security.

Natural gas as taken from nature varies chemically across production fields and needs to be processed before it can be distributed. Methane and ethane, the shortest and lightest of the hydrocarbons, constitute nearly 90% of natural gas. Since their boiling points are well below 0°C, they are present in gaseous forms under standard conditions. Clouds form when moisture at or below the dew point temperature of water causes condensation. The same is true for hydrocarbons such as butane, with a boiling point of ~0°C or higher, that can be in the liquid phase under some or all pipeline conditions. Gas pipelines are engineered to move gases not liquids, so these larger hydrocarbons need to be isolated. We are familiar with propane used in backyard grills and butane lighters. They are sourced from natural gas. Contaminants that can cause corrosion are removed, and these include CO_2, hydrogen sulfide, and water. Helium and nitrogen and other trace molecules are also isolated.

At this point the natural gas can be liquefied for transport or directly distributed as a gas.

As for the other fossil fuels, surface supplies of natural gas have been exploited for quite some time.[5] Around 500 B.C., the Chinese were building bamboo pipelines to transport natural gas that was used to boil seawater to obtain drinking water. This was likely the world's earliest gas pipeline and desalinization plant. More recently, natural gas was used as an illuminant in the eighteenth century, and you will remember the Bunsen burner from high school chemistry classes. Today natural gas is used in home heating and cooking, industrial processes, and power plants, to name just a few applications.

Climate change concerns and fossil fuel reserves are inversely related.

In 2013, worldwide annual use of natural gas was 120 trillion ft^3, and the consumption rate is steadily increasing. Imagine a 12-mi-diameter sphere: this would hold the world's annual use of natural gas, and this sphere is getting bigger. Predictably the Middle East, which has the highest reserves of petroleum (crude oil), also has the highest share of natural gas, at 41%. Here, though, Iran is the wealthiest in the Middle East, with 17% of worldwide reserves, followed by Qatar and Saudi Arabia. Eurasia is the next-wealthiest region, with a 32% share almost entirely in Russia, followed by Turkmenistan. Asia and Oceana have less than 8% of the world's deposits, with China, Indonesia, and Malaysia holding double-digit portions of this share. The North American share is just below 6%, found mainly in the United States. Europe once again gets the short end of the dipstick, holding about 2% of the world's natural gas reserves.

Although we tend to think of fossil fuels as energy sources, they also serve many other vital uses. The United States used 191 million barrels of liquid petroleum gases and natural gas liquids in the production of plastics materials, lubricants, and resins during 2010 and another 412 billion ft^3 of natural gas.

Whether as a fuel for our engines or feedstock for plastics, petroleum (crude oil and natural gas) is vital to our society, and countries with little or no resources have been challenged in their trade balances. Importing countries need to establish and maintain economic ties with exporting countries, and undoubtedly there is stress and political angles to each of these relationships. Japan suddenly incurred its first trade deficit in 30 years when it had to negotiate increased fossil fuel supplies in the wake of the Fukushima nuclear power plant calamity, a disadvantaged position from which to broker a necessity.

The concentration of fossil fuel wealth in such a few countries gnaws at our global security. Later in the book when we examine countries, we will see a wide diversity in energy consumption patterns and planning priorities related to the use of fossil fuels and renewable sources. We will also observe that real climate change concerns and fossil fuel reserves are unfortunately inversely related. Endowment of fossil fuel resources will prove to be highly predictive of a country's energy policy. We would expect resource-poor countries, particularly those in developed Western Europe, to embrace renewables in order to achieve energy self-sufficiency. Fuel-wealthy countries, on the other hand, will continue to produce and supply demand. They will play the hand they were dealt with gusto. An ideal hand versus the one you're dealt is the stuff of dreams and misplaced wagers. There are more than 200 countries, with just 10% having most of the fossil fuel wealth and the others awakening to the possibility of cutting the umbilical cord.

Notes

1. American Museum of Natural History, "The Origins of Coal," http://www.amnh.org/exhibitions/climate-change/climate-change-today/the-origins-of-coal/.
2. U.S. Energy Information Administration, "International Energy Statistics," accessed February 26, 2015, http://www.eia.gov/cfapps/ipdbproject/IEDIndex3.cfm?tid=6&pid=29&aid=12.
3. Gordon I. Atwater, "Petroleum," Encyclopaedia Britannica, July 1, 2013, https://www.britannica.com/science/petroleum.
4. Thomas Gold, *The Deep Hot Biosphere: The Myth of Fossil Fuels* (New York: Copernicus, 2001).
5. American Public Gas Association, "A Brief History of Natural Gas" (2015), accessed April 9, 2015, http://www.apga.org/apgamainsite/aboutus/facts/history-of-natural-gas.

3

Nuclear Energy and the Mass Defect

If one way be better than another,
that you may be sure is nature's way.
—Aristotle

As with fossil fuels, there are different types of nuclear reactions and fuels. Civil nuclear power has employed the fission reaction since the mid-twentieth century. Unlike the environmental disasters from fossil fuel production and development, mistakes with nuclear power implant themselves deeper into the psyche of the public. And like the old saying "You never get a second chance to create a first impression," all nuclear power confronts barriers in gaining the public's acceptance despite the potency of the energy source and the emergence of better solutions. Nature stashes enormous amounts of energy inside atoms, and human-kind's first efforts to harvest this energy have been primitive. New technologies developed over the course of the last half century can change the risk profile for practicing nuclear power. Had we known in the mid-twentieth century what we know now, nuclear would likely be the core piece of our energy puzzle, and climate change wouldn't be such a large global concern, since we wouldn't have used as much fossil fuels. We will discuss the technologies in subsequent chapters, but first we should understand the nature of nuclear energy.

Einstein's famous revelation in 1905 that $E = mc^2$ sheds understanding on the enormous amount of energy resident in atomic nuclei.[1] Who would have expected such large forces necessary in binding nucleons?

Who, beyond Stephen Hawking or a modern-day Galilei, would have deduced that the speed of light is intertwined with energy and mass as in Einstein's brilliant equation? The element iron (Fe) sits at a crossroads in chemistry's periodic table; it has the highest binding energy and lowest mass per nucleon. Elements lighter and heavier than iron have less binding energy per nucleon, while their mass per nucleon is higher. This is explained by Einstein's equation. Atomic energy is released as lighter elements to the left of iron combine (fusion) and heavier elements to the right break into fragments (fission). We are all familiar with the conservation of energy. There is a caveat, however—the conversion of mass into energy—and this is called the mass defect.

Fission: Cleaving the Nuclei of Large Atoms

Uranium, the predominant fuel source in mankind's practice of nuclear fission, is older than either the earth or the sun. Theories suggest that it was formed during the explosion of one or more supernovas. Estimates based on isotopic half-life and radiometric clocks indicate that earth's uranium was created 6.5 billion years ago.[2] One may think that nuclear fission is man's invention, but geologists studying the Oklo uranium deposits in the central African state of Gabon found evidence that about two billion years ago spontaneous natural fission "reactors" appeared and operated continuously for about a million years. The contemporary German chemist Professor Klaproth discovered uranium in 1879, where it was applied in a coloring process for glass and ceramics.[3] The world will look with mixed feelings upon the insightful observation made by another German chemist, Ida Noddack, in 1934. She discerned that elements in the middle of the periodic table were formed when a nucleus of uranium absorbed a neutron and fragmented. When a uranium atom splits, a surprisingly large amount of energy is released and converted primarily into the kinetic energy of the fragments (including neutrons). In the case of power plants, the released energy is converted into thermal and electrical energy. In the case of nuclear weapons, the pressure and kinetic energy of the blast winds combined with the thermal and nuclear radiation of the explosion wreak most of the destruction. The first nuclear weapons resulting from this discovery were used by the United States in World War II. Civilian applications for electricity generation came on line shortly thereafter in the late 1950s.

A fission reaction occurs when the nucleus of a heavy atom (i.e., uranium-235) absorbs a neutron and splits into lighter nuclei. As background, the higher the atomic number of an element, the higher the ratio of neutrons to protons. So, when a large nucleus splits into smaller fragments, the by-products are inherently neutron-rich and unstable. Radioactive decay is a process where the fragments transition to stable atomic forms. The term *radioactivity* refers to the energies and/or mass that are emitted as these nervous atoms settle. Ironically and unfortunately, transuranic elements, located above uranium in the periodic table, are formed when absorbed neutrons fail to cause a fission event— creating long-lived radioactive by-products.

All the world's commercial nuclear plants engage the use of fission, and the vast majority deploy thermal-neutron reactors. Although they have been potent sources of electricity, their spent fuel needs to be secured for thousands of centuries. This is a huge flaw and completely avoidable. According to the International Panel on Fissile Materials, it takes several hundred thousand years for the ingestion radiotoxicity of the spent fuel to fall below that of natural uranium, and even then it is still harmful. Their ionizing radiation can directly kill human cells or damage a cell's DNA. Every day we are exposed to natural and man-made sources of lower-energy nonionizing radiation. Natural sunlight and cell phones are examples of sources of nonionizing radiation, and although the energy is lower, overexposure can still be harmful.

The radioactive constituents of thermal nuclear spent fuel vary, so Table 3.1 is offered to convey the time scale for nuclear waste management. The right-hand column provides the percentage by weight for the specific radioactive material within the spent fuel. You can see that the radioactivity of spent fuel is glacially slow to change, particularly after the decay of the short-lived fission products. Another important takeaway is that greater than 96% of the spent fuel is still usable. Perhaps the unspoken plan for spent fuel is to someday soon utilize the waste. The technology section describes reactor designs that consume natural

Table 3.1. Nuclear Spent Fuel Content by Weight with Half-Life.

RADIOACTIVE ELEMENT	HALF-LIFE (YEARS)	WEIGHT PERCENTAGE
Uranium		95.48%
U-235	704,000,000	
U-238	4,470,000,000	
Plutonium		0.85%
Pu-238	88	
Pu-239	24,100	
Pu-240	6,560	
Pu-241	14	
Pu-242	373,300	
Minor actinides		0.11%
Am-241	432	
Am-243	7,370	
Np-237	2,140,000	
Long-lived fission products		0.20%
Iodine-129	15,700,000	
Technetium-99	211,000	
Zirconium-93	1,530,000	
Cesium-135	2,300,000	
Short-lived fission products		0.17%
Strontium-90	29	
Cesium-137	30	
Stable isotopes		3.19%

Source: Japan Atomic Industrial Forum, "Spent Fuel and Radiotoxicity," accessed September 30, 2015, http://www.jaif.or.jp/ja/wnu__si__intro/document/2009/m__salvatores__advanced__nfc.pdf.

uranium and plutonium, which, per Table 3.1, would dramatically reduce the amount of waste. Beyond managing the waste, of course, we are well acquainted with recent nuclear power plant accidents and remain concerned that other operational sites may be just as vulnerable. Our sentiments about nuclear energy are further complicated by the existence of nuclear weapons that can destroy the living planet.

Worldwide, 68,000 tons per year of uranium ore are used annually, a mere 600 railcars to use the earlier analogy with coal.[4] Proven reserves are sufficient for another 100 years at current rates with the technology as practiced. Australia, Kazakhstan, and Russia hold 50% of the world's uranium reserves. Another seven countries hold 40%, so, like fossil fuels, uranium is disproportionately distributed around the planet. As with petroleum, investment in exploration and development could extend uranium forward supplies by perhaps decades. Beyond uranium, thorium (fertile) is a potential fuel that could add substantially to forward supplies. It is estimated that thorium ore exists in the earth's crust at a quantity three times greater than that of uranium ore. Nevertheless, nuclear energy, as deployed, shares the same time clock as fossil fuels: economical depletion by roughly the end of the twenty-first century. Differently than with fossil fuels, the limits with nuclear energy are the technologies, not the supply of fuels, and there are solutions.

Readers might be surprised to learn how many countries use civilian nuclear power. In 2015 the following 30 countries generated electricity from nuclear power (listed from highest to lowest production): the United States, Russia, China, South Korea, Canada, Germany, Ukraine, The United Kingdom, Spain, Sweden, India, The Czech Republic, Belgium, Finland, Switzerland, Hungary, Bulgaria, Slovakia, Brazil, Mexico, South Africa, Romania, Argentina, Slovenia, Japan, Pakistan, the Netherlands, Iran, and Armenia.

Worldwide nuclear power derived from 430 reactors has been somewhat flat since 2000 and actually declined between 2011 and 2012. However, if worldwide planned and proposed nuclear reactor projects materialize, nuclear power will more than double by the middle of the twenty-first century, and new technologies will become the market standard beyond 2050.

A Neutron-Induced Windfall of Energy

Unlike fossil fuels, where thermal energy is generated by the energy released from chemical reactions, atomic energy is released when nuclei are combined in the case of fusion or broken apart in the case of fission.

Early in the twentieth century, scientists discovered that when a heavy nucleus absorbs a neutron, fission could occur.[5] It was further discovered that uranium (235) and plutonium (239), which contain odd numbers of neutrons, have high incidences of fission; they are called fissile. The uranium reaction is associated with the release of an incommensurate amount of energy. There are many fragmentation possibilities; a typical one is uranium-235 absorbing a neutron and breaking into barium-144, krypton-90, two neutrons, and the release of 200 MeV of energy. To provide some context for 200 MeV, during the combustion of coal, 5 eV of energy are released, while the uranium–neutron reaction releases 40 million times that amount. To help further appreciate the inordinate amount of energy released during a typical U-235 fission event, consider that a fully inflated car tire stores energy as compressed air. Depress the valve of an inflated tire and you get a sense of the stored energy. A single teaspoon of U-235 will release energy equivalent to the amount stored in 60 million fully inflated car tires.[6]

With Such Potency Come Hazards

When the public discusses nuclear energy, the topic of safety predominates the discourse and is centered on two major concerns. The first worry regards the risk of environmental contamination with radioactive material if a power plant fails. Fukushima is a recent example. A 6.6 M_w earthquake disrupted power to the plant, and the subsequent 45-ft tsunami knocked out the backup diesel generators. Although the nuclear reaction was successfully shut down, the reactor naturally continued to generate heat. The reactor's cooling systems failed, and this created a tremendous accumulation of heat. Confronting this catastrophic condition, the operators flooded the plant with seawater to cool the reactor. Failure of the redundant systems, most notably the backup diesel generators, created a cascading set of events that culminated in an environment contamination. Why a nuclear plant would be situated at sea level in a part of the world particularly vulnerable to earthquakes is another topic, but it is indicative of the choices countries are willing to make when they lack domestic supplies of fossil fuels.

The second concern is the viability of safe long-term storage solutions for spent fuel. Today in the United States, for instance, spent fuel resides in pools of water for up to 20 years where the radiation is shielded and the initially high decay heat is absorbed. To further safeguard the radioactivity and facilitate secure transport, the spent fuel is then placed in canisters and sealed. Today, though, the canisters remain unmoved, and this lack of a "permanent" plan unnerves many people, leaving them uneasy with the application of nuclear power.

Fusion: Following Nature's Lead

The sun and the stars are powered by nuclear fusion. A series of reactions fuse light elements and release energy, which is emitted throughout the solar system and beyond in electromagnetic waves. We perceive a portion of the electromagnetic radiation as visible light and warmth. Clean and renewable solar energy on earth is the by-product of nuclear fusion. We can be thankful for many things, but the sun shines most conspicuously among them. Nature is displaying every moment what may be our next "fossil fuel." The sun has sufficient fuel to burn for another five billion years, and securing energy from fusion on earth has long-term potential if we can only master an imitation.

Unlike fission, which harvests atomic energy by breaking nuclei into fragments, fusion combines the nuclei of light matter (less than iron in the periodic table) and captures a portion of the released energy. Researchers around the world are seeking to unlock the potential of several light matter reactions. For instance, our solar system's principal reaction is referred to as the proton-proton reaction. The complete cycle involves several intermediate reactions, but in essence the "fusion" of hydrogen produces helium and the release of 25 MeV of energy, which travels outward through our solar system and beyond.

Another promising reaction under research combines deuterium and tritium, which are heavy isotopes of hydrogen. As you might expect, the nuclei of deuterium and tritium are fused. When combined they also produce helium and energy. Tritium is radioactive; the atom contains one proton and two neutrons, while deuterium contains a single proton

The world's practitioners of civil nuclear power leaped before they looked. Ongoing research has produced vastly safer reactor designs, better fuel formats, and superior reaction types.

and neutron. This reaction produces 17.6 MeV of energy. Deuterium is quite abundant (30 g per m³ of seawater), but tritium is extremely rare in nature. So again, our pursuit of new energy sources has a hitch. Interestingly, tritium has been found in measurable quantities in the spent fuel rods from nuclear fission reactors. Scientists and engineers are experimenting with designs of fusion plants that simultaneously breed tritium from lithium.

Researchers are also considering the potential of helium-based fusion reaction, referred to as aneutronic because there are no neutrons emitted. In this reaction deuterium fuses with helium-3 to produce helium-4, a proton, and energy. Specifically, it produces 18.3 MeV of energy with no radioactive by-products. The glitch with this reaction is finding supplies of the helium-3 isotope, as the earth's supplies are extremely limited. The moon's surface soil, however, is laced with helium-3. Lacking an atmosphere, the lunar surface absorbs helium-3 from solar winds. If the helium-based fusion reaction proves viable, we could well be visiting the moon dressed in spacesuits and armed with shovels and pails.

What the sun does naturally, researchers hope to mimic. What we refer to as clean solar is the beneficiary of nuclear fusion, and we cannot ignore nuclear as a natural energy source. To underscore the vast potential of fusion, consider that the surface of the sun powers 63 million W/m^2 of electromagnetic radiation. Ninety-two million miles away, the earth's average surface radiation is less than 200 W/m^2, and on a clear day this still feels fine. Fusion is a potent energy source, and its "domestic" production promises thousands of years of supply.

But Can We Tame the Beast?

The enormous potential of nuclear energy is cloaked in doubts that sufficiently safe technology can be developed. Although transmutation, the changing of one element into another, occurs naturally throughout the universe, the feeling is that humankind thus far hasn't the means to safely practice the art. Fusion is attractive in its reaction simplicity; the combination of two light elements produces one and only one element. Fission reactions, meanwhile, can create many combinations of fragments, each with its own decay pattern and many of which require long-term isolation. Some fusion reactions produce relatively short-lived radioactive waste, while others do not. Some fusion fuels are abundant, while others will need to be secured from other celestial bodies.

To some extent the world's practitioners of civil nuclear power leaped before they looked. Ongoing research has produced vastly safer reactor designs, better fuel formats, and superior reaction types. Given the potency of nuclear energy, the burden to test and validate these advancements takes decades. Consequently, the civil nuclear industry is understandably slow to change. Meanwhile, the public's perception of nuclear energy is taken from experiences with the existing fleet of nuclear reactors. This brings us to the challenge with nuclear power: How does public opinion change when the topic itself is so radioactive?

If a more urgent response to climate change is called for in the future, nuclear energy, with its low full-cycle CO_2 emissions and high energy potential, is the well-positioned alternative energy source to meet this challenge. The technology section describes advances and research that strive to deliver safe possibilities for the future or at least the capability to properly dispose of the world's stockpile of spent fuel.

Notes

1. World Nuclear Association, "Nuclear Fusion Power," February 22, 2016, accessed April 9, 2016, http://www.world-nuclear.org/information-library /current-and-future-generation/nuclear-fusion-power.aspx.
2. World Nuclear Association, "The Cosmic Origins of Uranium," November 2006, accessed April 9, 2015, http://www.world-nuclear.org/information-library /nuclear-fuel-cycle/uranium-resources/the-cosmic-origins-of-uranium.aspx.
3. Bertrand Goldschmidt, "Uranium's Scientific History 1789–1939" (paper presented at the 14th International Symposium, Uranium Institute, London, September 1, 1989), http://ist-socrates.berkeley.edu/~rochlin/ushist.html.
4. U.S. Energy Information Administration, "International Energy Statistics."
5. World Nuclear Association, "Physics of Uranium and Nuclear Energy," September 2014, accessed April 9, 2016, http://www.world-nuclear.org/information-library /nuclear-fuel-cycle/introduction/physics-of-nuclear-energy.aspx.
6. Assuming a car tire of 1.0 ft3 and 35 psi results in 5,040 ft lb or 6.833 kJ per tire.

4

The Power behind Renewables

Waiho wale kahiko. [Old secrets are now revealed.]
—Hawaiian proverb

A look back through history reveals examples from around the world of humankind's creative endeavors to harness energy. We have milled grains, powered ships, and heated water from hydro, wind, and geothermal sources since ancient times. Unfortunately, the creativity witnessed in some early renewable applications went dormant with the advent of prevalent, transportable, and cheap fossil fuels. Milling no longer required access to a river, and work was no longer hostage to the fickle winds. Now 150 years later, we are awakening to fossil fuels' not-too-distant end and the need to transition to alternatives.

Most fundamentally, a renewable energy source isn't depleted when used. We can harvest their energy, and the sun still shines and the winds still blow. Given that our sun will perish, albeit five billion years from now, the renewable energy that the sun bestows, solar energy, will be depleted. It too is finite in time. The distinction between finite and renewable, then, is perhaps better defined in the context of the gap between the natural production and consumption of a resource. If the earth were producing fossil fuels as fast as we are consuming them, we would call them renewable, but alas, as of today we have already consumed as much as 40% of the economically recoverable resources that the earth produced millions of years ago. Happily, there is more than enough wind, solar, hydro, geothermal, and tidal energy than we can see ourselves consuming. Let's take a brief look at each.

From Whence the Wind Blows

Are there climates that feature prevalent wind conditions? Fortunately, they exist, and we know where and why. Temperature differences are the primary driver of all weather patterns; they cause atmospheric pressure gradients. When the sun heats the earth's surface, whether on land or at sea, the warming air rises, creating a surface low-pressure area. High pressure is created when air is cooled in the upper atmosphere and descends, spreading outward in all directions at the surface. Given a choice we may prefer the weather of a high-pressure system. As moist air rises and is cooled to the dew point temperature of water, clouds are formed. The opposite happens in a high-pressure area; the temperature increases above the dew point, and the skies clear. Air moves from high to low pressure, and voilà, you have wind. This is part of a never-ending process driven by the imbalances of temperature between the earth's equatorial and polar regions. There is a migration of energy and mass caused by this imbalance. Thermal energy wants to travel toward the poles. Since the imbalance is perpetual, we have an earth surface every day displaying a unique but somewhat predictable weather pattern. Equilibrium, thankfully, will never be achieved. Water plays a leading role in our climate as an energy-transfer agent. For instance, evaporation of equatorial waters removes heat, while rain showers release this heat into the atmosphere, the latent heat of vaporization. Water's latent heat of melting and freezing importantly contributes to the redistribution of energy. So, where are the wind-wealthy climates found?

The earth has primary high- and low-pressure areas, four in each hemisphere. Moving north from the equator there is the equatorial low-pressure trough, the influential subtropical high-pressure cells between 20- and 35-degree latitudes, the subpolar low-pressure cells around 60 degrees, and finally the polar high-pressure cells. Similar to the case of fossil fuels but less pronounced, there are parts of the world that are rich in wind resources and others that are poor. The descending

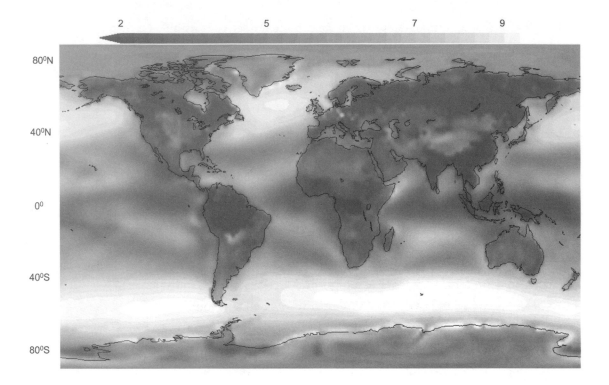

Figure 4.1. Annual average wind speeds (m/s) at 50 m. *Source:* NASA, Surface Meteorology and Solar Energy, July 1983–June 1993 (SSE Release 6).

air from the subtropical high-pressure cells feeds the two most potent manifestations of surface winds, the trade winds and the westerlies. Looking at Figure 4.1, the lighter bands in both hemispheres above 40 degrees are where we find the westerlies. You've heard of seafaring explorers in centuries past caught in the doldrums. Areas near the equator often have little or no wind for weeks, and yet blustery trade winds are just north and south. Beyond these two affluent regions other favorable locations exist from which to harvest wind energy. Where land meets the oceans, especially agreeable conditions exist for wind generation. When the air above the land heats faster than the air above the ocean, this difference conjures the delightful sea breezes we enjoy. In some locations the opposite happens in the winter months. The light colors seen in the North American interior, along Germany's North Sea coast, and across China are examples of locations particularly conducive to harnessing energy from the winds. Much of the world's high winds occur far offshore, and fortunately wind travels, which allows strategically placed nearshore locations an opportunity to harvest some of this energy.

Wind of course was applied to sailing vessels many thousands of years ago. The first documented designs of capturing wind energy for the application of milling are attributed to the Persians 2,000 years ago.[1] In this ingenious design, wind was directed through an opening in a building where vertical sheets were made to spin. A grinding stone was rotated over a fixed stone, milling the grain. Over centuries, the innovation spread and was improved. The Dutch refined the technology with rotating caps that could be directed into the wind. These enterprising engineers in Holland used the energy for pumping water to recover land. Windmills were used to pump water from wells across the western United States in the nineteenth century. All of these applications were dependent on a prevailing wind. Then, as today, people were confronted with the frustration of a fluctuating energy source. So, when fossil fuels delivered reliable on-demand energy through the naturally occurring stored-energy forms of oil, gas, and coal, wind power fell into disuse.

Today's wind power is based on absorbing the kinetic energy of air and converting this into electricity. Wind is affected by surface features as it moves. The closer to the surface, the more turbulent the flow as friction impedes its path. Even a flat surface such as the ocean depresses the power of the wind, causing the formation of waves. Consequently, wind speeds increase as you climb above the surface. High winds can be found in wide-open spaces with unimpeded flow and in mountain passes that funnel the air.

Figure 4.2 provides estimates of wind potential for most of the countries we examine in later chapters. To help provide scale to the onshore and offshore numbers, a potential factor is provided (Wind Potential/Total Electricity Generation), and the higher the number, the better. These are huge figures representing theoretical limits, not to be confused with the practical collection of wind energy, unless we are all prepared to don pinwheels. With the exception of Japan, Korea, and the United Kingdom, most of a country's practical wind potential is found onshore. This explains why Japan is experimenting with floating wind

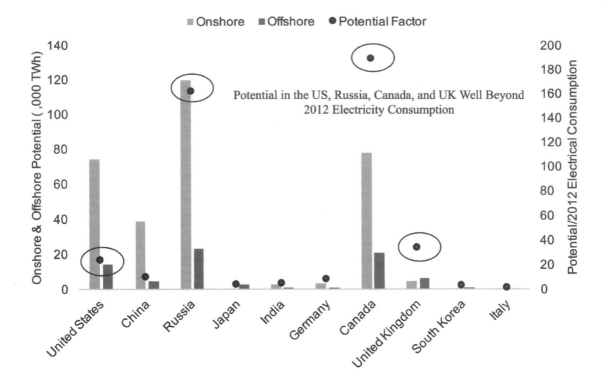

Figure 4.2. Potential wind energy and potential factor. *Source Data:* Xi Lu, Michael B. McElroy, and Juha Kiviluoma, "Global Potential for Wind-Generated Electricity," *Proceedings of the National Academy of Sciences of the United States of America* 106, no. 27 (2009): 10933–10938, doi:10.1073/pnas.0904101106.

platforms. Russia is wind-wealthiest by far, followed by Canada and the United States, but has the lowest level of wind generation within this top-tier group. Electricity generation from U.S. wind farms was 141 TWh in 2012, only 0.16% of the potential estimates in Figure 4.2. There is still a lot more wind energy to capture in the United States, though practically not 600-fold. We examine theoretical potential and apply value judgments in estimating practical potentials later in this section.

Every year the world over the wind blows with a certain consistency, while any specific location may experience high daily and seasonal variation. This is the nature of wind, not its fault. Managing and coping with the intermittent qualities of wind will be difficult but surmountable.

Solar Power: So Far and Yet So Dear

Introducing the star of the show for earth and her inhabitants, the sun. The sun is the engine that creates our weather and the conditions for life, not to mention positive attitudes. Burning every hour every day, the sun converts and shares her energy. No wonder solar power has the greatest potential by far of all the renewable sources. The sun burns with a steady consistency, and the earth receives as much energy in roughly one hour as the world consumes in a year.

The magnifying lens, invented in the early seventh century B.C., is likely the first recorded example of using solar power. It concentrated the sunlight and was useful in starting fires.[2] Stories are told of Archimedes, the Greek scientist, directing the use of bronze shields to focus light and setting afire Roman ships that were besieging Syracuse in the second century B.C. Modern-era recognition goes to a French scientist, Edmond Becquerel, who in 1839 discovered the photovoltaic (PV) effect, where electrical current increases in an electrolytic cell when exposed to light. In 1954 Bell Labs developed the first PV cell that was capable of running electrical equipment.

Amazingly and fortunately the sun's luminosity is extremely consistent, with variation of well less than 1% over its 11-year solar cycle. This consistency favors a system slow to change, providing a wonderfully pleasant and stable environment. The earth's surface already has enough variation caused by her daily rotation, the annual cycle of the axis of rotation, the solar orbit, and the lunar effect. If we had widely fluctuating solar energy patterns, the planet would be hostile to our survival.

The "Energy Budget" describes the inflows and outflows of the sun's radiation to the earth's surface and atmosphere.[3] Solar radiation is deflected outward by clouds and other molecules in the atmosphere and surface reflections. On average around the world, about 45% of the incoming solar insolation is transmitted through to the earth's surface as direct or diffuse shortwave radiation. *Insolation* is a term used to describe the amount of solar energy absorbed by a given surface location and is typically expressed as $kWh/m^2/day$. Insolation in turn powers subsequent natural processes. About 19% of the incoming solar radiation causes latent heat transfer (evaporation), 4% stirs convective heat transfer (turbulent), and 22% is net thermal radiation. So, when we calculate solar energy potential, surface insolation is the figure we use. If the plan were to deploy solar collection in space, as some countries are pondering, a figure roughly twice as high would be used.

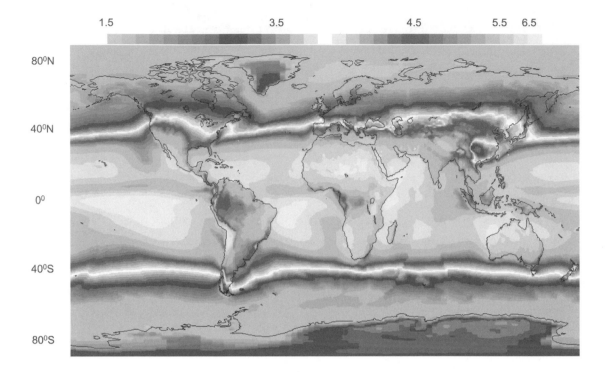

1.5   3.5   4.5   5.5   6.5

Figure 4.3. Average global horizontal insolation (kWh/m²/day).
Source: NASA, Surface Meteorology and Solar Energy, July 1983–June 2005 (SSE Release 6).

Practically, the world's demands could be met by collecting solar energy from less than 1% of land surface depending on the technology mix. As a comparison of land allocation, the World Bank publishes road density figures, which can be used to approximate the amount of land dedicated to roadways. The United States has a road density figure of 67 (km per 100 km²), which translates into more than 4 million mi of roads and up to 1% of the U.S. land surface when all is considered. To add additional perspective, farmland occupies 38% of U.S. land space, and the National Park System is allotted 3.5%. Providing an estimate of earthbound solar energy depends on how much surface area we choose to dedicate to collection. Ideally we would repatriate paved or otherwise altered land in equal measure to the undeveloped land we dedicate to sourcing energy.

The map in Figure 4.3 shows the average daily horizontal insolation (kWh/m²/day). Horizontal insolation includes both the direct normal and the dispersed radiation that strikes a horizontal surface. Technologies vary in their ability to harness these types of radiation. For instance, PV panels are installed with angles based on their location's latitude in order to capture the maximum amount of direct normal radiation. Looking at Figure 4.3, locations on and around the equator and generally at latitudes less than 40 degrees are ideal for collecting solar energy. The white bands farther north and south form somewhat of a boundary between high and low regions of solar potential because of the large swings in seasonal productivity. Noteworthy, though, Germany, located in the boundary zone, ranks first in global solar PV capacity despite a relatively low average daily insolation. Reflecting back on wind potential, to some extent wind and solar complement each other; wind speeds are generally higher north of 40-degree latitudes, while solar insolation decreases.

By location there is seasonal variation in addition to the expected variation throughout the day. Figure 4.4 plots the total surface area that receives more than 2,000 kWh/m²/y for selected regions and/or countries. Australia leads this group, with 79% of its land surface at or above this mark. Consider the vertical axis as a measure of a country's "solar fertility" and the horizontal axis as a measure of "solar wealth." Even though India appears relatively poor in solar fertility, at just 3%, that can still translate into a lot of solar energy.

Of all the renewables, solar technologies offer the largest diversity of methods for harvesting the sun's radiation. In contrast, wind is captured in fundamentally one way, converting kinetic energy into electrical. In the case of solar, technologies can convert electromagnetic energy into thermal, electrical, or chemical forms. Each of these methods is discussed further in the technology section. Throughout the book there will be references to fluctuating renewables, which include solar, wind, and tides, as opposed to large hydro, geothermal, and biomass, which by contrast are relatively steady sources. This subclass of fluctuating renewables will be referred to as temporals. The codevelopment of economical storage technologies with temporals will be crucial as the energy transition progresses.

Hydropower: The Modern World's First Major Renewable Energy Source

Once again, the all-powerful sun is responsible for the rivers that tumble their way to the sea. Solar energy heats the planet's surface, causing evaporation. When the heating is sufficient to overcome the water's latent

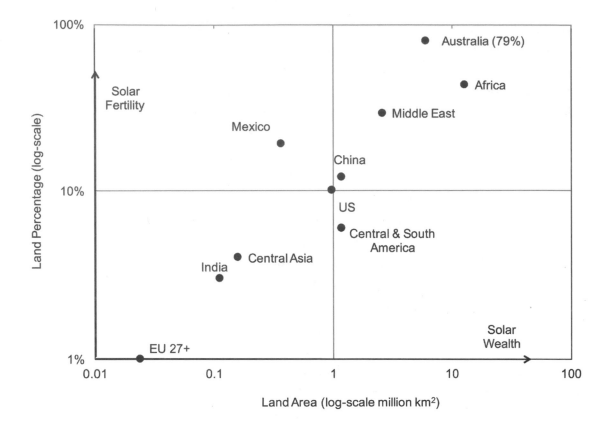

Figure 4.4. Land area with annual solar insolation above 2,000 kWh/m²/y.
Source Data: Greenbang, "What Parts of the World Get the Most Sunlight?" Sustainable Technology Forum, April 4, 2012, http://sustainabletechnology forum.com/what-parts -of-the-world-get-the -most-sunlight_21712 .html.

heat of vaporization, we have the magic of clouds, moisture. The warmer the air temperature, the more moisture it can hold. As the moist warm air rises, it expands and cools, and when the dew point temperature is reached the water condenses and releases its latent heat of vaporization. You've seen the awe-inspiring formation of cumulonimbus clouds where this is occurring. It is like the sky is boiling. These clouds very often dissipate in thundershowers. Less spectacular but more influential are large low-pressure systems that create snowpacks and rain showers that fill the rivers and reservoirs. Air is rising in a low-pressure system. Condensation occurs as the air rises. Often, sustained rain or snow showers occur within the low-pressure systems. A river's potential energy is translated into kinetic energy as it makes its way downward. Hydropower is the conversion of water's kinetic energy into electrical energy. Reservoirs are modern man's means of accumulating water's potential energy so that it can be metered out as needed.

The first water mill is credited to the Greek engineer Philo of Byzantium, in the third century B.C.[4] It comprised a waterwheel, the Perachora wheel, and toothed gearing. This invention rapidly spread and was improved upon as it traveled. Those fortunate enough to have access to moving streams and rivers used this methodology for centuries. More recently, two hydro demonstration projects in 1880 and 1881 provided direct-current lighting in Grand Rapids, Michigan, and Niagara Falls, New York. The world's first hydroelectric dam went into operation on the Fox River near Appleton, Wisconsin, in 1882.[5] By 2012, hydropower generated 3,600 TWh of electricity, representing 17% of the world's total generation. This percentage has been dropping, however, as hydroelectric potential is no match for the world's growth in electricity consumption.

While hydropower is a renewable source and one that does not contribute to greenhouse gases, it does come with a price. Rising waters held in huge reservoirs have displaced millions of people around the world. Ecosystem impacts have altered the benefit equations such that some dams are being removed to restore lost nature. River salmon migration that was made impossible by some dam projects is perhaps the most well-known example of ecosystem disruption. Many dams have formed mammoth reservoirs in arid climates, exposing precious freshwater to evaporation. In some cases, more water is lost to evaporation than is used. We are also discovering that the reservoirs accumulate silt year in and year out, which will ultimately render the dams incapable of holding water over time.

Geothermal Power: From Within

We may think that the sun supplies all the warmth we enjoy here on the earth's surface, and that is mostly true. Take away the sun, and the earth would be a cold and inhospitable rock floating through space. Beneath the surface, though, remnant heat from the massive collision of space debris as the earth was formed is still migrating to the surface, and the decay of radioactive isotopes provides a constant and ostensibly eternal source of thermal energy. Estimates place surface-level geothermal power at 42 million MW; solar power is 2,000 times this amount. So, yes, the earth would be cold but not completely without a source of heat.

Although the preponderance of the geothermal energy that makes its way to the surface is at too low of a temperature to be captured efficiently, there are good opportunities for this renewable resource. Fortunately, as with solar insolation, there are provinces where geothermal energy is concentrated. These higher-potential locations are for the most part located along the boundaries of the earth's tectonic plates. One type of concentrated source occurs where plates are moving apart and large volcanoes have formed, the globe-encircling mid-ocean ridge. A second source is where plates are colliding and one is thrust underneath the other in a process called subduction. Above the descending plate the temperatures can increase sufficiently to create molten rock. The density of the melted rock is less, and it ascends toward the surface. No matter how this is manifested on the surface, the temperatures below the surface are hotter than typical. Figure 4.5 shows the global geothermal provinces where favorable conditions can exist for the capture of concentrated supplies of thermal energy; they are located along the borders of the earth's major plates. A well-known region called the "Ring of Fire" borders the Pacific Plate, stretching from New Zealand through Indonesia and northward through Japan to Russia, eastward to North America, and then southward. Many of the top deployments of geothermal technology have occurred in countries bordering the Ring of Fire. Elsewhere, along the Western European portions of the Eurasian Plate, Italy and Spain are also leveraging this natural energy source. While many of the locations occur where plates are colliding, Iceland is located on the mid-Atlantic ridge, where plates are diverging.

Surface clues for geothermal potential can range from hot springs to thermal pools, boiling mud pots, geysers, and volcanoes. Within these geothermal provinces, when certain favorable conditions exist, harvesting thermal energy becomes technologically possible. A large water source in a network of fractured and permeable heated rock all sealed with a cap rock constitutes a sustainable environment ideal for today's technology. The network may contain tens or hundreds of cubic kilometers of water. The hydrothermal system within the network is stable and can endure for hundreds of thousands of years or more, a perfect quality for an energy

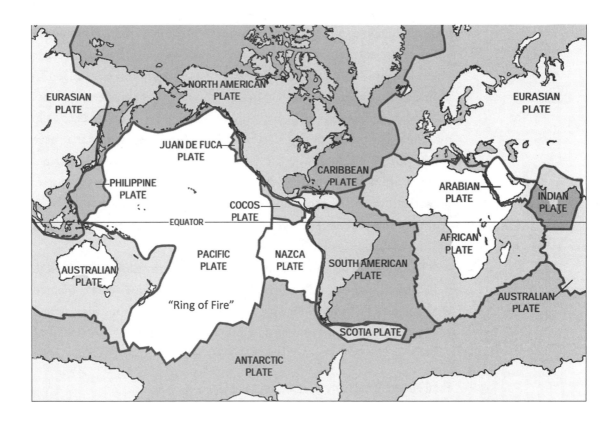

Figure 4.5. World geothermal provinces. *Source:* U.S. Geological Survey, "A simplified map of the earth's crustal plates," accessed June 21, 2014. http://pubs.usgs.gov /gip/earthq3/what .html.

source, persistent and round-the-clock steady. Geothermal wells drilled kilometers deep into these hydrothermal systems utilize the water as a heat-transfer fluid to drive steam turbines to generate electricity or the delivery of direct heat. And importantly, after sharing its thermal bounty, the water is returned to the well.

There are several methods used in the generation of electricity, and they depend on the temperature within the hydrothermal source. First and ideally, dry steams at temperatures above 235°C are fed to turbines where electricity is generated, and the condensate is returned to the underground reservoir. Prince Piero Ginori Conti is credited with inventing the first geothermal power plant in 1904 in Larderello, Italy, where the dry steam plant still operates. Flash steam is employed at lower temperatures (>150°C), where 30% to 40% of the water boils as it makes its way to the surface. The steam and water are separated, and the pure steam is again fed to the turbines to generate electricity, with the condensate and water returned to the well. The final method, binary cycle, is employed when temperatures are between 100°C and 150°C. The hot water is passed through a heat exchanger where a working fluid is caused to boil, and this steam is utilized to drive turbines. Some geothermal plants are designed to combine flash steam and binary cycle methods to help maximize the production of electricity. There are many places that have hot resources but lack a hydrothermal system. Enhanced geothermal systems are being developed to tap into this attractive energy source. As with all the other renewable energy sources, researchers are at work trying to innovate sustainable methods to gather as much of the 42 million MW as possible.

The merits of geothermal power are many for those fortunate enough to be located in the world's geothermal provinces. Unique when compared with solar and wind, geothermal is continuous throughout the day and year, able to contribute energy as a dependable baseload.

Tidal Power: Where the Oceans Meet the Shore

In 1687 Sir Isaac Newton published his famous *Principia*, within which he attributed the earth's tides to gravitational forces caused by the moon and the sun.[6] Newton discerned that the strength of the tidal forces was inversely related to distance (squared) and directly related to mass. As a consequence, the moon's influence is twice that of the sun's. To the business of harnessing renewable energy, the gravitational physics alone don't convey potential. The earth's surface is 70% water, with 870,000 mi of shoreline, so one would reasonably expect tidal power to be a potent and renewable source of energy. Quite surprisingly, though, practical tidal power is relatively small and concentrated. Why is this?

Think of the tides as a manifestation of the ocean's attraction to the moon. It causes the formation of a big wave known as the tidal bulge, with a wave length on the order of one-half of the circumference of the earth (~20,000 km). For comparison, an ocean wave in the western Atlantic has a length of just 120 m. The tidal bulge, then, is an entirely different type of ocean wave, best appreciated with time-lapse photography. As the moon travels around the earth, the tidal bulge follows. A second bulge on the opposite side of our planet is caused by forces of inertia, which is why most locations experience two high and low tides per cycle. The sun can amplify or weaken the effect of the moon. If the world's entire surface were deep water, a uniform pair of tidal bulges would be continuously circulating, and most everywhere the tide pattern would be the same. Introduce our continents, and the bulge now has stumbling blocks to satiating its gravitational desires. To follow the moon, the water moves as if through an obstacle course. The northern hemisphere more so than the southern hemisphere places obstructions in front of the tidal bulge because the North American, European, and Asian continents merge with higher latitudes. This causes a constriction in the flow of seawater, and this is influential in the size of tidal ranges.

There is another factor that further concentrates tidal power. The pattern of water flow into and out of any given bay or inlet (resonant frequency) is determined by many factors. When the flow pattern matches the tidal period, large tidal ranges can occur. Figure 4.6 displays a few locations around the world with tidal ranges greater than 7 m. Since tidal power varies with the square of the tidal range, these are prime locations for harnessing tidal power. The Bay of Fundy on Canada's North Atlantic coast experiences the world's largest tidal range. More water flows into the Bay of Fundy in one tide cycle than all the world's freshwaters. The tidal range can be as large as 50 ft, the equivalent of a five-story building.

Two principal mechanisms in the harvesting of tidal energy include streams and ranges. Tidal streams occur in coastal estuaries or bays and are associated with a landscape that funnels the flow, where water turbines can convert the energy. Tidal range technologies utilize the difference in height between the high and low tides. Tides can produce twice daily what dams are built to create, a reservoir of water at a height from which kinetic energy can be converted. The earliest recorded use of tidal energy is from around A.D. 800, when tide mills were erected along the Spanish, French, and British coasts. The concept is simple and ingenious. The incoming water from the rising tide was directed through a gate and into a pond. As the tidewaters peaked, the sluice gate was lowered, retaining the water. To harness the energy, water was directed through a waterwheel that powered a mill. The world's first and largest tidal-electric power

We look outside and see nature as erratic and fickle and conclude that an energy supply based on the earth's winds and the sun's radiation makes us vulnerable. Actually there is no greater consistency in our world than the power of the sun.

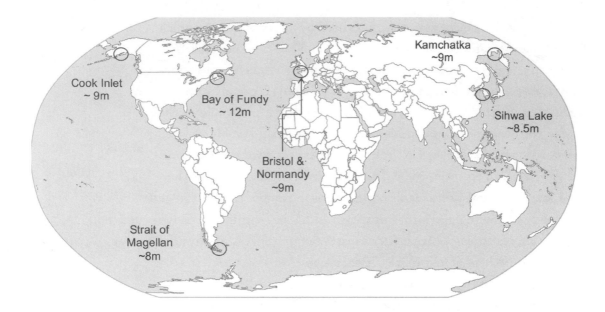

Figure 4.6. Locations with large tidal ranges.

facility was built on the Rance estuary at Ille-et-Vilaine in Brittany, France. Commissioned in 1966, the 240 MW power plant remains in operation.

Other Ways to Tap the Ocean's Energy

Ocean thermal energy conversion remains an elusive but attractive renewable energy source. It uses a heat pump to extract the ocean's thermal energy. Ocean thermal energy conversion has the potential to serve as a baseload because of its consistency, but temperature differences of only 20°C–25°C limit the efficiency of its harvest.

There are many research projects attempting to gather energy from the earth's oceans using waves, surface currents, and the salinity gradient. Although the sources of energy are attractive, current technology offers low extraction potential. But let's not despair; efforts today are still at an early stage.

Dormant Seeds of Invention Are Now Germinating

In contrast to the renewable sources, fossil fuels have provided a ready-to-serve source of energy. It is as if the world has been invited to an all-you-can-eat buffet, where patrons eat more than their required caloric intake for no better reason than they can. Complacency and waste set in as we had ample access to energy and our creative energies pointed elsewhere. With the end of fossil fuels in sight, however, invention is once again devising ways to harvest renewable energies. Our toolbox is so much larger than it was just two decades ago, and it is growing quickly. This is an exciting period, with practical products already being deployed and concepts such as solar space stations in early stages. The potential of renewables and humankind's stirring history of invention present the next several generations with the opportunity to participate in an exciting energy renaissance. Will we invest in our creative genius to usher in a post–fossil fuel era, or will we lock ourselves into inertia and linger too long before acting?

When we imagine a future relying on renewable energy sources, we look through glasses optically corrected for the use of fossil fuels. Since Drake's first oil well 150 years ago, fossil fuels have been abundant, relatively cheap, and always there when we needed them. We look outside and see nature as erratic and fickle and conclude that an energy supply based on the earth's winds and the sun's radiation makes us vulnerable. Actually there is no greater consistency in our world than the power of the sun. Our sun is as steady and consistent an energy partner as we could ever hope to find. Our grandmothers told us as children that we couldn't have our cake and eat it too. Once you eat that cake, you no longer have it. We are waiting for the alternative solutions as if we are unconstrained. But meanwhile, we continue to eat our cake, consuming more than our share of a resource we deplete when we use it.

Vignette: The Becquerels

Later I conclude that beyond solar, the other renewables have insufficient practical capacity to meet our twenty-first-century energy projections. We will depend on a combination of solar and nuclear power to close the gap, and we owe a notable debt to a single family, the Becquerels of France, whose momentous discoveries contributed to both our solar and nuclear tool kits. It is remarkable given our long history of scientific research that a single family provided critical insights in two major areas related to energy.

Born March 24, 1820, in Paris, Edmond Becquerel belonged to the second of four generations of scientists. His father, Antoine, who studied electric and luminescent phenomena, exposed Edmond to science at an early age. He was growing up in a time of discoveries on multiple fronts. In 1838, Charles Darwin was publishing controversial essays on his theory of evolution. Volta's battery had been invented 40 years earlier, and it would be a critical tool in helping Edmond uncover his most famous discovery. At the age of 19, experimenting with an electrolytic cell, Edmond observed that the electric current increased in the presence of light. He coined his observation the "photovoltaic effect," light current. At the time it was also referred to as the "Becquerel effect." The cause of that observed phenomenon has led to the development of our modern PV technologies. Our solar energy prospects would be discouragingly dimmed without this understanding.

Henri Becquerel, Edmond's son, was born in Paris on December 15, 1852. He made an eye-popping observation in 1896, opening the world's view to the property of nuclear radioactivity, the atomic release of energy. Photographic plates covered in thick black paper to protect them were nevertheless exposed in the presence of uranium salts. Henri concluded that the penetrating radiation (energy emissions) came from the uranium without need of any external excitation, "spontaneous radiation." The Curies followed this discovery with additional research that uncovered other elements with radioactive properties including polonium and radium. Together Henri Becquerel and the Curies shared the 1903 Nobel Prize in Physics. Sadly, Marie Curie died in 1934 from leukemia, likely resulting from her exposure to radioactive material. The nuclear age was dawning, and the twentieth century followed with rapid new discoveries that have led to the present day's nuclear tool kit. The seeds of the Becquerels' discoveries have taken more than a century to rise to a relevance so critical to the world's energy future.

Notes

1. Amin Saeidian, Mojtaba Gholi, and Ehsan Zamani, "Windmills (ASBADS): Remarkable Example of Iranian Sustainable Architecture," *Architecture–Civil Engineering–Environment* (Silesian University of Technology) 5, no. 3 (2012): 19–30.

2. U.S. Department of Energy, Energy Efficiency and Renewable Energy, "The History of Solar," https://www1.eere.energy.gov/solar/pdfs/solar_timeline.pdf.

3. Rebecca Lindsey, "Climate and Earth's Energy Budget," NASA Earth Observatory, January 14, 2009, accessed April 11, 2016, http://earthobservatory.nasa.gov/Features/EnergyBalance/.

4. Jeremy Norman, "The Earliest Evidence of a Water-Driven Wheel (circa 250 BCE)," HistoryofInformation.com, March 19, 2015, accessed April 10, 2016. http://www.historyofinformation.com/expanded.php?id=3509.

5. Energy.gov, Office of Energy Efficiency and Renewable Energy, "History of Hydropower," accessed April 9, 2016, http://energy.gov/eere/water/history-hydropower.

6. National Oceanic and Atmospheric Administration National Ocean Service, "Tides and Water Levels," accessed May 1, 2015, http://oceanservice.noaa.gov/education/tutorial_tides/welcome.html.

5

The Potential and Toll of Alternative Energy Sources

Life is pleasant. Death is peaceful.
It's the transition that's troublesome.
—Isaac Asimov

If global fossil fuel reserves are the means to assess their inventory, how do we measure the potential of renewables? Renewables aren't depleted like petroleum; they are power sources, and we create technologies to harvest, store, and deliver their energy. By examining the various renewables and the nature of their power, we can gauge their relative magnitude.

Delivering Energy Is Like Origami: It's an Art of Form Changes

Before we consider our future energy requirements and the potential of alternative sources to satisfy them, it is important first to distinguish primary from secondary energy. Primary energy is taken directly from nature without conversion. Fossil and nuclear fuels are examples of primary energy sources. Electricity is a secondary form, converted from a primary energy source such as coal. Hydrogen, which, as we will see, plays an important role in the future, is another example of a secondary energy because, like electricity, primary energy is used to produce it.

A second factor to keep in mind is energy efficiency. When we steal a relaxing sail or a pleasant dip in a natural hot spring, we are exploiting primary energy from nature. Most often, though, energy must go through a surprising number of conversions from its raw form before we use it in our daily lives. Depending on the type of conversion, the useful output energy is just a fraction of the input energy. Like with a leaky water pipe, there are energy leaks with every conversion.

Let's start with electricity. A coal power plant manages three distinctive energy conversions before it delivers its electricity into the grid. First, the stored chemical energy of coal is converted into thermal energy, and the efficiency is nearly 90%. Next the thermal energy is converted into mechanical energy inside the turbines, and this efficiency is only 40%. Finally, a generator efficiently converts the mechanical energy into electricity. Together these conversions leak 65% of the original chemical energy. The weak link in most of the world's fossil or nuclear power plants is the conversion of thermal energy into mechanical energy. It is an unavoidable price we pay for converting these fuel sources into electricity.

But electricity is only a portion of the energy we use. There are many applications that utilize thermal energy, and since this involves just a single conversion with fuels, the efficiency is much better. Fossil fuels in particular are ideal energy sources for thermal applications.

Now let's consider some of the renewables and the conversions necessary to deliver the different types of secondary energy. A hydroelectric plant is simplistic compared with a coal plant; it delivers electricity after just a single conversion. The mechanical energy of a water source is converted into electricity with an efficiency of 90%. This makes hydropower the most efficient method of converting nature's primary energy into electricity. Too bad its raw potential is way below what the world requires. A wind turbine, too, generates electricity after just a single conversion, kinetic to electrical, but it does so with an efficiency of 35%. Compared with hydroelectric this seems low, but it compares favorably with the fuel-based efficiencies and comes with many other advantages. Solar photovoltaic panels convert electromagnetic energy into electricity with an efficiency of only 23%. Still this is remarkable considering it is a renewable resource. Beyond electricity, only solar, geothermal, and biomass deliver thermal energy. So overall, renewable technologies are particularly suited to the task of replacing fuels in the power sector. One can also readily see the application of renewables within the transportation sector with electric vehicles. However, it is much harder to see the full-scale application of renewables for thermal applications without a renewable fuel.

So there are two main points to take away from this discussion. The first is to realize the number and type of conversions that occur with our use of energy, from driving our cars to lighting our homes and even storing energy, and to understand that these conversions are leaky, thus increasing the amount of energy we need to harvest to meet our demands. The second point is that our primary energy requirements in the future will depend upon the sources we choose, and here we must apply value judgments.

Just because a technology can unlock a supply of energy does not mean we should pursue it.

Nature, Technology, and Value Judgments

Now let's talk about primary energy and the potential of renewables to serve our appetites. We've looked at the natural phenomena behind the renewables and gained an appreciation and sense of their potency. It was nice to learn that the earth receives in one hour enough solar energy to power our world, but now we need to understand (1) Do we have the technology to capture and convert sufficient quantities? and (2) Can we live with the ramifications of those decisions?

The most difficult aspect of assigning potential is defining how far we bend nature to meet our palate. Just as we impose limits on the collection and use of fossil fuels, there will be limits on extracting renewable energy. Renewable technologies discovered long ago are now advancing quickly as they make their way to center stage for deployment. But as we explore what renewable sources have the power to satiate our thirst, it is equally important that we assess the "costs" of extracting renewable energy. Unfortunately, every energy source charges an extraction fee. The term *extraction* is commonly used to refer to the recovery of fossil fuels and is often viewed negatively. But it equally applies to renewable energy—extracting energy from renewable sources also takes a toll. We know what we sacrifice with fossil fuels. We need to be just as cognizant of what we sacrifice with renewables. It would be a mistake to turn against fossil fuels as environmentally unhealthy and embrace alternative sources without exorcising our rapacious use of energy. Although the "all you can eat" smorgasbord seems like a good value, the excess leaves the consumer unnecessarily overfed. Every choice of energy comes with negative impacts that we should strive to minimize. Planning not only should describe methods to harvest the necessary supplies of energy but should articulate preferences that reduce detrimental effects wherever possible. Our long-term survival depends on a diverse, healthy, and robust nature around us. Not only are we changing sources of energy, we are also transitioning to a sustainable system of supply and demand. In certain contexts, the use of the term *sustainable* stems from the desire to leave the world as we found it, so that we have not pillaged and plundered the planet.

Cautions in the Assignment of Renewable Potentials

The renewable potentials this book assigns are taken with the certainty that values will be the strongest determinant, that human-imposed limits will be as hard and fast as technological limits, that technology-oriented estimates will significantly overstate what localities around the world will collectively choose to collect, and that we are fortunate to have the power of choice and consequently the opportunity to tune our collection of energy. The following cautions stem from our tendency to lead with technology, and this has pitfalls.

First, just because a technology can unlock a supply of energy doesn't mean we turn the key. As the world adopts renewables, every locality will be unique in the choices it makes. Localities vary not only by the types of energy they can access but by the value systems they will exercise in making decisions. If a community is technology-driven and goes by numbers, it will select one deployment mix, while another community may apply values of land use, scenic quality, and environmental stewardship to arrive at a completely different deployment mix. Both communities will make trade-offs in securing their energy requirements.

Second, large potential estimates driven primarily by technology assessments obscure the pressing need to adjust our energy consumption downward. Bad habits are the hardest things to change, and we often apply technology to avoid the unpleasantness! Every source of energy takes a toll, and we take more than we need, so we should work hard to conserve and lower our footprints even with the prospects of abundant alternative potentials.

Third, large estimates for renewable power can underestimate the resistance to hundreds more megadams, tens of thousands of poor-land-use solar plants, and millions of wind and wave turbines dotting our landscapes and seashores. There are no passes in the name of renewables. Sensitive areas will be off-limits, obstacles to undermining scenic quality in the case of wind farms will develop, projects too close to population centers will be contested, and allocations of huge swaths of land for (concentrated solar power) low-density new-age renewables will be mightily resisted. It is conceivable that even before we can replace fossil fuels, resistance will limit renewable capacity.

Finally, unrealistically high potentials for renewables make it easy to dismiss the role of nuclear energy in the future, but that dismissal discounts the advances and relative merits nuclear technology offers in ending the use of fossil fuels this century. Germany, for instance, has chosen to discontinue the use of nuclear power and is retaining at least for the interim its dependence on lignite, the lowest-quality form of coal. Oppositely, China is placing a high value on nuclear power in the ultimate displacement of fossil fuels, particularly to reduce its dependence on coal because of poor air quality. When we look around the world we see that nuclear power remains an attractive option for many and that there are many examples of early deployment of far superior and safer nuclear technologies. Considering both uranium and thorium, nuclear fission fuels have centuries of potential, while fusion fuels have potential for thousands of years.

Placing too much credence on the potentials by proposing a precise mix at such an early stage in the world's transition is like the feeling of weightlessness. You don't feel the pressure of when and where to place a foot down. What the potentials do represent, when nuclear energy is considered, is a menu from which localities can plan to build nonfossil energy sources. Everywhere around the world the debate should be less about the continued use of fossil fuels and more about how we choose our alternative sources. Now for a little detail on the potential estimates for each of the renewables.

Solar: The Main Character in Renewables

Solar power has the largest potential of the renewables, and unlike with hydro, we have barely begun to deploy technologies to gather the energy. Although there are concepts for deploying floating platforms and space stations, we use a land-based assumption to frame solar potential without excluding these other options. Average surface solar radiation is approximately 160 W/m². As we've learned, the closer to the equator, the higher a location's potential. If we assume that 1% of land surface is used to collect energy, then solar potential would be approximately 200 TW. This is sixfold the world's projected consumption, which is good, but tough choices of where and how to gather the energy lie ahead. Collecting solar energy from a single percent of the earth's surface is a big deal. Can most of our solar energy be recovered from developed spaces? That seems like a good place to start. France is ahead of the game. In March 2015 legislators passed a law requiring all new buildings in commercial zones to be partially covered with either solar panels or plants.

Wind Power: From Moving Ships to Moving Turbines

Wind is solar energy transformed. What portion of solar energy finds its way into the formation of wind? A fair estimate is that 1% of the solar energy absorbed by the earth's surface is converted into wind, a two-magnitude difference. Across the earth's entire surface there is roughly 1,000 TW of natural wind potential. Fortunately, unlike location-specific solar insolation, the wind travels, and consequently a single point can capture a disproportionately high share of this potential. When we apply value assumptions on the spaces we use to collect wind energy and consider the concentrating effect due to the movement of air, we arrive at a 10 TW estimate. This represents about a third of our total projected mid-century requirement.

Hydropower: From Cutting Canyons to Spinning Turbines

Worldwide, in 2012, approximately 0.44 TWy (90% efficiency) of primary hydro kinetic energy generated 0.40 TWy of hydroelectricity, constituting

77% of the amount from all renewable sources. The world has been building hydropower capacity for more than a century. The projects have been some of the greatest engineering feats in our history, and yet hydro still only generates 17% of the world's electricity. This informs us of the scale and time frame required to replace fossil fuels, which provide 67% of the world's electricity.

Like every renewable, a hydroelectric dam from which we harness kinetic energy comes with a price tag; ecosystems are absorbed, and people are displaced. Harnessing hydropower doesn't necessarily imply the construction of a dam; many small hydro projects collect energy from the run of a river. Although these projects are smaller, typically less than 5 MW, they are inherently less environmentally intrusive and offer an attractive option for those communities with favorable natural resources. Even so, there are just so many rivers, flowing with just so much water, with elevation changes that support hydroelectric production. So it is unwise to project that hydro will play a significantly larger role. In the best case, we can project efficiency gains that will deliver more energy at existing sites and marginal net new capacity from both large and small projects. A generous estimate of 0.8 TW for hydropower represents a nearly 100% increase above 2012 levels and assumes a mix of improved efficiency at existing sites with the balance from new projects.

Geothermal: The Earth's Contribution

Geothermal energy, as discussed, is used directly in heat applications or utilized to generate electricity. According to REN21, worldwide investment in geothermal power in 2014 was US$2.5 billion, with most of this occurring in developing countries.[1] The report shared that through the end of 2014 there was 20.4 GWth of thermal capacity and 12.8 GWe of generating capacity. Considering potential estimates by the USGS and Western Governors' Association for just the western 13 states and the share of the world's geothermal provinces this represents, the book

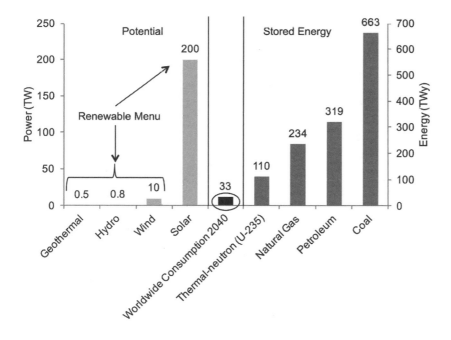

assigned a potential of 0.5 TW—a small number from a global perspective but an attractive source for the world's geothermal provinces.

Figure 5.1. Global renewable potential and stored energy in fuels.

Energy Is All around Us

So, can renewables deliver sufficient quantities of energy? The answer is a both resounding and cautious yes. Figure 5.1 summarizes the potential for four of the renewables. For reference, the energy potential for the world's fuels are also provided. There is an important distinction between renewables and fuels: the unit for renewables is power (TW), while that for fuels is energy (TWy). From the figure it's obvious that solar has the most practical potential of all the renewables by a wide margin. It is also noteworthy that the other three renewable potentials together are insufficient to power the world at forecasted consumption rates. Wind has been the fastest-developing renewable source, with a potential 12-fold that of hydro. Geothermal is site-specific to a high degree and globally limited. We will need a lot of solar along with the necessary help from nuclear if we are to meet the mid-century's expected demand.

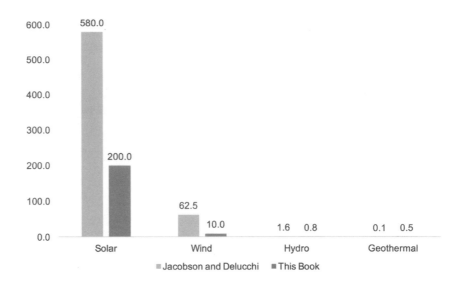

Figure 5.2. Potential mix of renewable energy sources (TW).

A Comparison with a Reputable Source

Alternative energy potentials are critical to energy planning. Figure 5.2 compares this book's more conservative estimates of renewable potential with an often-cited essay from Stanford University and the University of California at Davis, "Providing All Global Energy with Wind, Water, and Solar Power."[2] Jacobson and Delucchi's development of raw potential is an encouraging and important first step to acknowledging that we can move beyond the use of fossil fuels, and it generally aligns with this book's estimations. The differences between the report's estimates and this book's estimates are related to the cautions noted earlier in this chapter, that human-imposed limits will draw hard boundaries around the deployment of alternative technology. Both estimates, though, tell us that solar and wind will be the twenty-first century's renewables, while hydro and geothermal will obviously be valuable contributions where conditions and values permit.

We Really Don't Need to Set Fossil Fuels Afire to Thrive

Summarizing, Figure 5.3 shows the worldwide primary energy harnessed for each of the renewables compared with the potential expressed as a

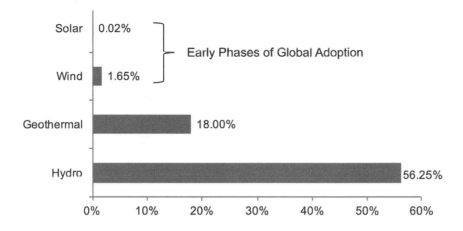

percentage. As you would expect, hydropower is nearest to reaching its full potential—we've been at that the longest. Wind is in the early stages of market penetration and should really whip up in the next decades. However, our global renewable energy future absent fossil fuels will predominantly depend on harvesting solar, and that is largely unrealized thus far. There are rapid advances in efficiency, manufacturability, and cost bringing us closer to deploying solar on a larger scale. It is comforting knowing that the potential is there. Thoughtful planning should establish a goal for an energy mix that comprehends technology but also natural resources and values. It took a century to realize ~56% potential in hydropower, and we should expect a similar time scale to build capacity with the other renewables.

A final point on the potential of renewables to replace fossil fuels involves storage. With the exception of large hydro, geothermal, and biomass, renewables are derived from temporal forces, implying that a strategy for the combination of direct use and storage will need to be determined. Thermal applications will need a fuel. Entirely replacing fossil fuels requires that a fair portion of renewable energy be stored. Storage can be long-term in the case of a new-age fuel and shorter-term in the case of batteries or pumped storage. Either way, the round-trip conversion of energy to storage and back incurs losses. Our total energy requirements in the future will be affected both by our storage strategy and by consumer behavior.

Figure 5.3. Renewable energy potential capacity vs. actual utilization.

Notes

1. REN21 Renewable Energy Policy Network for the 21st Century, *Renewables 2015 Global Status Report*.
2. Mark Z. Jacobson and Mark A. Delucchi, "Providing All Global Energy with Wind, Water, and Solar Power, Part I: Technologies, Energy Resources, Quantities and Areas of Infrastructure, and Materials," *Energy Policy* 39, no. 3 (2011): 1154–1169. doi:10.1016/j.enpol.2010.11.040.

PART II

Age-Old Technologies Find Application

Seeds of Innovation, Dormant for Centuries, Are Now Germinating

6

Technology Introduction

Our technological powers increase, but the side effects and
potential hazards also escalate.
—Alvin Toffler

Most new products offer marginal improvements within an established
category, but a few hatch an entirely new class, seminal advancements
that depending on your perspective can either enable or unleash changes
in our lifestyles. Often these big advances aren't recognized until the
space of time exposes their impact. Henry Ford's affordable personal
transportation led to the construction of 40 million mi of roadway worldwide, no doubt unfathomable to those technology pioneers of the early
twentieth century. The distance and time equation was changed; we can
travel 30 mi in the time we once spent to move a couple of miles. The car
ushered societal advancements, and as with all good things, too much
of anything can spawn negative outcomes. Our personalized vehicles
offered an alternative to the horse and forms of mass transportation,
allowed escapes from dense urban living, and brought access to goods
and services and markets. In the United States, the average light-duty
truck or car logs 11,000 mi annually, and there are more than 250 million vehicles on the road. Surprisingly, studies have shown urban pop-
ulations to be healthier than their suburban counterparts, who have
to jump into their cars to go most places. Simultaneously large thermal power plants able to serve customers far and wide were extending
access to energy. While personal vehicles were expanding our habitats,

our infant power sector was threading power lines alongside our roadways. These two synergistic advances together have altered where and how we live and work.

As we consider alternative energy technology, we should be alert not just to the new capabilities but also to the implications for the world's energy transition. Apart from hydropower the world is largely powered by fuels, and we've built our systems on their merits. We've learned that solar energy has the greatest potential of all the alternatives, and yet, as an energy source it could not be more different than our fuels. Where fuels are concentrated, solar insolation is diffused across the earth's surface. Where fuels are forms of stored energy, solar radiation is daily undulating between on and off. Where fuels are finite, solar is renewable. These differences pose implications for where we collect solar energy, how we accumulate it, and how we may shift consumption behavior. Distributed and renewable sources of energy along with storage technologies will change our world. How we deploy and even develop renewable technologies can be clouded by our current fuel paradigm. The ideal solutions will best leverage the merits of the renewable source and avoid a mindless imitation of our current marketplace.

We are at the outset of a foundational and fundamental transformation, and some specifics are already apparent, while others will need the same space of time and the insights of forward-looking planners. The power sector, once confined to large single points, is destined to decentralize as we collect our energy from temporals. Single points will remain but cede a large role to distributed sources of energy. Our demand-driven supply of electricity will morph into an equitable and efficient supply and demand system. Carbon-based fuels in the transportation sector will give way to alternatives, and more electricity will be generated to power these vehicles. Inefficient practices that are accepted today will be fertile ground for innovation. Less clear is how we displace fossil fuels for industrial heat processes and space heating and cooling. Options are emerging that presently lack full-scale commercial readiness, but they do provide a glimpse of a future when we can depart from the use of fossil fuels. As we examine the technologies in the chapters ahead we need to consider both the behaviors they enable and those they may unwittingly

Distributed and renewable sources of energy along with storage technologies will change our world. How we deploy and even develop renewable technologies can be clouded by our current fuel paradigm.

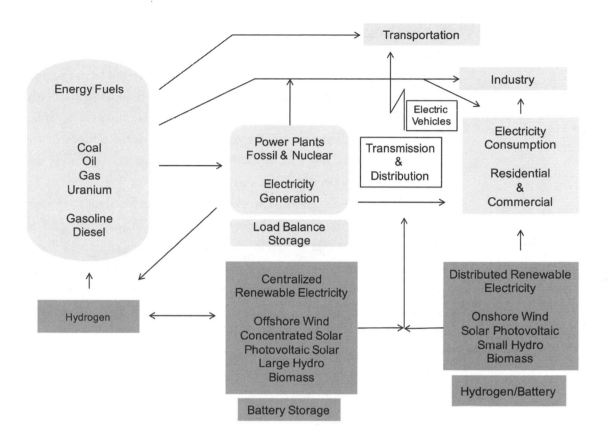

Figure 6.1.
Transitionary-phase
energy architecture.

unleash. This is an exciting period in our history as we contemplate the opportunities in our power, transportation, industrial, and residential and commercial energy sectors. We begin this part of the book by examining our current baseline.

As Things Are

Certainly there are exceptions, but for most of the world, electricity generation is powered by fossil fuels, as shown in the lighter color of Figure 6.1. Behind the scenes a network of power plants orchestrates an instantaneous response of supply to meet demand. Power plants burn fuels to generate electricity, and since the energy source can be stockpiled, it can be converted as needed.

Wind and solar power, shown inside the darker boxes, are at a distinct disadvantage when integrated into the current power infrastructure. Their supply is fluctuating. You cannot turn up a wind farm and pull more energy than nature is prepared to share when demand suddenly spikes. Today when we integrate renewables into the utility energy mix, fossil fuels provide the baseload to consumers, while the temporals are used opportunistically, and this will need to change. A baseload is the minimum level of electrical demand over a 24-hour period.

To contemplate a power network entirely without fossil fuels, storage technology and infrastructure become the key enabling capabilities. The cardinal argument against temporals is their fluctuating character. There are, however, solutions. Hydrogen is a fuel (energy carrier) that will support the accumulation of energy and subsequent generation of electricity, as shown in Figure 6.1. Also noteworthy is the distributed character of renewable sources and the requirements it imposes on infrastructure. Renewable capacity will be built in a combination of locations: both large-scale centralized and distributed. This will prove to be a great advantage in allowing an energy mix that minimizes environmental disruption.

Entirely departing from the use of fossil fuels is a monumental endeavor, and the power sector is the "easiest" place to start. The transition becomes much more difficult for heat generation and transportation applications. Globally, more than half of all final energy consumption is used for heat generation in buildings and industry. In the United States, the transportation sector uses about 25% of total primary energy. We are early in the energy transition, and most of our new technologies offer alternative means of generating electricity, but soon economically viable solutions need to emerge for these other applications. It's like a mechanic with a toolbox full of half-inch wrenches; it limits the scope and type of work. We need more tools that support high-temperature industrial processes, alternatives across all of the transportation categories, and a means to heat and cool the places in which we live and work.

This section's organization is a bit of a hybrid in order to help keep the material directed, so it warrants a brief explanation. Solar, wind, and nuclear are sources of energy, and in their chapters I describe specific technologies for electricity generation. Heat generation and transportation will have dedicated chapters. Finally, the role, status, and prospects of support technologies including advanced storage systems and smart grids will be explored.

It takes energy to build products such as cars, planes, phones, and renewable items such as solar panels. A viable renewable technology needs to generate more energy over its lifetime than that required to manufacture it; the energy yield ratio needs to exceed 1.00 by a lot. The more net energy produced, the better. The implication is that renewable technologies should be durable and built for a long productive life, delivering much more energy than it takes to produce them. The sustainability of collecting energy comes with a caveat: that manufacturing employs a robust and thorough recycling program for the critical raw materials, or we will be back to burning the remaining fossil fuels.

For readers who remain skeptical that alternative sources can replace fossil fuels, this section will showcase the breadth and readiness of technologies. Research reveals over and over that many of the scientific principles for the technologies were born long ago and now are finding application. Translating technical capability into economically attractive solutions is the critical work that is now occurring. Some of the technologies are already commercially competitive, while others are works in progress. Piece by piece the economics of a new energy system are falling into place.

7

Solar

Give me the splendid, silent sun with all
his beams full-dazzling.
—Walt Whitman

Before describing modern solar technology, it is worthwhile recognizing the sun's essential role in our food supplies and its influence on our dwelling designs and construction—that long-practiced applications in agriculture and architecture will continue to be vital in our future. Fossil fuels have increased our energy appetite, and as they lose appeal, we will capture much more of our supplies directly from the sun.

Tide, wind-, and water mills that harness energy are noticeable icons of early human ingenuity. Less obvious but much more influential in our flourishing has been our advancements in securing food. The earth's food chain is powered by the sun, and over time we've honed agricultural technologies to where we can now feed more than seven billion people, the energy equivalent of more than three million tons of coal per day in food calories. Tilling the land changed our societal structures, and inventions such as irrigation helped mitigate deadly seasonal variations in rainfall and prevent famines. Still powered by the sun, the food industry is a global enterprise applying an impressive ordinance of tools to feed the planet.

And visible architectural innovations remain from centuries past that integrate intelligent passive designs for heating and cooling, showcasing the rich possibilities for tapping natural processes. Hawa Mahal, Jaipur, which translates as Palace of the Winds, is such an example. Located in India's Rajasthan State and built in 1799, the palace incorporated an ingenious cooling system that delivered gentle breezes during the region's oppressively hot summers. And throughout the northern hemisphere homes purposely situated with southern exposure capture the winter sun and provide shade during the summer. Long before the development of the triple-pane window, cooling towers and solar chimneys provided dwellings comfort from the outdoors. Now, on to our modern technologies.

There are no free passes in harnessing energy.

First, Concentrated Solar Power

As the name implies, the technology category of concentrated solar power (CSP) concentrates the solar insolation striking a large parcel of land—typically several thousand acres—to heat a medium that powers the generation of electricity. Worldwide, in 2014 CSP capacity reached 4.4 GW, the equivalent of five Grand Coulee Dams. Because CSP plants have such relatively low power densities, we are trading lower carbon emissions for two magnitudes the land space used by today's nuclear and fossil power plants. There are no free passes in harnessing energy, and a thorough evaluation of the relative ecosystem disruptions of renewable and fossil fuels is beyond the scope of this book.

The four basic CSP designs include the parabolic trough, the linear Fresnel reflector, the power tower, and the dish engine. The parabolic trough system reflects solar energy from a parabolic mirror to a focal point where piping contains a heat-exchange fluid (i.e., synthetic oil). The fluid can reach temperatures of 750°F. This superheated fluid flows through a heat exchanger, producing steam that powers a turbine. The linear Fresnel reflector is principally the same as a trough design, but it uses parallel rather than parabolic mirrors to reflect the solar energy. Instead of using synthetic oil, the Fresnel system utilizes water, which is converted to high-pressure steam that drives electricity-generating turbines.

A third design, the power tower, looks like something out of J. R. R. Tolkien's *Lord of the Rings* trilogy. Mirrors referred to as heliostats surround a tower where they focus the solar energy. These heliostats are constantly adjusted during the day, turning like the sunflower, to ensure that the energy is focused at a fixed point on the tower. Inside the tower, a medium such as a molten salt absorbs the reflected solar radiation and follows the familiar path to a heat exchanger, with water to produce steam, which drives turbines.

As noted, these first three designs are suited for large-scale electricity production. The fourth design, referred to as the dish engine, is intended for smaller applications. It is another reminder that discoveries from long ago are often at the root of today's renewable technologies. Our vehicles' internal combustion engine depends on many things, but a fuel tank with gasoline is essential. The Stirling engine, first invented in the early nineteenth century, requires no fuel—just the heat conveyed by the sun's radiation. Setting fossil fuels afire isn't the only way to gather and apply thermal energy. A parabolic dish focuses the radiation to a point where a Stirling engine is mounted.[1] The thermal engine contains a closed working fluid (i.e., gaseous hydrogen), which expands when exposed to heat and is compressed again in a continuous cycle. The emissions-free engine spins quietly, and the motion is used to generate electricity. Stirling's invention exemplifies the difference between the depletion of fossil fuels and the sustainable qualities of renewable energies. Although it is not the answer for the world's energy future, it epitomizes a revival in applying old methods in harvesting energy, showcasing options beyond extracting the chemical energy of fossil fuels. For readers curious to see the engine in operation, numerous videos are available online.

CSP Is a Landgrab with Poor Production Economics

Large-scale parabolic systems net around 12% efficiency in converting the sun's electromagnetic energy into electricity. The mirrors used for solar collection typically only occupy 40% of the plant's property. That results in a paltry 5% land-use efficiency, a measure of the percentage of the solar energy that hits the plant's property that is converted into electricity. To compare the parabolic land-use efficiency with the power tower design, we can use Ivanpah as a representative plant.[2] Ivanpah was the first renewable plant to receive *POWER*'s Plant of the Year Award. Located on 3,500 ac in California, Ivanpah has a power rating of 372 MW and a projected 30% capacity factor. Given that the average annual insolation is 2,000 kWh/m^2/y for the Mojave Desert, it results in a land-use efficiency of 3.5%. Ivanpah's first-year capacity factor was even less—resulting in a 2% land

If we were to depend on CSP technology alone to deliver the world's energy demands in 2040, we would need to dedicate fully 1% of the world's land surface.

efficiency. From these two examples we learn that CSP has a land-use efficiency between 2% and 5%. Energy crops have solar-to-biomass efficiencies of 2% or more and nowhere near the capital investment. It seems like a lot of technology, land, and capital investment for marginal improvements over what nature recovers. Ivanpah, though, aligns with today's centralized power paradigm, so we are force-fitting this technology to mimic our current power industry structure. Surrendering large tracts of land and disrupting even more ecosystems for the sake of centralized energy may not be the optimum choice. At the least it should be limited.

The Ivanpah tower plant (372 MW) cost $2.2 billion to construct, translating to $5.91 million per MW, while the parabolic plant (75 MW) in Martin, Florida, cost $476 million to construct, translating to $6.35 million per MW. For comparison, the capital cost for a large wind turbine is one-third this figure. When capacity factors, operating costs, land use, and plant lifetimes are considered, the cost of CSP-generated electricity is no better than $0.10/kWh, making this class of renewable at least twice and perhaps as much as four times the cost of wind power. Looking ahead a decade, the cost of energy for CSP may hope to approach $0.06/kWh, but there is no escaping the poor use of land with this technology.

Several large CSP projects are incorporating thermal storage capability, hoping to improve their overall value mix.[3] Storing heat will allow these plants to continue to deliver electricity on cloudy days and into the night. A CSP plant's ability to store thermal energy introduces a new value. Gemasolar, a CSP plant in Spain with thermal storage, was able to generate electricity around the clock for 36 straight days. Solana, in Arizona, is a 280 MW parabolic plant with six hours of storage capacity. Several other new megaprojects due to come on line will couple thermal storage with CSP, breaking the notion that the supply of solar energy is limited to daylight hours.

The Centralization of Power Is Hard to Break

If we were to depend on CSP technology alone to deliver the world's energy demands in 2040, we would need to dedicate fully 1% of the world's land surface. It may offer a means for capturing and storing energy thermally for use when the sun goes down, but it is hard to imagine this technology surviving the advent of other storage solutions. At the end of 2014, world-wide CSP capacity was 4.4 GW, compared with a capacity of 177 GW for solar photovoltaic. Looking forward, according to REN21, 90% of world-wide investment in solar capacity is directed toward solar PV, reinforcing the assessment that CSP will be a niche solution as compared with PV.

Solar PV: The Becquerel Effect

Material science researchers have developed many types of photovol-taic cell constructions.[4] All photovoltaic technologies absorb the sun's electromagnetic radiation and convert a portion of this energy into an electrical current. Inside a PV cell, electricity is generated when electrons in the cell's PV material (i.e., silicon) absorb sufficient energy from a pho-ton strike to break their valence bond. This free electron, with higher energy, is now able to move as current. Stacking cells made of different PV materials, each with unique spectral qualities, in multicell configura-tions has set efficiency records for the portion of solar energy converted into electricity, reaching 44%.

No fewer than four different categories are finding their way into today's PV market: crystalline silicon, thin-film, multijunction, and con-centrated PV. Crystalline silicon panels are the rectangular panels we commonly see on rooftops, while flexible thin-film technology can be incorporated into building materials, becoming less apparent. Multijunc-tion technologies stack different cell types like pancakes to improve effi-ciency, allowing them to set world records. Finally, concentrating solar power does what you would expect; it focuses the solar radiation into a cell to maximize performance. These systems are deployed with track mechanisms in order to follow the movement of the sun.

Several specifications are important in comparing solar PV products: module efficiency, power rating, and fade rates. As an example, a typical warranty would guarantee a minimum of 80% of the panels' rated out-put for 25 years. Locations vary by air moisture, snow and ice, and solar intensity; consequently site-specific warranted life is most important.

Crystalline silicon modules, with 90% market share, have seen steady improvements in efficiencies, going from 14% to 22% between 2009 and 2016. Still, this is a surprisingly low figure when headlines talk of breaking the 40% barrier and targeting the 50% level. The waterfall chart in Figure 7.1 depicts the factors that contribute to the loss of efficiency in PV technology. Pre-photovoltaic losses occur due to shading, snow, dirt, and reflection. System losses occur in the wiring, inverters, and other components of the system. The largest losses occur when photons on either side of the cell's spectral sweet spot fail to cause a PV effect. Thermal losses increase with higher operating temperatures—an unfortunate relationship given that high solar radiation implies higher temperatures.

Worldwide solar PV capacity reached 177 GW in 2014, up from 138 GW in 2013 and nearly 80% above capacity levels just two years before. According to a REN21 report, 55% of the world's 2014 investment in renewable energy was spent on solar, followed by wind at 36%. China, Japan, and the United States are leading the world in solar investments, while the historically active markets in Western Europe have recently slowed. Encouragingly, developing countries are starting to make significant solar investments.

Even more encouragingly, subsidy-free solar PV is competitive with retail electricity prices in 19 markets and below retail prices in Australia, Brazil, Denmark, Germany, and Italy. Whereas the costs of CSP are best compared with wholesale energy costs, distributed PV is best compared with retail pricing to gauge its competitiveness. The vast majority of the capacity is in off-grid and rural areas, but as of 2014 there were 53 solar PV plants (>50 MW each), representing a total capacity of 5.1 GW (<4% of PV market). The largest PV plant is a 320 MW facility in China colocated with a large 1,280 MW hydro plant. Where the economics of solar competes favorably with conventional sources, impressive contributions are beginning to occur. Solar PV generated 7% of Italy's electricity and 5.3% of Germany's, well above the worldwide average of less than 0.5% in 2012.

Thin film, considered second- and third-generation PV, has a small share in today's market for a number of reasons but most notably because of its lower efficiency. However, thin film partially offsets these disadvantages with a superior temperature coefficient, allowing it to operate more efficiently in higher temperatures. Also on the plus side, thin films made of flexible semiconductor material broaden the range of installations and simplify the manufacture for this class of PV. Recently the solar efficiency gap has been closing, and thin film may be poised to grow in market share

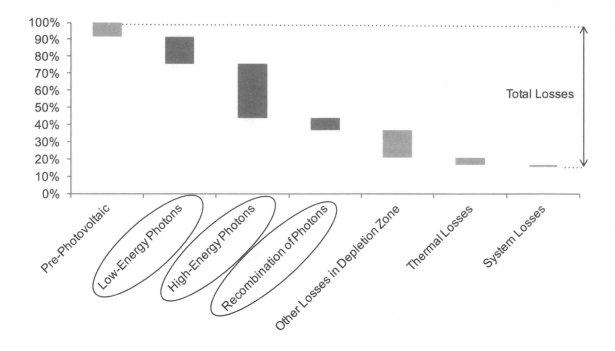

Figure 7.1. Sources of energy losses with solar photovoltaic systems. *Source Data:* Green Rhino Energy, "Energy Yield and Performance Ratio of Photovoltaic Systems," 2013, http://www.greenrhinoenergy.com/solar/technologies/pv_energy_yield.php.

on the basis of some of these unique advantages. Versatile thin-film technology can be aesthetically integrated into building designs, becoming both an architectural element and a source of electricity. This will offer more surface options, and most importantly it will utilize existing spaces.

Innovation is advancing a whole new category referred to as biological PV, a microbial fuel cell. Colonies of specialized bacteria convert sunlight into an electrical current. This technology is early stage, but it could unlock whole new types of applications.

Solar PV: A Scalable Technology

Figure 7.2 shows solar PV system pricing, which includes modules, installation, and a balance of system components for selected countries. The worldwide weighted system average is just over $2.00/W and is expected to fall to $1.00/W by 2050. A typical residence would require somewhere between 2,000 and 10,000 W of capacity. The pricing in Italy, Australia, and Germany is stimulating installations, while higher pricing elsewhere poses challenges for broad adoption absent financial incentives. China has the lowest solar PV pricing but thus far has prioritized

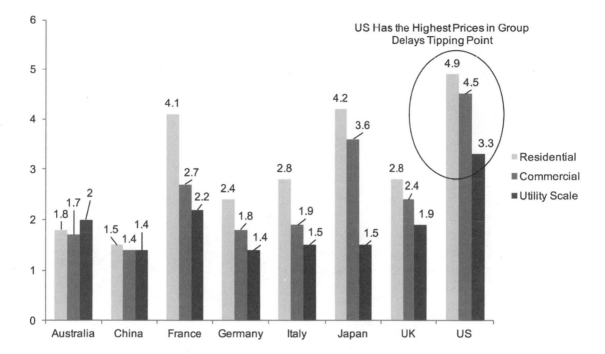

Figure 7.2. Solar photovoltaic system prices (US$/W), 2013. *Source Data:* International Energy Agency, *Technology Roadmap: Solar Photovoltaic Energy—2014 Edition* (Paris: Organisation for Economic Co-operation and Development/ International Energy Agency, September 2014), accessed February 27, 2015, https://www.iea .org/publications /freepublications /publication/Technol ogyRoadmapSolarPhoto voltaicEnergy_2014ed ition.pdf.

other renewable technologies. Japan has addressed the challenge with feed-in tariffs designed to stimulate residential and commercial installations, and builders include solar power to differentiate their product. Encouragingly, worldwide PV module pricing that was slightly less than $1.00/W in the past couple of years is now projected to be in the range of $0.55/W–$0.65/W. The majority of the system costs will be installation labor, warranties, and a balance of system components such as inverters. Utility-scale solar PV installation pricing is marginally lower than residential pricing, underscoring this technology's scalable quality.

For reference, the average retail prices in the United States for electricity were 12.5¢/kWh in November 2014.[5] Examining the rooftop solar PV levelized cost projections (in Figure 7.3) between now and 2040, solar PV around the world will provide savings over conventional sources. Similarly at the utility scale, solar PV systems will provide savings by 2040, or earlier if conventional costs increase. Solar PV pricing comes down when manufacturing is able to scale up, while most adoption only occurs with demonstrated savings. Activation energy is needed to spur change, and absent sustained policies including financial incentives, adoption will be deferred.

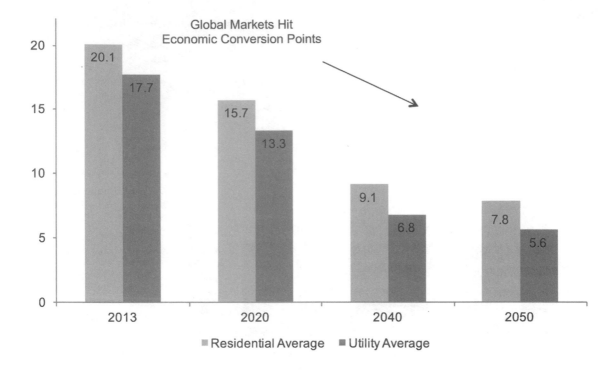

Figure 7.3. Average levelized cost of electricity (US¢/kWh) for rooftop and utility-scale photovoltaic systems. *Source Data:* International Energy Agency, *Technology Roadmap: Solar Photovoltaic Energy–2014 Edition* (Paris: Organisation for Economic Co-operation and Development/ International Energy Agency, September 2014), accessed February 27, 2015, https://www .iea.org/publications /freepublications /publication/Technology RoadmapSolarPhotovoltai cEnergy_2014edition.pdf.

The economic return of a new rooftop installation will depend on many factors: where you are located, your site's orientation and quality, the utility cost of electricity, and the total installation cost of your system. There are many online calculators that can estimate your location's average daily insolation (kWh/m²/day) for each month of the year. For instance, in Salt Lake City, Utah, the amount of solar energy that on average hits the surface is 2.3 kWh/m²/day and 6.3 kWh/m²/day for the months of January and June, respectively. A solar module that is 22% efficient, then, will generate 0.5 kWh of electricity per day per square meter for the month of January. Determining the economic sense of installing solar panels depends on the electricity it generates, the local price of electricity, and the cost of installing the system. Location by location around the world, the return economics of PV are turning positive.

Capturing Energy from Developed Spaces

As discussed previously, land use should be an important consideration in the solar technology we choose to deploy. Will the installations be on multi- or single-purpose surfaces? Solar PV is more attractive than CSP for at least two reasons; it can be deployed in multipurpose surfaces, and it converts higher percentages of solar radiation. How we mix the multiuse and single-use surface deployments of solar will reflect both economic and value judgments. The tendency to refer to renewables as green technology may obscure the need to apply value judgments to their deployment. In early 2016, a project was announced in France that will resurface 1,000 km of an existing roadway with purpose-built PV panels called Wattways, the essence of multiuse. This is the world's most ambitious plan to harvest solar energy from a highway, and the project hopes to demonstrate the large-scale practicality of the application in terms of both durability and efficiency. The expectation is that for every 4 m, the solar highway will generate the annual electricity demand of a typical French residence excluding heating.

If we are ultimately replacing fossil fuel, there will need to be upward of 1 million km^2 of solar PV collection. That is a lot of material. Are the raw materials in the collection technology plentiful and recyclable? Current thin-film technologies employ critical raw materials in their semiconductors, which limits their capacity to the gigawatt scale. Later in chapter 27 I note that as we confront the depletion of economical fossil fuels, we need to shift attention to establishing comprehensive recycling programs to secure sustainable alternative products.

Distributed PV: Bringing Energy Sourcing Closer to the Consumer

When people say that solar PV energy costs too much, it is important to know the context of their position. Are they taking the perspective of the power generator, the utility, or the consumer? Distributed sources, such as solar PV energy, are disruptive to the power sector's business model. A residentially installed and owned solar PV system cuts out middlemen and their margins on the electricity they generate. In the long term, utilities are going to adapt to the opportunities of distributed energy sources or participate in a shrinking market.

If you are concerned about climate change and air quality, you may be motivated to install solar panels to help reduce CO_2 emissions, and economics may play a secondary role in your decision making. Roughly, for every 1,000 kWh of solar PV electricity generated, the consumer avoids a half ton of coal, which would have produced a ton of CO_2. This may be startling, but remember from chemistry class that the atomic weight of carbon is 12, while the molecular weight of molecular CO_2 is 44, and the carbon content of coal is generally around 80%.

Looking forward to 2050, solar PV capacity is expected to reach 4,600 GW, up from the 177 GW base we have today. That will require worldwide investments of upward of 200 GW of capacity per year. Even at this impressive pace, in 2050 solar PV systems will still only generate 16% of the world's electricity.

No Limits to the Imagination

Harnessing solar energy with our earthbound technologies is just the beginning. Innovators are already envisioning space-based outposts that capture and deliver the sun's offerings. Concepts range from orbiting to moon-based stations that escape the temporal qualities imposed here on the earth's surface. We are unnecessarily bound to fossil fuels when we defer acting on these realizable alternatives for sustainable energy collection.

Notes

1. Energy.gov, Office of Energy Efficiency and Renewable Energy, "Dish/Engine System Concentrating Solar Power Basics," August 20, 2013, accessed April 13, 2015, http://energy.gov/eere/energybasics/articles/dishengine-system-concentrating-solar-power-basics.
2. National Renewable Energy Laboratory, Concentrating Solar Power Projects, "Ivanpah Solar Electric Generating System," November 20, 2014, accessed April 13, 2016, http://www.nrel.gov/csp/solarpaces/project__detail.cfm/projectID=62.
3. Kent Harrington, "Molten Salt Gives Concentrated Solar a Unique Advantage," American Institute of Chemical Engineers, October 25, 2013, accessed April 13, 2015, http://www.aiche.org/chenected/2013/10/molten-salt-gives-concentrated-solar-unique-advantage.
4. International Energy Agency, *Technology Roadmap: Solar Photovoltaic Energy—2014 Edition* (Paris: Organisation for Economic Co-operation and Development/International Energy Agency, September 2014), accessed February 27, 2015, https://www.iea.org/publications/freepublications/publication/TechnologyRoadmapSolarPhotovoltaicEnergy__2014edition.pdf.
5. Ibid.

8

Wind

Pray look better, Sir . . . those things yonder are no giants,
but windmills.
—Miguel de Cervantes

We've come a long way since the Persians first conspired to harvest energy from nature's winds some 2,000 years ago. Modern wind turbines appear as enormous flying machines firmly tethered to the ground, tirelessly grinding out electricity instead of taking flight. Although a few turbines point away from the wind, most are designed to point directly into the wind. Like birds, they are smart machines that constantly measure the wind, adjust their angle when the wind direction changes, and even alter the pitch of their wings to optimize performance for a given wind condition. Imitating birds that won't fly in extreme conditions, wind turbines turn off to avoid damage.

They are massive machines with hub heights (distance from the ground to the turbine's center) as high as 90 m, blade diameters as wide as 100 m, and power capacities up to 8 MW. To get a sense for the production potential of the world's largest wind turbines, let's compare them to a representative nuclear power plant. Considering actual production, it would take roughly 1,500 2-MW wind turbines to match the annual electricity generated from a single nuclear plant with a power rating of 1,000 MW. These immense turbines are most often installed in groups, both on- and nearshore, on wind farms in the business of selling electricity. Much smaller models, typically less than 100 kW, can be used to generate electricity in isolated environments such as on ships at sea and can directly feed electricity into homes, commercial buildings, farms, and schools. In 2013, 2,700 of these smaller turbines were installed in the United States at an average cost of $7.00/W, as compared with U.S. installed solar PV costs of $5.00/W.

Perhaps you can recall playing with pinwheels as a child. We formed wind when we exhaled and watched the pinwheel spin furiously with a good puff. It was amusing how much of our breath the pinwheel could translate into an artful spin. A perfectly designed field of wind turbines could theoretically still the wind. Fortunately, turbines are less than perfect. Nevertheless, every unit of energy gleaned dampens a portion of the wind as it passes. We locate the wind farms where the wind likes to run, and the wind exits a little less potent.

If you've ever noticed a wind turbine in operation, you've probably wondered how a blade moving so slowly could produce much in the way of electricity. A gearbox inside the turbine converts a blade rpm of 30 to 60 to a high-speed shaft rpm of 1,000 to 1,800. The high-speed shaft drives a generator that produces the electricity. Historical data collected from wind farm installations report average capacity factors of around 30%, and this figure is improving. The capacity factor is the percentage of the actual energy produced compared with a turbine's full energy potential (per nameplate power rating). A fossil fuel power plant's capacity factors can be dialed in, but nature determines the wind farm's capacity factor. A 30% factor means that on a typical day, a turbine is producing electricity roughly eight of every 24 hours.

By the end of 2014, wind farm installations were producing significant portions of several countries' total electricity. In Europe, for instance, Denmark, home to the world's leading turbine manufacturer, was producing 39% of its electricity from wind, while Portugal, Spain, and Ireland were producing 15% or more. Although the United States and China have large shares of the world's installed wind capacity, they generate only 3.5% and 2.0%, respectively, of their total electricity. Through the end of 2014 there were 370 GW of worldwide wind capacity, with 66 GW in the United States.

By the end of 2014, wind farm installations were producing significant portions of several countries' total electricity. Denmark was producing 39% of its electricity from wind.

There are many components to a modern wind turbine: yaw motors that keep the turbine pointed into the wind, a pitch system that adjusts the angle of the blades to keep rotation speeds within tolerances, and a gearbox that drives an electrical generator. Of course, all of these components take energy to keep the wind turbine operating.

Key wind turbine specifications include nameplate power (in MW), hub height (in meters), rotor diameter (in meters), and specific power (in W/m²), and these specifications have changed in the past several decades. Fifteen years ago in the United States, for example, the average nameplate capacity, hub height, and rotor diameter were 0.7 MW, 55 m, and 48 m, respectively. By 2014, average wind turbine specifications had increased to 1.9 MW, 83 m, and 99 m. Since surface resistance diminishes the wind, higher hub heights are able to tap higher power potentials at a given location. Wind power varies directly with the cube of the wind's speed; a doubling of wind speed results in an eightfold increase in power. So, hub heights became influential purchasing factors for wind farms as well as the nameplate power ratings.

Recently, however, wind turbines with lower specific power (W/m²) ratings have taken increased market share. Originally, lower-specific-power turbines were designed for lower-quality wind locations. New wind farm construction is now purchasing these low-specific-power turbines for all qualities of wind locations. While higher power ratings require high wind speeds, the occurrence of high wind is less frequent. Wind farmers are seeking to maximize energy production, optimizing between producing energy at lower power at lesser wind speeds on more days and producing more energy at higher power on fewer days.

Any increase in a wind farm's production directly reduces the cost per unit of energy. The market is striving to optimize the hours of operation and energy production to maximize the net electricity generation. How many times have you driven by a wind farm whose giant turbines stand still despite a fair amount of wind? A lower-specific-power turbine at that site would perhaps be generating energy under those conditions. Wind turbine technology has advanced, and we are witnessing the results of the learning curve.

Unlike solar technology, which is rapidly changing, modern wind tur-bines are further along in their development, with commercially viable technology and economically competitive energy production. The prob-lem is that wind energy is temporal and not predictable—thus the value (not the cost) to purchasers of electricity is lower since they also need to secure baseload supplies (such as coal) for the wind farm's rendition of sailors caught in the doldrums.

Try as we may to limit liabilities, every energy source has them. Being classified as a renewable energy source doesn't provide immunity against having negative impacts.

Because the Winds Can Be Stilled, Wind Energy Has a Lower Value

Wind-generated electricity is becoming one of the world's lowest-cost supplies, but this comes with a caveat: that we use it directly. As a fluc-tuating source its value is less because it isn't necessarily there when we need it. Given its low cost but lower relative value, wind capacity will hit a limit without parallel investment in storage and the continuation of financial incentives. Examples of financial incentives in the United States include production tax credits and the investment tax credit. According to the U.S. Department of Energy 2013 *Wind Technology Market Report*,[1] the value of these financial incentives is about $15/MWh. On the reve-nue side a wind farm negotiates a power purchase agreement (PPA) with an electricity purchaser for a certain dollar amount per MWh of energy. These agreements then allow the developer to concentrate on the pro-duction side of the project, while the purchaser gains compliance with state-level renewable portfolio standards.

Figure 8.1 from the same report shows the wholesale price ranges and the long-term 2012–2014 levelized PPAs for wind projects across regions of the United States. The percentages next to the region names represent the portion of total U.S. wind capacity through the end of 2014. Back in chapter 4, a map of wind speeds at 50 m revealed that the interior of the United States has the greatest onshore potential. So, it is not surprising to observe in Figure 8.1 that the interior region (on the left) has the lowest average levelized wind PPA prices. That the PPAs are below the wholesale price range bodes well for further market adoption. This is not the case with the other regions. Thus far financial incen-tives have been crucial in the development of wind capacity, and absent

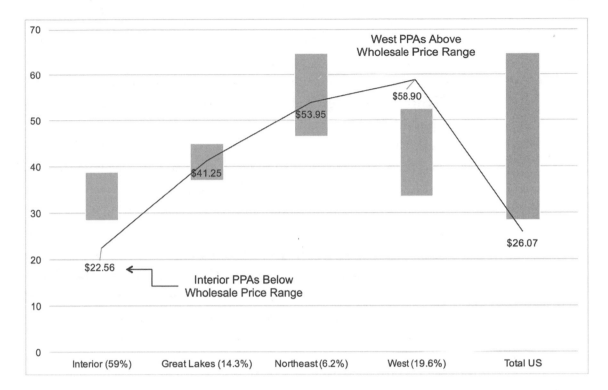

these incentives, PPAs would need to be even higher for continued investment, and that seems unlikely. So, sustained investment depends on the belief that conventional prices will increase, economical storage capacity will emerge, and incentives will remain in place. Investors, for now, are embracing that belief!

A Look Ahead for Wind Power

Turbine technology is maturing; efforts are now focusing on improving capacity factors, reliability, and reducing costs. Wind energy has already established remarkable production levels in early-adopter nations. Offshore wind energy production comes with increased costs related to construction, transmission, and susceptibility to harsher operating environments, which affects reliability. The first large-scale offshore wind farms are moving into production (i.e., Germany), and their performance will be informative for gauging the commercial readiness of the current

Figure 8.1. Levelized long-term wind power purchase agreements (PPAs) in 2012–2014 and wholesale electricity price ranges (US$) by region.
Source Data: Ryan Wiser and Mark Bolinger, *2013 Wind Technologies Market Report*, U.S. Department of Energy, Energy Efficiency and Renewable Energy, August 2014, accessed February 26, 2015, http://emp.lbl.gov/sites/all/files/2013_Wind_Technologies_Market_Report_Final3.pdf.

technology. Around the world several projects are implementing floating wind platforms for nearshore locations.

Wind farms are confronting opposition as they extend their placements, just as nuclear and fossil fuel projects face resistance. Wind farm opposition cites damage to wildlife, the landscape, and even human health. One oft-cited concern, for example, involves reported bird kills in certain locations along migration pathways. By far the largest source of human-caused bird kills is the buildings we occupy. Necessity is the mother of invention, and now there is a patterned coating for glass that is transparent to the human eye but visible to birds. Similarly, with wind turbines, there is an alternative approach that may resolve the problem in these vulnerable locations. To avoid bird kills, a new-concept tower design captures and concentrates air through a funnel, where isolated internal turbines collect the energy. Try as we may to limit liabilities, every energy source has them. Being classified as a renewable energy source doesn't provide immunity against having negative impacts.

A heads-up going forward is the use of rare earth elements such as neodymium and dysprosium in turbine systems. They improve both reliability and the generation of electricity, but they are on the European Union's critical raw materials watch list.[2] Neodymium is characterized as limited availability with future supply risk. Chapter 27 will discuss this further in the context of scarce materials and the need to build robust and sustainable recycling capabilities.

Finally, at times peak renewable production and peak consumer demand are in synch, but that is a coincidence. If ever there was an independent variable, it is energy production with temporals. Until our power networks can be improved to store and retrieve energy to meet demand throughout the day, the value paid for temporals will be depressed, and conventional sources will be preferred. Establishing higher and higher levels of wind energy is intertwined with storage technology, grid technology, and changes in demand-side behavior. Wind is already bumping up against these constraints by virtue of curtailment. Curtailment is manifest when our systems are not keeping pace with temporal capacity. Lagging infrastructure, storage capacity, and smart grid investments is the Achilles' heel to wind power achieving both its full economic value and its capacity potential.

Notes

1. Ryan Wiser and Mark Bolinger, *2013 Wind Technologies Market Report*, U.S. Department of Energy, Energy Efficiency and Renewable Energy, August 2014, accessed February 26, 2015, http://emp.lbl.gov/sites/all/files/2013__Wind __Technologies__Market__Report__Final3.pdf.
2. European Commission, "The European Critical Raw Materials Review," press release, May 26, 2014, http://europa.eu/rapid/press-release__MEMO-14-377__en.htm.

9

Nuclear

When the mind is burdened with a conclusion, a formulation,
there is the cessation of inquiry.
—J. Krishnamurti

First, as Things Are

I remember as a university student in the mid-70s learning about a plan to safely dispose of spent nuclear fuel. A surprisingly small amount of waste resulting from the conversion of a huge amount of nuclear energy, what looked like a glass pellet, would be dropped into a deep layer of stone, where its initial temperature would liquefy the stone. Within a few years, however, the radioactive decay would be diminished enough for the stone to return to a solid, "forever" encapsulating the waste. At the time it sounded reasonable, at least to me. I left the class as a would-be engineer, secure in a belief that mankind's ingenuity could solve anything.

Of course all these years later the problem isn't solved. Our nuclear waste is stored in stockpiles all around the world. If one is concerned with safety from either unintended human error, malicious, or natural disasters, it seems we are taking the greatest possible risk as things are today. We can think about the globally sprinkled stockpiles of spent fuel as analogous to the Deepwater Horizon oil spill, but with a major difference. BP was assigned responsibility to clean that spill up, while nuclear power still accumulates its waste at reactor sites, with no apparent plan to guard against environmental exposure. In the United States, nuclear power plants are charged fees for every unit of generated electricity. The revenues are theoretically there to fund a viable long-term spent fuel management plan. So, we build the nuclear plants, we generate electricity from these plants, and we collect money from these plants, yet we have no evident pathway for the waste, and leadership shrinks from engaging the public in establishing a plan. Even the use of the term *radioactive* has moved beyond its atomic roots to describe topics that are untouchable, which explains politicians' aversion to discussing nuclear power. Thus, many of us who might be open to nuclear power remain conflicted on operational nuclear due to the lack of an actionable plan for spent fuel management. We are storing as opposed to disposing of spent fuel as though it will one day be repurposed—which it quite likely will be.

Another concern is the risk of a nuclear plant event, and Fukushima in Japan is the most recent reminder of this concern. Unlike the images of oil-smothered birds and plumes of oil spilling into the Gulf of Mexico, the toxic radioactive releases from the Fukushima plant disaster were invisible. It could have been a lot worse, and it is a wake-up call regarding the potential of other catastrophic events at nuclear facilities where flaws in design and/or operation lie in wait. There is a lack of standardization in plant designs where maximum safety requirements are needed.

Civil nuclear power based on fission will either alter its decadently poor use of fuel or fold its operations this century and leave behind spent fuel stockpiles that will require safeguarding for thousands of centuries. Let's now imagine hearing about a newfound discovery from which the world could safely harvest vast amounts of energy stashed inside atomic nuclei. Secure atomic batteries would deliver energy for 30 years with no air pollutants or greenhouse gas emissions, and securing the radioactive waste stream would be a matter for 200 years rather than for thousands of centuries. Imagine also that we could discuss this new energy source outside of the context of cataclysmic nuclear weaponry. Wouldn't we consider this an exciting area of research and hope for the future? Fortunately, some intriguing work in alternative fission has been and is continuing to be done. This work is referred to as Generation IV (Fast Reactor), and when delivered in a small modular reactor it gives nuclear fission a pathway to relevance.

Thermal-neutron reactors' most glaring deficiency beyond the appallingly inefficient use of uranium fuel is the consequential accumulation of waste that is seemingly toxic for an "eternity."

Let's take a cursory look at how today's thermal-neutron reactors work before we review the principles, benefits, and potential application of this fast reactor technology. Most of the world's nuclear reactors are referred to as thermal-neutron because the expelled neutron (central to the chain reaction) is slowed to achieve thermal equilibrium with the nuclear fuel. A moderating material (i.e., water) absorbs some of the neutron's kinetic energy. Unfortunately, slowing the neutrons comes with consequences that either were not fully understood or were not appreciated by the inventors. The neutron dispatched from a fission event is tuned to the use of fissile uranium (U-235), which makes up only 0.72% of naturally occurring uranium ore. The thermal-neutron reactor's most glaring deficiency beyond the appallingly inefficient use of uranium fuel is the consequential accumulation of waste that is seemingly toxic for an "eternity." The pioneers of nuclear, however, were not as concerned about fuel efficiency or the management of spent fuel as they were about the ability to breed weapons material. This is quite understandable since a nuclear weapon had recently been used, to the world's dismay. In a "fast" reactor the neutron is not slowed, and thus it can breed fissile plutonium from fertile uranium (U-238). So, a superior technology that is 60-fold more efficient in the use of uranium and produces a fractionally small percentage of waste was potentially tabled because it could breed fissile material.

With thermal reactors the naturally occurring uranium is a sideshow, whereas it becomes a centerpiece with plutonium in the fast reactor, and this delivers the higher fuel efficiency and the dramatic reduction in the waste stream. Guarding against the production of weapons material is a legitimate concern, but it's also time to manage civil nuclear power's waste stream. Today there are advances that significantly mitigate both the concern with breeding fissile material and the accruing of radioactive waste.

Radioactive transuranic elements apart from plutonium (minor actinides) are also part of the reason we need to manage nuclear waste. Generally, the transuranic elements with atomic numbers from 93 through 96 are produced from fission reactions, while the larger elements are created using particle accelerators. Transuranic waste is

> *The majority of waste from all thermal-neutron reactors would be fuel to a fast reactor.*

produced when neutrons are absorbed without a subsequent fission reaction. These transuranic elements can be consumed as well within a fast reactor.

In chapter 3 we reviewed an example composition of spent fuel from the Japanese Atomic Industry Forum (Table 3.1). Radioactive uranium made up 95% of the waste, and another 1.5% comprised long-lived by-products. The half-life of U-238 is 4.5 billion years, while some of the long-lived elements are in the millions. Together they ostensibly pose health risks forever. It makes one wonder whether or not a viable solution can ever be found with things as they are; nobody wants spent nuclear fuel (SNF) in his or her backyard. But the lack of a solution doesn't seem to be slowing most countries down. Sooner or later these stockpiles will contaminate our environment. Given our use of nuclear power over the past 50 years, there is an obligation to deal with the waste, past, present, and future. The promising news is that with fast reactors, we can trade a liability for an advantage. Fast reactors consume uranium and the transuranic elements as fuel, and that significantly reduces both the quantity and the long-term radioactivity of the nuclear waste. The vast majority of what is waste from the world's thermal-neutron reactors would be fuel to a fast reactor. If nuclear fission is to play a relevant role in the world's energy mix beyond the twenty-first century, the fast reactor technology must be successfully commercialized.

Fast Reactors: The Outlook

Generation IV (Fast Reactor) research and development has been undertaken for decades and now focuses on various designs for cooling the reactor. In January 2014, the Organisation for Economic Co-operation and Development's Nuclear Energy Agency published a status report on Generation IV projects conducted by 12 participating countries.[1] This report includes an update on the progress of six different designs for fast reactors and reveals that three of the six designs are fairly well advanced into the performance testing phase. Performance testing means that the technology is at an engineering scale where there is both verification and optimization of design.

There are two attractive benefits to fast reactor designs that happen to address both of the concerns stated earlier: plant safety and spent fuel management. Argonne National Laboratory demonstrated in 1986 that its fast reactor fuel format naturally shut down the reaction when the cooling systems failed.[2] As the temperature increased in the reactor, the fission reaction stopped. This was due to the design of the metallic fuel format, where the heat expansion increased the distance between nuclei of the fuel to such an extent that the neutrons no longer hit their targets and the reaction stopped.

Complementing the enhanced fuel formats are different methods for removing heat from the reaction chamber, for example, liquid metals. One of the designs uses sodium to cool the reactor. Sodium remains a liquid across nearly 800 degrees (C), as opposed to water's range of 100 degrees (C), and this difference expands a plant's operating temperature range and confers a measure of safety. The thermal conductivity of sodium is substantially higher than that of water, and this improves the coolant's ability to remove heat. However, liquid sodium explodes when exposed to water and spontaneously ignites with air, so again, every good idea has its risks. Lead-cooled fast reactors have some technological challenges related to refueling and servicing but offer attractive heat management properties. Validating, selecting, and standardizing a reactor coolant is the crux of the work necessary prior to the broad commercialization of Generation IV fission.

Arguably, the biggest advantage of fast neutron reaction is its ability to consume our existing stockpiles of spent fuel waste. A fast reactor's fission event (Pu-239) generally releases more neutrons, and these extra neutrons not only support subsequent fission but also convert what are called fertile elements into fissile elements. A fertile element can absorb a neutron and transmute to become fissile. Through this process, fast reactors actually consume both uranium-238 and transuranic elements. Sellafield, in the United Kingdom, stores 120 tons of plutonium from spent fuel. GE Hitachi Nuclear Energy is proposing to deploy its PRISM fast reactor to deplete plutonium stockpiles while also producing enough energy to supply electricity for 500 years![3]

Summarizing, fast reactors with mixed oxides of uranium and pluto-nium fuel, metallic fuel formats, superior fuel efficiency, and liquid metal coolants address the principal concerns with thermal-neutron reactors, and the technical capability seems close at hand for the world's nuclear innovators. Globally, small-scale fast reactors have a combined 400 reactor-years of operation and are a high priority in many countries. Russia, a member of the Organisation for Economic Co-operation and Development Generation IV team, has recently started the world's larg-est fast reactor, the BN-800. Imagine a nuclear plant without nuclear waste! The plant has a power rating of 780 MWe, and in December 2015 it was brought on line with the service grid at 35% capacity. The plan was to ramp the capacity to 100% over the course of the first few months. Less advanced, the U.S. Department of Energy is soliciting fast reactor design proposals, with prototypes anticipated by the 2030s. China has recently joined Russia in performance testing a smaller-scale fast reactor. Finally, France generates an amazing 75% of its electricity with nuclear and has set goals to replace 50% of its thermal-neutron reactor fleet with fast reactors by 2050. As a large practitioner of civilian nuclear, France has plenty of spent fuel it can repurpose for fast reactors.

Even if you don't think we should be using nuclear energy, we should pursue fast reactor technology research to responsibly deal with spent fuel stockpiles from power, military, and research applications. Some esti-mates predict that using existing stockpiles of spent fuel in fast reactor plants could supply hundreds of years of nuclear power.

If we had known in the mid-twentieth century what we know now, we might have deployed fast reactors as opposed to thermal-neutron reactors from the start, and "nuclear" would not have its current neg-ative reputation. We would have far fewer stockpiles of waste, posing risks for hundreds of years as opposed to thousands of centuries. We would have less need for uranium ore because the reaction is nearly two magnitudes more fuel-efficient. We'd have no weapons-capable stockpiles that pose national security risks and probably wouldn't be burning as much fossil fuel. Can we, though, mitigate the valid risk of weapons material production with this technology? Some believe that deployment of the small modular reactor, a nuclear battery, can play a role in addressing this concern.

A Nuclear Battery

A joint report from contributors in the United States, France, and Japan describes in great detail a design for a 50-MW modular fast reactor.[4] The benefits of the design include the ability to consume spent fuel, incorporate the passive safety benefits of metallic fuels, and significantly increase the energy potential of nuclear fuels. Think of the small modular reactor as a nuclear battery with a 30-year lifetime.

The team's goal was to simplify and improve the operational and safety measures of nuclear power generation. The initial fuel cartridge sourced from radioactive waste would provide energy for 30 years without refueling. Nuclear plants that routinely handle spent fuel combined with on-site storage are points of exposure to diversion risk. The small modular fast reactor avoids this risk while reducing worldwide inventories of spent fuel. The cost of nuclear power is relatively expensive, and left unaddressed this alone will limit its role in the future. The other benefits of a small modular reactor include the standardization of design and construction and the ability to reduce time to commercialization, and this may help improve the economics of nuclear power.

Thorium: A Fertile Fuel

An alternative nuclear reaction receiving attention uses thorium as a fuel. Thorium is fertile and needs to be used in conjunction with fissile material to operate. Three times more abundant than uranium, thorium can work in many types of reactors including fast reactors. This can substantially increase the energy potential of nuclear fuel. Beyond extending the energetic supply of nuclear fuels, the use of thorium has less risk of nuclear weapons proliferation than uranium reactions and produces less long-lived radioactive waste. Funding for the feasibility of a thorium reaction within a fast reactor environment would make a lot of sense, but we have plenty of time for that work, assuming we begin depleting the world's SNF stockpiles.

With Nuclear Technology, the Past Should Not Hold the Future Hostage

With all the promise and advantages of the fast reactor, the question you may be asking is, Why the continued use of thermal-neutron reactors? It is hard not to conclude that we intentionally de-optimized civil nuclear power to avoid proliferation of weapons-capable plutonium. Civil nuclear power and nuclear weapons development were born at the same time, and one strongly influenced the other. We deploy thermal-neutron reactors that utilize less than 1% of the natural uranium ore and produce long-lived radioactive waste as opposed to more capable designs that are safer and more fuel-efficient, though they can also be used to breed plutonium—central to bomb making. It seems that every choice in energy has a thorn, and this is a big one. According to the World Nuclear Association, the world's current fleet of 430 nuclear reactors in 30 countries is set to increase, with 63 reactors under construction, another 165 in planning, and finally another 316 under proposal. The number of countries practicing civil nuclear power will increase to 48, and while a portion of these reactors are fast reactors, most are still thermal-neutron reactors. The world is in a nuclear transition; many countries are building safer thermal-neutron plants, while others are just beginning to implement the more compelling fast reactor technology. Building a thermal reactor today would only make sense if the plan was to later migrate to a fast reactor.

Most would agree that we have an obligation to address nuclear waste, no matter one's opinion on nuclear energy. Lacking a collective plan to dispose of spent nuclear fuel, countries are storing radioactive waste at sites all over the world. This seems counterintuitive to minimizing the exposure risk. The continued commercial use of nuclear fission should be contingent upon a rational plan and action to manage its by-products. The man-made pyramids of Egypt are 5,000 years old, while SNF will require secure housing for thousands of centuries as things are.

Ideally, over 96% (by weight) of the waste from a standard thermal-neutron plant could be used to convert nuclear energy (uranium and plutonium).[5] After decades of accumulating spent fuel with no practical solution in sight, the fast reactor finally offers a two-for-one answer: the simultaneous consumption of long-lived radioactive elements while producing vast amounts of energy. While acknowledging that almost everywhere the discussion of nuclear power is politically radioactive, many folks when properly informed on energy mix pros and cons would support fast reactor research and modular deployment upon successful validation. This is where steadfast government funding and support is vital.

Fusion: Combining the Nuclei of Small Atoms

Fusion, the hunt for energy's most elusive source, has been researched for decades, and as with the pursuit of flight, it is impossible to know when success will occur. One day, though, the fusion that powers our solar system will become a primary energy source on earth. For now, we must plan the near-term energy future with what we know.

Research and development of fusion is happening on several fronts. The ITER project, a collaboration of seven countries, is combining technology and financial resources (greater than $15 billion and growing fast) to pursue fusion power. The effort is seeking to demonstrate the technical feasibility of applying commercial-scale fusion to the generation of electricity. *Iter* is the Latin word for "the way," and if the project successfully demonstrates fusion, it will pave the way toward a clean and long-term energy solution. The project is building a 500-MW power plant in Durance River valley, France. ITER members include China, the European Union, India, Japan, South Korea, Russia, and the United States.

Specifically, the goal is to create a self-sustaining deuterium-tritium-burning plasma reaction that generates at least 10× the input energy.[6] The plasma temperature must achieve and hold at 150 million °C in order to sustain fusion—the temperature of the stars. The tokamak machine, the core of the plant, is scheduled to begin commissioning in 2019, and the first plasma production is scheduled to begin in 2020. The deuterium-tritium fusion reaction is scheduled for 2027. The project is struggling with schedule delays and significant cost overruns and recently changed leadership in hopes of correcting course. This is not unusual for early-phase development projects and reinforces that fusion cannot be assumed in the world's near-term energy planning.

Another effort that has already achieved a critical milestone is occurring at the Lawrence Livermore National Laboratory in California. This project is taking a different approach by utilizing a set of high-powered lasers directed at isotopes of hydrogen to force fusion. Encouragingly, at Livermore's National Ignition Facility, the project has been able to generate net positive energy. The next milestone is to achieve fusion ignition and positive energy gain. Ignition means that the fusion reaction is limited by the amount of fuel and the sustained input of energy to hold the necessary conditions. There are other efforts directed at fusion (i.e., Germany's Wendelstein 7-X project and China's Experimental Advanced Superconducting Tokamak), but all of them are several decades at best from commercialization.

Nuclear power converts millions more energy units per nuclei than does fossil fuel from its chemical bonds. So, solving fusion will be one of humankind's most significant developments in energy sourcing. Since fusion occurs naturally, there is reason to be optimistic that someday soon success will be achieved, but it is too soon to establish commercial timing. Fusion is a little like the pursuit of flight. For thousands of years humankind watched birds and contemplated flight in much the same way as we feel the sun's warmth and want to mimic its fusion process. We toiled over the centuries in pursuit of flight, not knowing when the breakthrough would occur. We created fabulously inventive prototypes for flight, persevered, and suddenly the pieces (i.e., propulsion) that enabled flight came together and the dream was realized. A mere 110 years or so after Orville and Wilbur Wright's flight, we are landing on comets, planets, and moons, and launching explorers beyond our solar system. Still, air travel is in its embryonic phase. Fusion research will challenge researchers as flight has done, and one day it will succeed. How coincidental it seems that just as the Wright brothers in 1903 were launching themselves into the air, a young Albert Einstein in 1905 discerned a wonderfully simple relationship, $E = mc^2$. Flight may be a critical invention for enabling fusion, since we may need to fly to the moon to gather the helium-3.

Notes

1. Nuclear Energy Agency, *Technology Roadmap Update for Generation IV Nuclear Energy Systems* (Paris: Organisation for Economic Co-operation and Development Nuclear Energy Agency for the Generation IV International Forum, January 2014), accessed February 26, 2015, https://www.gen-4.org/gif/upload/docs/application/pdf/2014-03/gif-tru2014.pdf.

2. George S. Stanford, "What Is the IFR?" Nuclear Engineering Division, Argonne National Laboratory, May 2013, accessed January 1, 2015, http://www.ne.anl.gov/About/reactors/ifr/What%20Is%20the%20IFR.25.pdf.

3. Fred Pearce, "Are Fast-Breeder Reactors the Answer to Our Nuclear Waste Nightmare?" *Guardian*, July 30, 2012, accessed May 12, 2015, https://www.theguardian.com/environment/2012/jul/30/fast-breeder-reactors-nuclear-waste-nightmare.

4. Y. I. Chang, C. Grandy, P. Lo Pinto, and M. Konomura, Small Modular Fast Reactor Design Description, report no. ANL-SMFR-1, Argonne National Laboratory, Commissariat à l'Energie Atomique, and Japan Nuclear Cycle Development Institute, July 1, 2005, accessed June 4, 2015, http://www.ne.anl.gov/eda/Small__Modular__Fast__Reactor__ANL__SMFR__1.pdf.

5. Japan Atomic Industrial Forum, "Spent Fuel and Radiotoxicity," accessed September 30, 2015, http://www.jaif.or.jp/ja/wnu__si__intro/document/2009/m__salvatores__advanced__nfc.pdf.

6. Osamu Motojima, "Status of ITER" (paper presented at the 34th Annual Meeting and Symposium, Fusion Power Associates, Washington, D.C., December 10, 2013).

10

Advanced Energy Storage Solutions

Today's storage technology portfolio could be described as little
more than a pile of coal and wasteful levels of capacity.

The difficulty in parting ways with fossil fuels is neither in the gathering
nor in the available supply of sufficient energy. The challenge, instead, is
the advancement of supportive technologies and infrastructure to marry
supply and demand. If your big bold energy plan is based on an ongoing
mix of fossil fuels with minor temporal contributions (<20%), as with the
United States, then solving storage is not critical. But if you are urgently
driving toward the goal of terminating the use of fossil fuels, nothing is
more important. Not focusing on storage is either an indication of short-
term planning or an ill-advised comfort level with the continued long-
term use of fossil fuels.

Our current power paradigm relies almost entirely on stored energy
conveniently in the form of fossil fuels. In the United States as an
example, its 1,000-GW power sector has only 2 GW (0.2%) of storage capacity. Such a small amount of storage works because utilities
can dispatch fuel-based plants. It can be argued that fossil fuels have
become the centerpiece of the world's energy supply primarily because
they can be transported and used on demand, two seductively attractive traits that convince us to tolerate their negative health and environmental characteristics.

Grid reliability is a measure of the power system's ability to deliver quantity and quality electricity to all points of consumption. How does the power system maintain a reliability of 99.9% for the consumer? It does so through three tiers of plants—baseload, load-following, and peaker—which together supply round-the-clock electricity. Baseload power delivers the minimum grid demand within a 24-hour period. Plants of this category are set to operate at an output that permits the most efficient generation of electricity. Coal plants, used for baseload, have ideal levels at which they convert energy and emit the lowest possible greenhouse gas emissions. During the day and early evenings, when demand increases, load-following plants are dispatched. Load-following plants are turned up or down to shadow demand. The cost of electricity is naturally higher during these periods. It's as if you chose to alternately step on your car's accelerator and brake to meet a random speed target; your fuel mileage will suffer. Fuel-based plants are intrinsically capable of meeting a dispatcher's request for more or less power. Peaker plants are brought on line when demand spikes above the capacity to load-follow. We see in chapter 28 that the cost of electricity from a peaker plant is multiples of the cost from a baseload plant.

Our utilities sustain such a healthy and functioning grid primarily through supply-side adjustments. Over the last century, the power segment has become an expert player in a game of copycat. When the first light switch is flipped, a distant energy source instantly delivers the electricity. The consumer makes a move, and immediately the power network mimics the move. In this game the consumer is the puppeteer. The costs are higher than they really need to be, and the swings in production costs are largely invisible to the consumer. Since supplies of fossil fuels have been plentiful and their costs are relatively cheap, we are content with this three-tier plant system.

When we integrate large portions of temporals, the complexity of the power sector changes. The future will see the emergence of demand-side adjustments based on taking maximum advantage of low-cost energy corridors. The consumer will respond to supply, possibly lowering demand at times through a gained awareness of supply-side dynamics. When

Not focusing on storage either indicates short-term planning or reveals an ill-advised comfort level with the continued long-term use of fossil fuels.

the supply is low, storage systems will retrieve energy to provide the difference. This also works in reverse. Times when the supply is high, the user may schedule energy-intensive activities, with any balance diverted to storage. The grid caught between these pushes and pulls will play the role of peacemaker.

Even with changes in consumer behavior we need storage for two purposes: (1) to provide hour-by-hour real-time load balance and grid regulation and (2) as backup for low production periods with renewables. No matter how much we learn to shift demand, there will remain applications that cannot be stayed. Hospital emergency rooms, traffic lights, and communications cannot be turned on or off like a reading light; robust energy management systems become crucial for the future.

During these early decades of the twenty-first century it is vital that storage technologies and infrastructure be established with an eye toward completely weaning the world from its fossil fuel dependence during the second half of the century. To some extent our situation is analogous to personal computing devices; without connectivity and memory our smart electronics are less useful, as are the renewable technologies without smart grids and storage solutions. The controversy is not about what should be done but, rather, with what urgency. A colleague recalled the frustrations of a marketing director who kept asking a software engineer if he could develop certain applications. The response was always yes, but nothing ever materialized. As frustrations mounted, the coworker suggested that next time the marketer might ask the software engineer a simple follow-up question that would clarify expectations: "Can this task be accomplished in our lifetime?" We are facing a similar scenario—yes, the challenge of storage can be solved, but when will we begin to tackle it in earnest? If the world does not explicitly place a stake in the ground on when fossil fuels will be retired as an energy source, we won't place the right urgency on laying a new foundation. The world is quite conflicted on the timing of the fossil fuel depletion window. More urgently than debating climate change and CO_2 reduction commitments, the world needs to debate and agree on the timing for the decarbonization of our energy sourcing. We need to harmonize

We need to harmonize and work with exigency in the short term in paving the way for a wholesale energy transition.

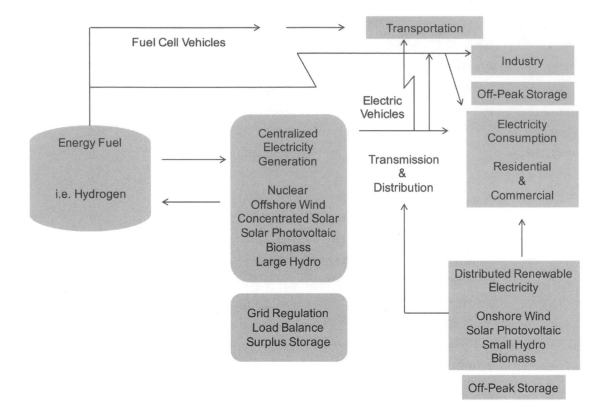

Figure 10.1. Energy architecture post–fossil fuels.

and work with exigency in the short term in paving the way for a wholesale energy transition.

Now let's look at Figure 10.1, which jumps to the future and depicts the energy sector absent fossil fuels—a picture both enlightening and somewhat intimidating. In this view, hydrogen emerges as a new fuel. Centralized power will utilize a combination of energy sources, large hydro, geothermal, nuclear, biomass, and hydrogen—as necessary to deliver electricity. Distributed locations that harvest energy will also integrate storage solutions. The figure depicts the vital role for products that store and retrieve energy; they are the threads that knit tomorrow's power sector together. Hydrogen plays a diverse role across the energy landscape—cooking, high-temp industrial processes, even as fuel in your car with fuel cells. In this chapter I will describe hydrogen's role in the power sector.

Storing a Lot of Energy in an Easily Retrievable Manner Is of Paramount Importance

Studies in Germany and California have shown that solar and wind technologies are often complementary in terms of when they produce energy. This is encouraging and suggests that when solar and wind capacities are coupled with smart grids, much of their energy could be used directly, thus avoiding the inherent inefficiency of storing and retrieving. As we finally depart from the use of fossil fuels there can be no doubt, though, that storage will be required. We've built societies incapable of living without energy, and temporals individually and in combination will always have downtimes.

Fortunately, many possibilities are on the horizon. Developing technologies will allow us to "make hay while the sun shines," so to speak, harnessing temporals' energy when it is available and setting it aside for later use to overcome their fluctuating character and seasonal variation. Storage also makes it possible to postpone capital investments in new capacity because stored energy can instead be called upon during peak demand periods.

Another argument against renewables is the inefficiencies of storage. But every storage method loses energy during the conversion process. Load-following and peaker operations with fossil fuels are inefficient. It is the cost we pay for our on-demand power sector. Anytime we use energy directly as opposed to recovering stored energy we take a savings. As discussed earlier, fossil fuels on average converted less than a percent of solar energy into chemical energy, and the average efficiency for converting that chemical energy into electricity is 36%. So when detractors comment on storage inefficiencies, it is relevant to keep in mind that our current fossil fuel economy is less than 1% efficient (0.36% [1% × 36%] efficient), far less efficient than the combination of temporals and storage technologies. Figure 10.2 shares the round-trip efficiencies for current energy storage technologies. Hydrogen has round-trip efficiencies between 40% and 50%, lower than that of the other technologies but offering other compelling advantages, including the useful employment of a portion of the "lost" energy within micro—combined heat and power

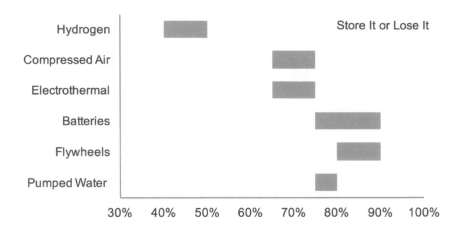

Figure 10.2. Round-trip efficiencies for energy storage. *Source Data: "Round Trip Efficiency," Energymag*, February 8, 2014, accessed August 31, 2015, http:// energymag.net /round-trip-efficiency/.

systems. The other technologies, with round-trip efficiencies up to 90%, will support shorter-term functions.

Storage Is the Key to Breaking the Hold of Fossil Fuels

As consumers we enjoy on-demand supplies of electricity. To deliver this experience the power sector has aggregated sources and overbuilt capacity to complement the inherent character of fuels. Will tomorrow's marketplace continue to deliver these consumer expectations at the cost of excess? Or will the market adopt a new approach where consumers shift some of their demand when supply is low? Renewables encourage the latter, and the next chapter discusses this further. Today's storage technology portfolio could be described as little more than a pile of coal and wasteful levels of capacity, and that would not be far off the mark. A new market is forming, though, where distributed energy sources will employ energy management solutions to gain grid autonomy—a market that today is already acquiring solutions from the fledgling storage industry.

There are many functions for storage technologies now and in the future. What is new is the hand-in-hand requirement for energy storage as more and more temporal sources come on line. When fossil fuels fall into disuse, the scale of baseload sources such as nuclear energy, large hydroplants, and biofuels in the power sector will dictate the requirements for long-term storage. The greater our ability to shift demand, the less the need for energy storage. But try as we may, we cannot adopt large portions of temporals without a lot of storage. The generation and transmission markets will be obvious end users, but distributed locations will most likely be the first adopters. Commercial and residential markets will couple renewable capacity with energy management solutions, which means storage, intelligent software systems, and electronic components.

There are at least seven classes of storage technologies, including pumped hydro, compressed air energy systems, thermal, mechanical (i.e., flywheel), electrochemical (i.e., conventional and flow batteries), phase change, and hydrogen. This chapter will investigate product categories with universal applicability, as they are so critical going forward. This includes both conventional and flow batteries along with hydrogen, a new-age fuel. The next chapter examines how smart grid technology can help balance supply and demand and coordinate the accumulation and retrieval of energy.

Hydrogen: A Recyclable Fuel

A complete hydrogen system requires technologies on four fronts—the ability to produce (i.e., electrolysis), transport (i.e., physical state and method), store (i.e., liquid, gas, organic compound), and retrieve (i.e., fuel cells or combustion) energy. The good news is that these technologies exist. Hydrogen is not the only fuel under investigation (i.e., ammonia), but it is the leading candidate.

In chapter 2 we discussed how each of the fossil fuels stored energy in a chemical form. But fossil fuels are not the only means to store energy chemically. Shortly after the modern battery was discovered, William Nicholson observed that an electric current could be used to separate oxygen and hydrogen from water. This electrically induced chemical reaction is referred to as electrolysis, the first step in a reversible hydrogen cycle. Unfortunately, only a portion of the electrical energy makes its way into the chemical bonds of the hydrogen gas. Like fossil fuels, hydrogen can be stored and transported. The future will need a way to accumulate energy, and the electrolysis of water in the production of hydrogen is an attractive option.

Not long after Nicholson discovered electrolysis, Sir William Robert Grove observed that hydrogen in the presence of a catalyst generated electricity. This is the retrieval side of the hydrogen cycle. Modern hydrogen fuel cell technology was inspired by Grove's discovery in 1839. Hydrogen gas is passed through the anode side of the fuel cell, and in the presence of platinum it ionizes. The hydrogen proton moves through a proton exchange medium, while the electron moves through an external circuit where it can perform work. On the cathode side of the fuel cell, oxygen and the hydrogen's proton and electron combine to form water. The round-trip efficiency of storing and retrieving electrical energy with hydrogen is less than 50% but very helpful if the alternative is the complete loss of energy.

Electrolysis and fuel cell technologies were invented in the early nineteenth century but had little practical application with such abundant supplies of fossil fuels. Now in the twenty-first century these scientific discoveries are quite relevant. NASA rejuvenated this technology with two practical applications. The space station utilizes electrolysis to provide oxygen inside the cabin, and its rocket propulsion system uses liquid hydrogen.

Closer to earth, we are seeing projects demonstrating the potential of hydrogen fuel in the transportation sector. In the home and office energy market too, stationary hydrogen fuel cell units are already in use in Japan, as we see in later chapters. Another promising application is the use of portable hydrogen. Today we use batteries and small tanks with propane or butane for stand-alone energy use. Hydrogen fuel cells can be available in a portable format for the same types of applications. Finally, power plants will retrieve energy from hydrogen in the generation of electricity.

Hydrogen does pose several problems in its transportation, though, and those challenges are driving some interesting innovations. The melting point of hydrogen is −259°C, while its boiling point is −253°C, so this molecule is naturally gaseous in the world we occupy. As a compressed gas it is inefficient because it has an extremely low energy density.

Alternatively, storing hydrogen cryogenically as a liquid is expensive and associated with loss due to evaporation. Finally, hydrogen is highly flammable, so alternative chemical formats that can overcome these disadvantages will be indispensable. One of many interesting storage methods is referred to as a liquid organic hydrogen carrier.[1] There are several liquid organic hydrogen carrier candidates, and they share a common approach: the reversible storage of hydrogen in a liquid-phase molecule at standard ambient temperatures and pressure. With one contender, hydrogen is fixed to toluene (C_7H_8) during a hydrogenation reaction to produce methylcyclohexane (MCH: C_7H_{14}). Both toluene and MCH are components of gasoline and are liquids at ambient temperature. MCH is stable and has a high energy density, making this an attractive option for long-term storage. Converting the MCH back into toluene releases the hydrogen, and the toluene is once again available for recycling. This method and other solutions to storing high-density hydrogen in a format conducive to long-term storage and transport are critical for the future.

Ammonia (NH_3), sometimes referred to as the other hydrogen, is a bit of a hybrid energy carrier. Renewable energy would be used to produce hydrogen through electrolysis. A secondary reaction between hydrogen and nitrogen produces ammonia. When called upon to deliver energy, an ammonia and oxygen reaction would take place, with nitrogen and water as the by-products. This cycle, like the hydrogen cycle, is reversible; energy can be captured and retrieved repeatedly.

Building a Full-Cycle Hydrogen System Will Take Decades

The task of replacing fossil fuels is less about alternatives and more about weaning ourselves from an utter dependence on a finite energy source. Still it is nice to know that chemistry offers a pathway. Hydrogen will be a vital component in the post–fossil fuel era. It can be stored and transported like fossil fuels and leverage some of our existing natural gas infrastructure. It can also be used to build strategic levels of storage to mitigate supply risk. If technologies converge, it may well be a fundamental fuel for the transportation sector, where it would compete with pure electric vehicles. The big challenge is to build a full-cycle hydrogen system. This includes creating hydrogen production facilities, determining the best means to store and transport it, building refueling structures, and adapting heat generation equipment. Even given the proper economic environment for investment, this will still take several decades.

Batteries: An Electrochemical Exchange

Batteries are yet one more example of a "new" technology that has been around for many years. Modern history dates the invention of batteries to the late eighteenth century, but the "Parthian Battery" discovered near Baghdad is estimated to be 2,000 years old. It includes iron and copper terminals built into a container that would presumably hold an electrolyte solution (i.e., vinegar). Tests conducted with reproductions measured the battery's voltage to be between 1.1 and 2.0 V. Opinions vary as to whether this device was used as a battery, medicinally, or for electroplating precious metals, but no matter; its discovery shows that advanced societies thousands of years ago understood the principles of the battery well before the modern-era scientists credited with its invention.

Alessandro Volta, an Italian physicist, is credited with inventing the "modern" battery. His battery used zinc and silver discs in pairs stacked into a "voltaic pile." Paper soaked in saltwater was placed between the zinc and silver discs, performing the function of an electrolyte. This device, capable of producing a continuous current, was the first in a new category, the chemical battery. As an aside, Volta is also credited with the discovery of methane, the main molecule of natural gas, a relic of history's multidisciplinary researchers. A half century earlier, Benjamin Franklin was experimenting with the nature of electricity and is attributed with coining the term *battery*. His invention from the "New World" describing electricity as a single fluid was in sharp contrast with the conventional understanding, giving credence to the adage that "seeds sown in virgin soil should produce minds of fresh understanding."[2] If Franklin and Volta could have spoken across the ages, they may have learned a few things from the Parthians.

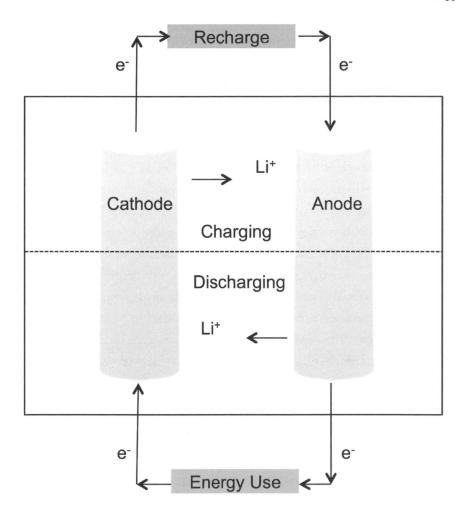

Figure 10.3. Lithium ion battery.

Take Lithium as an Example

A conventional battery consists of two electrodes made of metal or metallic compounds. In the case of the lithium battery, an electrolyte between the electrodes allows positive ions to move between terminals as the battery is charged and discharged. Although both electrodes can attract electrons, the cathode's attraction is greater. The difference in attractive properties of the electrodes determines the voltage of the battery. While discharging, the lithium ions are traveling from the anode to the cathode through the electrolyte, while the electrons move through an external circuit performing work (lower part of Figure 10.3). The reverse happens when the lithium-ion battery is charged. Lithium ions move from the cathode to the anode, and the electrons move externally from the cathode to the anode through a charger. Rechargeable batteries allow this process to happen repeatedly, absorbing and delivering electrical energy through their reversible chemical reactions.

Researchers around the world are racing furiously to design a better battery that will unleash applications in the transportation and power sectors. For instance, recent news from scientists at the Beijing Institute of Technology reports on a new construction for the lithium battery's anode, incorporating a regenerated silk fibroin material.[3] The material stores five times the charge of graphite, a common choice for the anode material. The University of Cambridge also reports progress on a next-generation lithium battery, the lithium air, portending similar advantages. When either of these or other research is translated into commercial products, electric cars, for instance, will overcome most range objections. We are likely within a decade of these advances, but new and improved chemistries for the modern battery will undoubtedly occur and unlock large commercial markets.

Flow Batteries: An Interesting Twist to an Old Idea

Flow batteries take a different approach than conventional batteries; the energy is stored outside in two electrolyte solutions. The electrolytes are pumped through a flow battery pack somewhat akin to hydrogen and oxygen moving through a fuel cell. When the electrolyte solutions travel through the flow battery pack, they deliver electrical energy just like a conventional battery. There are several compelling advantages with flow batteries. First, the energy and power ratings are separated. The amount of energy, from kWh to MWh scales, can be tailored to an application by sizing the electrolyte storage tanks, while the power is determined by the number of flow batteries stacked together. Second, the charging and discharging of the electrolytes can occur separately. One pair of electrolytes can be discharging while another pair is reenergized, thus allowing a virtually instant recharging capability by switching electrolyte feeds. Unlike some conventional batteries, flow batteries operate at ambient temperatures, adding to their attractiveness. As with all things, however, with advantages come disadvantages. Typically flow batteries have low energy densities that limit their appropriateness for certain applications, namely, transportation. They appear to have great potential for large energy storage applications where the size of storage tanks is not an obstacle.

Before we built our vast transmission and distribution infrastructure, batteries were our means of delivering electricity. They powered telephones in the early twentieth century, and they were used as the "igniter" in early automobiles before the emergence of the magneto and electric starter. Most often we think of batteries as supporting portable or off-grid energy applications such as cell phones, computers, cameras, and flashlights, to name a few.

Now batteries are about to enter a whole new market in the energy field. The marketplace potential for battery storage is estimated by some of the entrants to be in excess of a trillion dollars per year. This will include commercial and residential installations and temporal energy producers. Utilities that offer net metering (discussed in later chapters) will need storage systems as temporal capacities increase. Because they will be used daily, battery technologies will need to have low fade rates for this application. This means that the battery can be cycled many times over the course of decades with little degradation in potential. To be of practical value for many applications batteries will need high energy densities (Wh/L). Other important characteristics are the operating conditions, construction material, efficiency, grid integration software, and of course cost.

As you can see, then, battery storage systems are a large market opportunity and a critical capability as the world's power sector weans itself from fossil fuels. Companies ranging from large established organizations to start-ups are investing research and development toward storage solutions. The storage industry is somewhat unnerving in that there are so many product concepts that lack commercial readiness, more and more localities adopting high levels of electricity from temporal sources, and immediate decisions to be made. Having such a bounty of creative ideas is great as long as you have time. Our options include the tried-and-true pumped hydro, flywheels, and storing compressed air in vast underground caverns; we can also choose to move trains up and down big hills, a process that mimics the pumped hydro model; or we could inadvertently select a battery vulnerable to ignition (i.e., blowup). We have so many battery chemistries in development, and all of them are seeking the perfect combination of cost, longevity, raw material supply and safety, and energy density, which altogether portrays a market where the buyer should beware. Leaders are, however, emerging and delivering integrated solutions that will allow the world to break the storage barrier.

From Humble Beginnings to Vital Roles in the Future

Regardless of who emerge as the winners in advanced storage solutions, good technology exists, and more is on the way. Companies are building batteries from 1 kWh to 2 MWh, and early testing suggests that some of these batteries can be cycled daily for more than 20 years with minimal fade.

As temporals generate a higher and higher portion of our total energy, the cost per unit of energy from fossil fuel power plants will increase substantially because plants will need to recover their fixed costs across smaller and smaller amounts of electricity. It doesn't take a crystal ball to see that an extended transition period is an expensive way to make the energy switch.

It is also important to note that the direct production of electricity from renewables is quickly achieving parity with fossil fuel. However, when the need to accumulate energy is factored into relevant costs, it becomes clear that cost-effective energy storage is key to the retirement of fossil fuels.

Notes

1. Yoshimi Okada and Mitsunori Shimura, "Development of Large Scale H2 Storage and Transportation Technology with Liquid Organic Hydrogen Carrier (LOHC)," Chiyoda Corporation, February 5, 2013, https://www.jccp.or.jp/international

/conference/docs/15rev-chiyoda-mr-shimura-chiyoda-h2-sturage-and-transpor
.pdf.

2. Catherine Drinker Bowen, *The Most Dangerous Man in America: Scenes from the Life of Benjamin Franklin* (Boston: Little, Brown, 1974).

3. "Silk May Be the New 'Green' Ultra-High-Capacity Material for Batteries," Kurzweil Accelerating Intelligence, March 11, 2015, accessed April 12, 2015, http://www. kurzweilai.net/silk-may-be-the-new-green-ultra-high-capacity-material-for-batteries.

11

Smart Grids

I do not fear computers. I fear the lack of them.
—Isaac Asimov

As the twentieth century was dawning, so was the genesis of the world's electrical grids. Inventors had a blank sheet of paper during the advent of this entirely new industry. Fledgling utility companies were competing on the merits of two methods for transmitting electricity: direct current and alternating current.

There are two relationships that help us explain why the benefits of a higher voltage in electrical transmission factored into our choice of a grid standard. First, power is directly proportional to the voltage and current. A higher voltage can deliver the same power at a lower current. Second, the dissipation of power in the transmission of electricity is proportional to the square of the current, and lower is better. Direct current, limited at the time to lower voltages, was promoted by Thomas Edison and initially perceived as safer but restricted in transmission distances. Power plants needed to be within miles of the end users, and this offered no advantage in the age of fossil fuels. This was an impractical vision for a power grid, particularly when rural customers were considered. Nikola Tesla, perhaps the most overlooked genius of modern times, developed equipment for the generation, transformation, and transmission of alternating current. The capability to transmit alternating current at a high voltage opened the door to the long-distance transmission of electricity.

Centralized power stations could deliver electricity to urban and rural customers, and our modern power grid was born.

From Decentralized to Centralized and Back Again

In spite of Thomas Edison's fierce and controversial defense of direct current, alternating current won out and became the standard for the world's electrical grids. Today we have a grid built with alternating current serving as the foundation of our centralized architecture. Similarly, standards will need to be established for smart devices that operate within the grid. These smart devices will need to communicate and share information, and as with the Internet, their interoperability requires the use of standards.

The convenience of fossil fuels helped the decision to centralize, and now distributed renewables will point us back to a decentralization model. Unlike in the beginning of the twentieth century, we don't have that same blank sheet of paper to work from as we consider the future. The U.S. grid consists of 300,000 mi of transmission lines incorporating 9,200 electric generating plants with a million megawatts of power capacity. Although 9,200 generating plants seem like a lot, in the future we will see millions of locations gathering energy, three orders of magnitude more than today's centralized architecture, as we mount solar panels on rooftops and erect wind turbines across fields and near shorelines. Upgrading the grid is complicated because we need to use it simultaneously. Not a simple task, it will be like trying to repave a road while traffic whips along.

Figure 11.1, taken from the Danish Energy Agency, shows the degree to which decentralization has occurred in Denmark since the 1980s. This trend isn't complete; there will be additional dots of distributed generation appearing in the future. Properly managing these new sources requires a system with intelligent components that measure, communicate, and act in harmony to deliver energy where needed, to shift demand where possible, and to avoid waste. Many of the distributed sites are wind, but many others are combined heat and power, where a variety of energy sources are used on-site for both electricity and heat generation.

Smart grids will be as critical to transition as the coxswain is to the sculling team, managing millions of supply points with millions of demand points.

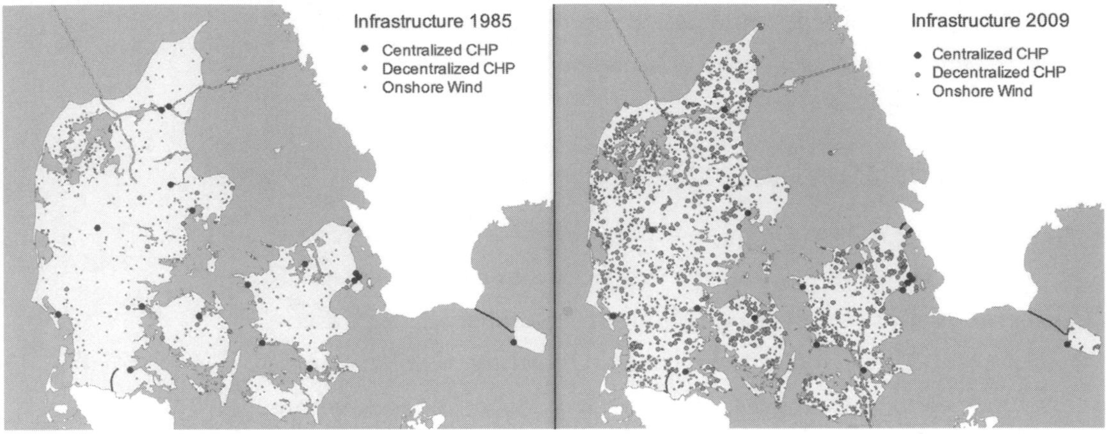

Figure 11.1.
Decentralization
of energy sources
in Denmark. CHP =
combined heat and
power.
Source: Danish Energy
Agency, "Oversigtskort
over Danmarks
El-infrastruktur
anno 1985 og 2009,"
accessed August 1,
2016, http://www
.ens.dk/info/tal-kort
/energikort/download
-faerdige-kort.

Electrical grids today are structured like our own circulatory system—the heart pounding out a steady beat as it delivers energy to our body. The voltage and the frequency on the grid are like our blood pressure and pulse, measures of our circulatory health. Decentralized renewable energy sources don't easily fit today's grid model. Instead, they are like sculling teams, where each member contributes his or her energy and collectively the boat is powered. Successful sculling teams coordinate their energies and maximize efficiency. Within our current system, single-point supply is delivered to multipoint consumers. Decentralized energy production will change that pattern to multipoint supply and demand. Smart grids will be as critical to this transition as the coxswain is to the sculling team, managing millions of supply points with millions of demand points.

Another factor pulling for smarter grids is found in the behind-the-scenes marketplace that sells and buys electricity in real time, insulating the consumer from the challenges with delivering their kilowatt-hours. The price moves up and down much like the stock market. The prices reflect many factors including supply, demand, what energy sources are on line, local congestion, and losses. Our retail utility bills in the United States largely mask this underlying market, and that isolates the consumer and to some extent unwittingly creates inefficiencies. Smarter grids will involve the consumers in demand shifting, and this will boost the overall efficiency of electricity generation in the following way.

Satisfying peak consumer demand is an extremely expensive endeavor for the utility companies. The U.S. Department of Energy estimates that 10% of all generation assets and 25% of distribution assets are only used 5% of the time! We have built a system that in an unconstrained fashion delivers what we want when we want it, regardless of waste. Having access to information through smart grids will allow consumers to make decisions that dampen peak loads, creating greater efficiency where everyone wins.

More specifically, a smart grid is an electrical network that incorporates computer-based equipment with two-way communications to measure, control, and automate the supply and demand of electricity. Today in the United States, according to the Bureau of Labor Statistics, there are still 38,000 utility employees who read meters, and these jobs won't survive the twenty-first century. In the future, smart systems (i.e., smart meters) will know in real time the demand of your home down to some of your appliances. That knowledge can reduce demand in peak periods and deliver energy when you select a smart system that can turn appliances on and off in an efficient manner. A smarter grid would deliver benefits in today's fossil fuel–dominated power sector, so we should be investing. However, tomorrow, when millions of alternative energy sources are interacting with a similar number of storage systems, an unintelligent power grid is unfathomable.

As noted earlier, much of what occurs within the power grid is taken for granted. We appreciate that our system works but really don't think we need to know the inner workings. Smart grids and devices that connect to the grid will involve us directly. Beyond smart heating and cooling systems, consumer appliances such as washers and dryers will be grid-configurable. Appliances with "set-it-and-forget-it" grid compatibility will allow consumers to actively participate in a market where price varies by the hour. Consumers will be able to set energy-intensive appliances to operate within low-cost time corridors. Generating and storage devices will be remotely monitored and controlled to collectively meet demand. Transmission and distribution lines will be monitored to improve reliability.

We benefit whenever we can avoid inefficient peaks. The utilities will improve capital utilization and defer capital investments with demand

Think of a smart grid as one that clips the highs and lows of both supply and demand to optimize the efficient generation and use of electricity.

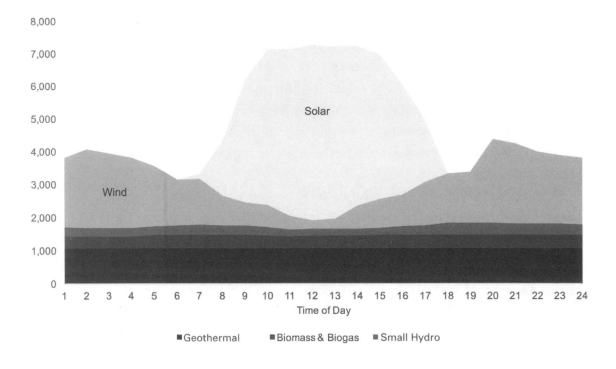

8,000 7,000 6,000 5,000 Solar 4,000 3,000 Wind 2,000 1,000 0

Figure 11.2. Hourly breakdown of average energy output from renewable sources (MW), November 6, 2015.
Source Data: California Independent System Operator, "Renewables Watch," accessed April 14, 2016, http://www. caiso.com/green/ renewableswatch.html.

shifting. Commonwealth Edison, the largest utility in Illinois, already has a Residential Real-Time Pricing Program.[1] It allows participating customers to at times pay the wholesale price for their electricity on the benefits of demand shifting where the utility saves many-fold more expensive peak production costs.

Successfully integrating large amounts of energy from temporal sources will depend on not only changing our demand-supply patterns but also managing smart storage systems. A smart grid will facilitate both of these capabilities. Figure 11.2, from California Independent System Operator, displays how renewable energy was generated for November 6, 2015, as an example. From top to bottom the renewable categories include solar PV, wind, small hydro, biomass and biogas, and geothermal. Solar PV stands out as the most potent and the most dynamic of the renewables. But how does one marry this production pattern with consumer demand? Today with relatively small portions of total electricity provided by renewables, the grid is manageable. But this will be entirely impossible without smart upgrades orchestrating the use of storage systems.

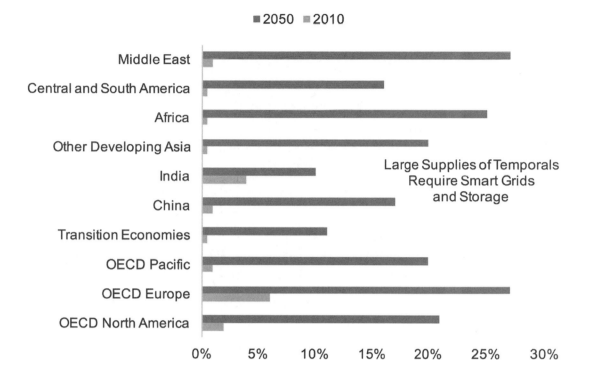

Large Supplies of Temporals
Require Smart Grids
and Storage

■ 2050 ■ 2010

Figure 11.3, from an International Energy Agency report, shows the projected portion of electricity delivered by variable (renewable) resources for a group of countries. For definition, "variable sources" includes wind, solar, run of the river, tidal, and combined heat and power. Not surprisingly, the nations of the world are moving collectively from less than 5% variables in 2010 to a projected 20% in 2050. Experience from localities with relatively large temporal capacity suggest that as we approach 2050 we will not be efficient (i.e., curtailment) without smart grids.

Currently it's a perfect storm, with increasing temporal resources, inadequate storage systems, and grids that are both blind and incapable of influencing decisions beyond a manually operated on/off switch. Exacerbating the situation, a utility is obligated to operate as though a temporal source will stop producing energy at any moment, significantly undermining efforts to be efficient. Astoundingly, worldwide electricity demand is forecasted to nearly double between 2015 and 2040.[2] In many parts of the world, investment in the grid has lagged behind increasing

Figure 11.3. Portion of electricity generated by variable resources, actual in 2010 versus projections for 2050. OECD = Organisation of Economic Co-operation and Development. *Source Data:* International Energy Agency, *Technology Roadmap: Smart Grids* (Paris: Organisation for Economic Co-operation and Development/International Energy Agency, April 2011), https://www.iea.org/publications/freepublications/publication/smartgrids_roadmap.pdf.

electricity demand for decades. The time is approaching when the grid's reliability will suffer. Investments are long overdue to secure dependable access as the energy mix changes. In the United States, 33%–50% of consumer utility bills are set aside to maintain and extend the grid infrastructure. Lacking alternative funding, that rate will need to rise during the transition period when the old system needs to be maintained while upgrades are implemented. The problem is exacerbated by the increasing use of electricity in the transportation sector. Internal combustion engines giving way to electric-powered vehicles is forecasted to consume 10% of total electricity generation by 2050, placing additional stress on the world's power grids.

Summarizing, think of a smart grid as one that clips the highs and lows of both supply and demand to optimize the efficient generation and use of electricity. A smart grid will automatically trigger diversion to storage devices during low-demand periods and retrieval during periods of high demand.

Micro-Grids and the Pursuit of Autonomy

Micro-grids are a promising reflection of the way we will transition energy sources. Localities will build "domestic" sources, and micro-grids will help them manage their investment. Like the main grid they will need to be smart as well. Where large power plants produce electricity in quantities able to serve large networks of communities, the micro-grid is powered by local renewable sources. The micro-grid will also incorporate storage capability to advance its independence from the main grid. A disruption of supply in the main grid won't imply a loss at local levels when distributed energy sources exist. Retaining connectivity to the main grid provides a means to mitigate outages when local supplies fail to meet local requirement. There will be localities simply unable to produce all the energy required, and thus a connection to the main grid will abate this shortfall. This will often be the case in high-population areas or locations with energy-intensive industries.

There are many advantages to the development of micro-grids. They become the integrators of local supply and demand of energy. They allow a natural means to transition to a decentralized model reflecting the best advantages of renewables. They offer the means to maximize the localities' potential and build a mix of alternative sources tuned to the setting and value systems. We've noted that the long-distance transmission of electricity wastes at least 6% of electricity through thermal loss; this can largely be avoided within micro-grids.

The path to replacing fossil fuels will include a smart grid but also a shift in the architecture. Government policies have helped build renewable capacity, but lagging investments in the electrical network threaten to limit the amount of temporal capacity. The technology is here. What we lack is timely investment.

Notes

1. ComEd: An Exelon Company, "ComEd's Hourly Pricing Program," accessed April 14, 2015, https://hourlypricing.comed.com/.
2. U.S. Energy Information Administration, International Energy Outlook 2016, 165, Table A2.

12

Transportation

The future has a way of repaying those who are patient with it.
—The Rev. Arthur Pringle

Today, sales of light-duty vehicles powered by the internal combustion engine outnumber those of electric vehicles by roughly 100 to 1, but way back in the 1890s the shoe was on the other foot. Electric vehicles outsold the internal combustion models by 10 to 1. Ultimately, limited range, poor electric infrastructure, and the invention of the electric starter to replace the hand-powered crank needed to start gas engines flipped consumers' choice in favor of internal combustion—powered vehicles. Now, consumers seem to be switching back again, returning to the electric vehicle of the past. Battery technology is much improved, promising to overcome most of the range constraints, and of course it avoids the undesirable tailpipe exhaust of gasoline- and diesel-powered vehicles linked to negative environmental and health effects. The really new transportation technology emerging is the hydrogen fuel cell, and that was invented in the nineteenth century. Consumers are no longer limited to choosing between vehicle models; they can now choose the type of energy they prefer. For some, reducing environmental and health impacts trumps other purchasing factors.

Given the substantial investment in our transportation infrastructure, our road networks and rail tracks will remain the paths we take, but how we propel ourselves from point A to point B will be altogether different. We will see more energy choices and greater efficiency in all classes of vehicles. Smart devices and mobile apps will keep us informed of our vehicle's energy status. Our pattern of refueling and/or recharging our vehicles will change as well. Leading contenders for alternatively powered vehicles are plug-in electric vehicles, hydrogen fuel cell vehicles, hybrid combinations, and flex-fuel vehicles. Even diesel-powered locomotives are really hybrids, with diesel engines driving generators that power huge electric motors. Electric power already drives intercity transit trains, subways, and trams and is also set to power new high-speed rail systems. Changes are already under way.

Today, though, the personal transportation sector is largely fueled by petroleum; the price per barrel directly correlates to the price per liter or gallon of gasoline. In the United States, petroleum, natural gas, and renewables account for 93%, 3%, and 4% of the transportation energy used, respectively. The U.S. transportation sector in 2012 consumed 28% of the total primary energy, ranking second behind the power sector. Replacing fossil fuel usage in transportation is just part of a broader migration to alternatives.

Two major reasons for switching to alternative vehicles are related to security and health. A country can gain a measure of petroleum independence if it can switch to alternative vehicles powered by domestic energy sources. Another obvious factor driving a desire to introduce alternative-powered vehicles is the role that the transportation sector plays in both poor air quality and greenhouse gas (GHG) emissions. According to a 2013 U.S. Environmental Protection Agency (EPA) report,[1] the transportation sector contributes 30% of U.S. emissions and 13% of global emissions of greenhouse gases. In the United States, cars and light trucks are responsible for 60% of the transportation GHG emissions.

Constant and intensifying pressures on reducing vehicle mass emissions and increasing fuel efficiency are improving air quality. Higher fuel efficiency inherently reduces CO_2 emissions, while anti-pollution technologies are employed to lessen the amount of toxic compounds in the exhaust. The revelation in 2015 that Volkswagen purposely designed, developed, and installed software to cheat emission testing exposes the strength of the regulatory forces. The vehicle's software was designed to sense the conditions of an emissions test and adjust operations in order to pass the test. The air ratios and exhaust flows were modified to temporarily satisfy the regulations.[2] Afterward the vehicle reverts back to normal operations, which have emissions up to 40 times the U.S. EPA limits for certain toxins. Presumably, the reduced horsepower from complying with the emissions regulations was thought to lessen the vehicle's appeal, so the company rigged the system. As this example so clearly demonstrates, increasing health and environmental pressures on petroleum-based transportation will open the door for alternative vehicles.

Turning back atmospheric greenhouse gas levels means adopting vehicles in transportation with lower net carbon emissions. Mass emission and fuel regulations have had a tremendous impact in reducing air pollutants, but there is no avoiding the CO_2 emissions, a natural by-product of burning gasoline. If the world is poised to double the number of vehicles and regulation reduces GHGs/vehicle by 50% because of better fuel economy, we have accomplished nothing toward reducing transportation-related GHG emissions. Providing alternatives for the consumer, then, is a necessary step to meaningful reductions in global transportation-related emissions. In the case of transportation, the goal is to reduce both toxic and greenhouse emissions, and that means a shift away from the use of fossil-based fuels. Yes, electric cars will continue to have carbon footprints as long as the electricity they use is sourced from coal or another fossil fuel. However, while gasoline vehicles have fixed emissions over their lifetime, electric vehicles will get cleaner as the power sector shifts to renewables. Ultimately the plan is to power alternative vehicles entirely from nonfossil sources, and that will take time.

Transportation is fundamentally decentralized, so decarbonizing the sector means the adoption of alternative vehicles. The predominately fossil-fueled power sector, on the other hand, can in theory employ carbon capture and sequestration at a single point and substantially reduce emissions. If our goal was only to reduce emissions, we could charge electric vehicles from lower-emitting fossil power plants. In this way, we reduce carbon emissions, but we are not addressing the goal of

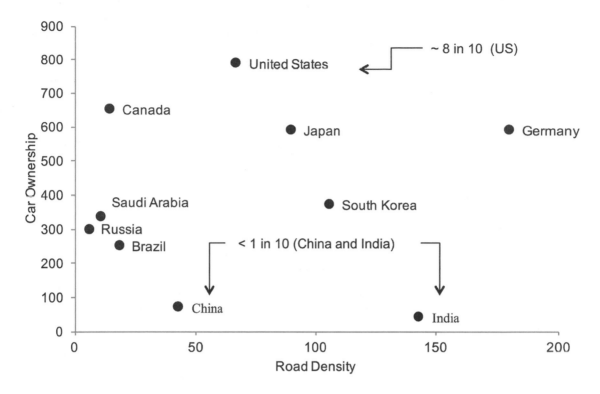

Figure 12.1. Car
ownership (per 1,000
people) and road
density (km per 100
km²).
Source Data: ChartsBin,
"Worldwide Total Motor
Vehicles (per 1,000
People)," accessed
April 14, 2015, http://
chartsbin.com
/view/1114.

transitioning away from the dwindling supply of economically accessible fossil fuels. This is an example of where the goal of reducing carbon emissions diverges from goals to transition to alternative energies.

Figure 12.1 shows the cars per 1,000 people for the countries we investigate later in this book. The United States is in a class by itself, with the world's highest rate. Canada, Germany, and Japan set the next-highest bar. Urbanization and access to mass transportation can mitigate the intensity of car ownership, but cars still represent personal freedom, and the hundreds of different makes and models reflect our continuing desire for personalized travel. As countries gain wealth, we can expect car ownership to rise substantially in places such as China, India, and Brazil. Figure 12.1 also shows the road density for the same countries, representing each country's investment in its road infrastructure.[3] Germany has the highest road density, followed by India, with a surprisingly high density for such a low car ownership level. When you have a population well over one billion in a relatively small space, road density increases. As noted earlier, although our vehicle's power plant will be different, we are not going to see new modes of personal transportation in the near future, because of the huge investments in our road and rail infrastructure.

So what are our near-term options in transportation for non-fossil-powered vehicles? The following section compares and contrasts electric and hydrogen fuel cell vehicles and their commercial readiness and considers the long-term relevance of biofueled vehicles.

From Four-Stroke Engines to Hydrogen Fuel Cells

It is a basic principle that, as with electric vehicles, if hydrogen is produced from fossil fuels, the claim of zero carbon is misleading. While most of today's hydrogen is produced utilizing fossil fuels, the full merits of hydrogen and electric will only occur when they are provided by non-carbon sources. When renewable sources are producing beyond demand, the energy will be diverted to charge batteries or generate hydrogen. The realities of our current energy mix shouldn't cloud the plan or dim the excitement for these new transportation technologies.

Although the internal combustion engine represents the majority of what is sold today, sales of alternative vehicles are starting to make inroads. Hydrogen fuel cell vehicles are one of the exciting new options entering the market, but interest is dampened by the lack of refueling infrastructure. They are, however, finding applications in niche markets. Warehouse forklifts are typically electric-powered. The weight, recharge time, and progressive loss of power with electric forklifts favor the use of hydrogen fuel cells. Fleets of quietly running hydrogen-fueled lawn mowers provide an attractive alternative to the noisy gasoline-powered versions for golf course maintenance. Neither of these markets suffers from the lack of refueling infrastructure; consequently they provide proving grounds for the refinement of fuel cell technology.

Major car manufacturers are launching compressed hydrogen gas vehicles with impressive fuel ranges. The vehicles utilize well-tested electric motors, fuel cells, and onboard batteries that provide supplemental power. These vehicles are really hybrids, with hydrogen fuel cells providing the base power and a battery capturing energy during deceleration and braking and boosting the fuel cell at other times to maximize efficiency.

The main advantage of hydrogen is the ability to scale the technology to all classes of vehicles, providing range and timely refueling. The technology exists now, but the lack of fueling station infrastructure is blocking the early adoption of this type of vehicle. In 2015, according to the Alternative Fuels Data Center,[4] there were only 12 fueling stations in the United States offering hydrogen, and those are mainly located in California. Germany is working to develop a network of 1,000 stations in the relatively near term. Decades from now hydrogen fueling stations may be located as conveniently as gasoline stations. For now, the lack of infrastructure provides electric vehicles a market window to establish share and improve their overall offering. There is, however, novel equipment that produces on-site and on-demand hydrogen that may speed the establishment of refueling stations.

Electric Is Here, Albeit with Limitations

Already in 2015, there were more than 20 alternative models from which the consumer could choose. They are available in pure electric and hybrid versions. Supporting the adoption of these vehicles are more than 10,000 recharging centers, with 26,000 outlets in the United States. Flexibly, electric cars can also be recharged at the owner's residence. For the present, electric technology is ideal for short-distance commutes and the light-duty vehicle class. Traveling longer distances and carrying heavy loads are impractical because a complete recharge takes hours, and even a quick charge that restores 80% of the battery can take 30 minutes. Extreme cold weather further reduces the vehicle's range from 25% to 50%, so drivers need to factor this into the vehicle's use. Looking into the future, a car battery with higher energy density would address most of these drawbacks. And given the size of the market opportunity and the level of research and development, a better battery is a good bet. Even better would be a technology that speeds the recharge time for purely electric vehicles.

U.S. electric vehicle sales increased to nearly 120,000 in 2014, posting an increase of 18% in an overall market growing in the vicinity of 3%. But electric vehicle sales are still less than 1% of total auto sales. So, although their economy per mile is attractive, many potential buyers are sidelined because of range limitations, recharge times, model selection, and convenient access to recharging outlets. As with many embryonic markets, these initial obstacles will be overcome.

To Many a False Choice: Hydrogen versus Electric

The current cost of electricity in the United States is around $0.12/kWh, making electric vehicles the most economical from a US$ per mile perspective versus both gasoline and hydrogen. Table 12.1 compares the range and fuel economy of the typical gasoline, electric, and hydrogen fuel celled vehicles. Mileage economy ranges from $0.04 a mile for electric vehicles to $0.17 per mile for hydrogen fuel cell vehicles, while conventional gasoline vehicles rate $0.10 a mile. The driving range varies from less than 100 mi to more than 200 mi for pure electric vehicles, to nearly 300 mi for hydrogen fuel celled vehicles, to more than 400 mi for some conventional models.

Both electric and hydrogen vehicles have disadvantages that limit their inroads, as summarized in Table 12.2. Nevertheless, the light-duty vehicle market has the necessary technology to begin the replacement

Table 12.1. Cost per Mile Comparison for Gasoline, Electric, and Hydrogen Vehicles.

COST FACTOR	GASOLINE VEHICLES	TYPICAL ELECTRIC VEHICLES	TYPICAL HYDROGEN FUEL CELL VEHICLES
Capacity	16 gal	65 kWh	5 kg
Mileage	30 mi/gal	3 mi/kWh	60 mi/kg
Range	480 mi	195 mi	300 mi
Cost per unit of fuel	US$3.00/gal	US$0.12/kWh	US$10.00/kg
Cost per mile	US$0.10/mi	US$0.04/mi	US$0.17/mi

Table 12.2. Advantages and Disadvantages of Electric and Hydrogen Vehicles.

VEHICLE TYPE	ADVANTAGES	DISADVANTAGES
Hydrogen fuel cell	Fuel cells weigh less than battery packs, thereby they are more readily scalable to larger classes of vehicles	Lack of refueling infrastructure: only 12 stations in the United States, mainly in California
	Refueling takes minutes	Expensive to build infrastructure
	Superior for long-range travel	Initially fuel more expensive per mile than gasoline
Electric	Relatively inexpensive to create recharging infrastructure	An 80% recharge takes 30 minutes, while a full recharge takes more than an hour (85 kWh); Tesla is floating concept of battery swap to address this disadvantage
	There are approximately 10,000 static stations, with 26,000 outlets in the United States	Category limited to smaller class of vehicles
	Assuming an electricity rate of $0.12/kWh, the cost per mile is $0.04	Weight of battery packs makes scaling to larger vehicles impractical (e.g., tractor trailers)

of internal combustion models. To accelerate adoption, though, electric vehicles need batteries with higher densities and/or faster recharging, while hydrogen fuel cell vehicles need refueling infrastructure and improved fuel prices.

A Renewable Fuel Additive

Biofuels are derived from living matter. As the world looks to replace fossil fuels in the transportation sector, biofuels have allowed markets to act with motor vehicles and refueling infrastructure as they exist. Biofuels are blended with gasoline and diesel, and, voilà, a small shift to renewables occurs. Worldwide biofuel production increased 16.1% since 2000, to reach 1.9 million barrels per day in 2011.[5] In the United States the growth over that same period was 20%, reaching 0.94 million barrels a

day. The United States and Brazil account for 70% of the world's ethanol production, while most of the rest of the world appears less inclined
to place land and food resources to this purpose. What the United States
and Brazil share is a history of biofuels that began in response to the
Organization of Petroleum Exporting Countries' 1973 oil embargo and
the subsequent quadrupling of the price of oil in 1974.

Bioethanol is principally derived from corn and wheat, but there are
many other energy crops. We understand that bioethanol is still carbon-based, but producers claim that it is carbon-neutral—the carbon
dioxide consumed during the growth of the crop offsets the CO_2 emitted during combustion. This can be debated if all carbon contributions
from agriculture are considered, but nevertheless, they remain less car-
bon-intensive than standard gasoline. The other trade-off of course is
directing agricultural products to fuel rather than to feeding communities. For these reasons it seems that biofuels are an interim solution for
at least the light-duty vehicle class until other renewables can establish
sufficient capacity. The United States has hit the E10 (fuel with 10% ethanol) blend limit, so continued industry growth will need to find export
markets or look to the EPA to increase blend allowances. Recently, the
U.S. EPA granted waivers that allow but do not require blends to E15 for
use in model years 2001 and newer light-duty motor vehicles.

A class of vehicles called flex-fuel vehicles uses modified engines that
operate with ethanol blends as high as 85% ethanol. Across the major car
manufacturers more than 100 different models are available. And there
are other benefits to flex-fuel vehicles, including higher horsepower, less
engine knocking, cleaner fuel-injection operation, and less exposure to
line freezing in cold weather. Since ethanol has less energy per volume
than gasoline, consumers should expect lower fuel mileage, in the 15%–
30% range. Biodiesel is blended with mineral diesel up to 5% without
affecting engine performance. The most common source for biodiesel is
oil seed crops, vegetable waste, and animal oils. It is a renewable by virtue of the crop production but also a recycled fuel by virtue of the vegetable waste.

As you can see in Table 12.3, of the alternatives, biofuel vehicles
are the most popular, followed by hybrids, purely electric vehicles, and

*Vehicle technology
is way ahead of
consumer refueling
or recharging
stations. Access to
recharging stations
where you work and
live determines your
ability, not your
desire, to purchase
an alternative
vehicle.*

Table 12.3. U.S. Alternative Vehicles by Class.

ENERGY SOURCE	NUMBER OF VEHICLES IN USE IN 2011
Ethanol flex-fuel (E85)	862,837
Liquid petroleum gas (propane)	139,477
Natural gas	121,650
Electric battery	67,295
Hydrogen	527
Ethanol flex-fuel (all)	9,946,091
Hybrid gas/diesel-electric	2,126,357

Source Data: U.S. Department of Energy, Alternative Fuels Data Center, "Alternative Fuel Vehicles in Use, 2011," accessed April 29, 2015, http://www.afdc.energy.gov/data/.

embryonic-stage hydrogen fuel models. Biofuels reduce the use of fossil fuel and reduce greenhouse emissions, and flex-fuel (85% ethanol) vehicles specifically come pretty close to fully displacing the use of fossil fuels. And there are already 2,500 refueling stations with E85 (85% ethanol fuel) in the United States.

With Choices Comes a New Mix in the Transportation Sector

Robust alternatively fueled ground transportation technology is already here, but since cars have a 10-year life span, it will take decades to replace the world's fleet of fossil fuel–based vehicles. The good news is that, unlike the power sector, the auto industry is providing consumers with choices and the infrastructure is developing. As renewable energy adoption in the power sector is slowed by lack of storage and grid infrastructure, so the transportation market's conversion to alternative-powered vehicles is slowed by infrastructure that is just emerging. Vehicle technology is way ahead of consumer refueling or recharging stations. Access to recharging stations where you work and live determines your ability, not your desire, to purchase an alternative vehicle.

California Setting the Pace in the United States

Not surprisingly, within the United States the state of California is leading the way toward the widespread adoption of alternative vehicles. California has also led the country in policies designed to reduce vehicle emissions, and its targets have influenced the national standards. Now, the state's leadership is setting an ambitious goal, via executive order: an 80% reduction in economy-wide greenhouse emissions below the 1990 baseline by 2050, far exceeding the national goal.[6]

Since transportation accounts for 30% of those emissions, California, which is home to some of the nation's worst smog zones, has established an Alternative and Renewable Fuel and Vehicle Technology Program. The plan includes mechanisms to foster the development of infrastructure for these new vehicles, reflecting a long-term determination to shift away from transportation's petroleum dependence. Impressive momentum is already evident. Electric recharging stations have increased 368%, while E85 stations have increased 412%—impressive growth, despite the fact that these percentages are based on small initial numbers. Unfortunately, a lofty goal of reducing the use of petroleum-powered cars and trucks by 50% by 2030 was abandoned in a broader piece of legislation, highlighting the resistance to regulating the consumer's choice in vehicles. California did, however, pass the Clean Energy and Pollution Reduction Act of 2015, which called for procurement of electricity from renewable sources to be increased from 33% to 50% by 2030. So indirectly, the more electricity generated from renewables, the more capacity there is for electric forms of transportation.

As the world's eighth-largest economy, California is operating on a planning horizon far longer than the national scale. It is making investments today that have soft benefits decades away. California believes that the long-term benefits of decarbonizing far exceed the shorter-term costs and understands that public investment speeds realization. Building refueling infrastructure is the only real means to encourage the public's support for alternative vehicles.

It's difficult to conceive of any transition, whether to a new home, a new job, or even a vacation, that doesn't involve an initial cost for a future benefit. California is accentuating the importance of investing in programs today in order to secure long-term economic benefits. Policies that are often built around the notion that there should be an immediate return on investment for such a massive enterprise as energy aren't just a bad idea in their own right. They wreak havoc on future generations that will lack adequate lead time for the task. We wait and wait for an economic trigger point that never comes, while the future consequences mount. Forward-looking leadership will have the courage to acknowledge and communicate the need to sacrifice in order to achieve longer-term benefits as the twenty-first century gives way to new sources of energy.

Freight: Moving People and Stuff

The leading freight energy consumers are medium and heavy-duty trucks, air, water, and rail at 22%, 8%, 4%, and 2% of total transportation, respectively. It is surprising to learn that rail consumes only 2% of the energy since it is responsible for the majority of freight ton-miles. Not surprisingly, data from the U.S. EPA report that medium and heavy-duty trucks contribute 53% of CO_2 emissions for the entire freight category, while air, rail, and water contribute 25%, 6%, and 8%, respectively. Trucking, the least efficient energetically and with the highest carbon dioxide emissions, will feel the greatest pressure to change. At this juncture, the trucking segment will need the scalability, power, and range bestowed by hydrogen fuel cell technology and consequently will not be converting from carbon-based fuels en masse for a decade or more, until both the technology and refueling infrastructure improve.

Railways: As Relevant as Ever

Railways serve two important functions: they efficiently move freight and people, often on the same set of tracks. Well into the twentieth century, steam engines were the principle means of drive. Steam gave way to the diesel engine on the merits of better performance and economics and remains the dominant energy plant on trains, driving generators that power electric motors. The U.S. Department of Transportation Federal Railway Administration's National Rail Plan in 2010 forecasted a growth of 1.2 billion (~40%) tons of freight over the next 40 years.[7] Ironically, a fair portion of this growth is the transport of more coal. Taking advantage of rail where possible can have a big impact on energy efficiency and lowering emissions, since rail averages 413 ton-miles per gallon to trucking's 110 ton-miles per gallon.

On a worldwide basis rail accounts for 39% of total freight ton-miles, followed by truck and water at 29% and 12%, respectively. Interestingly pipelines move 20% of worldwide ton-miles, highlighting how much petroleum is moving around the globe. Rail's continued investment in infrastructure is critical to move more freight on the merits of efficiency. The deregulation of rail in the United States was influential in the revitalization of the system. Today, the U.S. rail system is one of the most efficient in the world. Years ago, rail lost share to trucking with the building of the interstate highway system, but now congestion on interstates is bringing some of this business back to rail, as evidenced by more and more trailers loaded on flat wagons and moved by trains.

When it comes to moving people, rail transit in the United States uses 23% fewer BTUs per passenger mile than automobiles, so the advantage is partially energy efficiency but also reduced congestion on the highways.[8] In January 2015, California broke ground in Fresno on an all-electric $68 billion high-speed rail. The plan is to procure all of the train's energy from renewable sources. One of the ways California plans to meet this goal is through its Renewable Procurement Plan, which works with utilities along the route to deliver on-site energy. The train will achieve speeds of 220 mph and make the commute from L.A. to San Francisco in three hours. Breaking ground was the culmination of persistent leadership and farsighted planning.

Hydrogen fuel cell technology may also have a role in the future of rail. Already there have been successful hydrogen fuel cell demonstration projects; Germany is moving forward with plans to deploy 20 hydrogen-powered regional trains by 2020. This new class of train is referred to as hydrail. Unlike the switch from steam to diesel in the twentieth century, the transition to hydrail in the next decades will be driven by goals to reduce emissions while maintaining economic competitiveness within the sector.

Waterways: Slow, Steady, and Busy

The irony of moving fossil fuels is even more pronounced with maritime shipping. Forty-one percent of sea trade is oil, liquid gas, and coal, and ocean traffic has never been busier. Accepting this fact, though, water is able to move 576 ton-miles per gallon, the most efficient of all modes. Diesel engines are the predominant power plant on cargo ships. Near-term efforts to reduce emissions include liquid natural gas and lower-carbon-emitting biofuels. Long-term, as with rail, hydrogen fuel cells may emerge as the new power plant option. Recently a company by the name of Nuvera announced plans to power high-end marine vessels for the Italian shipbuilder Fincantieri using advanced fuel cell technology.[9] The power plants, rated 260 kW, will generate electricity driving the vessels' propulsion and charging onboard battery systems. Although military naval vessels routinely deploy nuclear power, there isn't discussion of broadening its application to the world's maritime trade.

Like rail, water has significant economic advantages over alternative modes. But just as a boat moves slowly across the sea, so will the shift in maritime dependence on fossil fuel. It will take many decades.

Trying to mitigate health, environmental, and climate impacts while remaining fossil fuel–dependent is like applying a new coat of paint to rotten wood.

From Crop to Fuel Additive and Electricity

Particularly in the United States and Brazil, biomass is refined into fuel additives for both gasoline- and diesel-powered vehicles. Of the methods for collecting solar energy, bioenergy is the least efficient and a relatively poor use of the world's precious land and food resources. Looking far ahead, it will be a niche renewable source in the transportation sector, still carbon-based but with a lower footprint compared with fossil fuels. Biomass, like hydropower, was one of the world's first forays into the use of renewables for electricity generation. And like hydro, biomass in the power and transportation sectors will play a small yet vital role in the wholesale replacement of fossil fuels.

Air travel doesn't forecast an alternative to liquid hydrocarbons for decades, and a practical electric-powered airplane is far into the future. However, airlines have been testing biofuel blends since 2008 and have reported positive results. More than 10 airlines have conducted tests with blends up to 50%, requiring no engine modifications. Biofuel performance requirements for airlines are more challenging, requiring much higher energy content. Consequently, different crops and sources are under investigation to fuel airplanes, including switchgrass and even algae. The aspirations are to develop high-performing blends that can reduce "life cycle" greenhouse emissions by up to 80%. These fuels are not available commercially at this time, but certification standards are being developed, paving the way for commercial use in the future.

Biofuels may be the midterm future for the airborne portion of air travel, but changes may come faster for moving planes around on the ground. The jet engines designed for air propulsion are a grossly wasteful means for taxiing a jet to and from the runway. A European carrier, easyJet, has announced plans to prototype a plane outfitted with an eco-friendly green system designed to perform these ground duties. The airline estimates that 4% of its fleet's jet fuel usage occurs on the ground, and this poses an opportunity to reduce fuel costs and emissions. At the heart of the system is a hydrogen fuel cell that is charged by recovering kinetic energy upon plane landings. Subsequently, pilots control the taxiing with electric motors mounted in the plane's main wheels that are powered by the fuel cell. Silencing the jet engines on the ground will provide a quiet, energy-efficient, and clean alternative to the jet engines even after the arrival of biofuel blends.

The *Solar Impulse*, a lightweight propeller plane that made a five-month round-the-world journey in March 2015, is a spectacular, if not practical, demonstration of human inventiveness. Amazingly the plane is completely powered by 17,000 solar cells integrated into its wings. The *Solar Impulse* is no model airplane; this is a 2.3 ton plane with a 72 m wingspan. The itinerary included crossing both the Atlantic and Pacific, which involved flying at night! Onboard lithium-ion batteries collected excess solar energy in order to power the overnight flights. The *Solar Impulse* project is an amazing display of ingenuity and a deliberate flash of technological prowess.

Readers will by now have identified a theme that is developing as the book examines alternative energy technology and deployment: a timely migration from fossil fuels is lacking. There is an imbalance in managing the priorities of extracting fossil fuels to power the world in the present against the opportunity to develop alternatives for our long future. Mitigating health, environmental, and climate impacts while remaining fossil fuel–dependent is like applying a new coat of paint to rotten wood.

Notes

1. U.S. Environmental Protection Agency, *Fast Facts: U.S. Transportation Sector Greenhouse Gas Emissions 1990–2011*, report no. EPA-420-F-13-033a, September 2013, accessed February 26, 2015, http://nepis.epa.gov/Exe/ZyPDF.cgi?Dockey=P100GYH6.pdf.

2. Gwyn Topham, Sean Clarke, Cath Levett, Paul Scruton, and Matt Fidler, "The Volkswagen Emissions Scandal Explained," *Guardian*, September 23, 2015, accessed October 28, 2015, http://www.theguardian.com/business/ng-interactive/2015/sep/23/volkswagen-emissions-scandal-explained-diesel-cars.

3. NationMaster, "Transport. Road Density. Km of Road per 100 Sq. Km of Land Area. Countries Compared," accessed April 17, 2015, http://www.nationmaster.com/country-info/stats/Transport/Road-density/Km-of-road-per-100-sq.-km-of-land-area.

4. U.S. Department of Energy, Energy Efficiency and Renewable Energy, "Alternative Fuels Data Center," accessed March 1, 2015, http://www.afdc.energy.gov/.

5. U.S. Energy Information Administration, "International Energy Statistics."

6. California Energy Commission, *2014 Integrated Energy Policy Report Update*, report no. CEC-100-2014-001-F, January 2015, http://www.energy.ca.gov/2014publications/CEC-100-2014-001/CEC-100-2014-001-F.pdf.

7. U.S. Department of Transportation, Federal Railroad Administration, "National Rail Plan: Moving Forward," September 2010, accessed February 26, 2015, https://www.fra.dot.gov/Elib/Document/1336.

8. C2ES Center for Climate and Energy Solutions, "Transportation Overview," accessed April 14, 2015, http://www.c2es.org/energy/use/transportation.

9. "Ahoy There! Fuel Cells on the High Seas," Nuvera Fuel Cells RSS, November 7, 2013, http://www.nuvera.com/blog/index.php/2013/11/07/ahoy-there-fuel-cells-on-the-high-seas/.

13

Heat Generation

Waste is inevitable with energy conversions,
but avoidable waste is not.

Wherever we live, the use of energy for hot water and cooking is fairly consistent. Our use of energy for space heating and cooling, however, depends much more on geography, so new technologies need to span the spectrum to serve high, nearly year-round heating/cooling as well as locales that seldom use heating or air conditioning. Today's range of products reflect this geographic influence. Our technologies include simple fans, swamp coolers, heat pumps, and dedicated furnaces and air conditioners. To power these products our energy is delivered either as electricity through the grid or via a tanker truck hauling fuels. As you can see in Figure 13.1, 72% of residential site energy consumption is related to heating and cooling.

As in the residential market, the major portion of industrial energy use is for heat generation. Beyond the employment of energy for space heating and cooling, there are industrial processes that require the direct use of heat. Temperatures for heating, drying, and curing processes can range from hundreds to thousands of degrees Fahrenheit. To understand the scale, in the United States, the demand for heat generation in the industrial sector is twice that of the residential and commercial sector.

The alternate technologies we share include solar thermal panels, ground- and water-source heat pumps, and certain combined heat

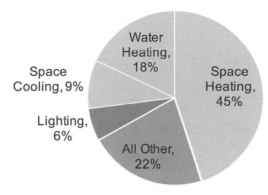

~ 72% for Heating and Cooling Applications

and power systems. But this collection of technologies has little to offer for high-temperature industrial processes. And that is why hydrogen becomes so important in the future. It will be used directly as fuel much like we use natural gas. It will also be used by fuel cells to generate electricity in decentralized environments. The criticality of infrastructure for the distribution of hydrogen will emerge and perhaps even eclipse that of today's power grid. So, we introduce the alternative fuel hydrogen. Technologies that generate electricity get all the press, but solutions for heat generation deserve some attention as well.

Figure 13.1. U.S. residential site energy consumption by end use, 2010.
Source Data: U.S. Department of Energy, *Buildings Energy Data Book* "Residential Site Energy Consumption by End Use," http://buildingsdatabook.eren.doe.gov/ChapterIntro2.aspx?2#1.

Combined Heat and Power

Images of steam locomotives and steamboats recall the dawn of the nineteenth century, when there was a certain innocence and bravado on display as the world charted new frontiers. Today we can still visit these marvels of mechanical motion, but they have given way to better methods. Later in the nineteenth century the steam turbine was applied to the generation of electricity, the origins of our centralized thermoelectric power system. Surprisingly, though, the steam turbine remains the primary means of generating electricity with fuels; even "advanced" nuclear power plants built a century later continue to employ them. The underlying flaw with centralized thermoelectric power plants is not the use of steam but that the majority of energy is squandered. Generally, these isolated plants don't have a practical application for thermal energy. Consequently, a

Most of the energy from thermoelectric plants goes into the pointless formation of clouds, not electricity. Tall cooling towers spewing puffy white vapor are the escape routes the energy takes.

Waste is inevitable with energy conversions, but micro-combined heat and power systems offer much more efficient use of energy.

traditional coal power plant delivers less than 40% of the fuel's energy to the consumer in the form of electricity, and the rest is thermal pollution. Where does it go? Most of the energy from thermoelectric plants goes into the pointless formation of clouds, not electricity. Tall cooling towers spewing puffy white vapor into the skies are the escape routes the thermal energies take. Abundant supplies of affordable and acceptable conventional fuels garnered little sympathy for the waste. Now, it is compelling that the quest for alternatives has the world examining the inefficiencies of the power sector and the idea that these thermal energy losses can contribute to the world's demands for heat generation. This deficit in the age of conservation and efficiency is stimulating some very creative products to harness this otherwise lost thermal energy. Some of the solutions can be applied in centralized locations, while others can only be realized through decentralization, bringing power generation closer to the demand for heat, and this technology is referred to as combined heat and power (CHP).

Very simply, combined heat and power systems package the delivery of electricity and heat, colocating electricity generation where the thermal energy is required and not in some faraway power plant. Large industrial customers have already adopted CHP systems powered by conventional fuels. What is new is that CHP systems are becoming smaller and less expensive, so that residential and commercial customers can imagine their adoption. Fuel cells incorporated into CHP systems will hasten the decentralization of power generation. Hydrogen in particular is best positioned to power these fuel cells. We've discussed hydrogen fuel cells, so I will be very brief here. The conversion efficiency of a hydrogen fuel cell is approximately 50%, so the balance of the energy can be directed to heat applications.[1] In actuality 90% efficiency can be achieved with hydrogen fuel celled CHP, a dramatic increase above the coal plant's 40%. The thermal energy is used for hot water heating, space heating, and even space cooling. Japan, Korea, and Germany are leading the world in the development of what are referred to as micro-CHP systems. A decade ago they were deployed in early demonstration projects. Japan's micro-CHP is the furthest advanced commercially, with 120,000 systems installed residentially as of 2015. Government subsidies have been essential in fostering early adoptions, and this has allowed suppliers to gain market experience and lower costs. Because of this, capital acquisition costs are dropping, and sales projections are in the millions over the next decades.

When the use of electricity and heat is in balance with the fuel cell's efficiency, the overall use of energy is best. This is not necessarily in line with the consumer's pattern though. To broaden the applicability of the CHP system, backup heating units are also installed. The same supply of hydrogen that feeds the fuel cell can be used directly just like we use natural gas today. There are differences between natural gas and hydrogen, but more importantly, they are sufficiently similar to allow functional substitution. The early demonstration and commercialization of these micro-CHP systems have provided market innovators a big head start on refining their products. These units pose attractive export opportunities for these countries.

An obvious question is, Where does the hydrogen production take place? Today most of the world's hydrogen production is sourced from hydrocarbons via steam reforming. In the world of alternatives there are several pathways. Existing nuclear power plants can apply otherwise wasted thermal energy to steam electrolysis; and certain biological material (i.e., algae) can be used in a process called photobiological water splitting. Finally, whenever the supply of electricity exceeds demand, the excess can be directed toward the production of hydrogen.

The vision for the energy sector is evolving as we begin to embrace new sources. We will find a new balance between the centralized and decentralized generation of electricity, and now a subtle shift in philosophy is guiding our pursuit of solutions. Waste is inevitable with energy conversions, but avoidable waste is not. Micro-CHP systems offer the promise of much more efficient use of energy. Large companies are refining the costs, and soon these "appliances" will displace energy-inefficient stand-alone heating and cooling systems.

Hydrogen: An Alternative Fuel Similar to Natural Gas

So how much fuel storage will we need when we transition to alternatives? Like many countries, the United States maintains strategic reserves of petroleum. Its Strategic Petroleum Reserve is approximately 700 million barrels, or 30 days at current consumption rates in the transportation and industrial sectors. Additionally, we maintain stockpiles of fuels (i.e., coal) for the power sector, and nuclear energy can continue to serve a role here. When we transition from fossil fuels to alternatives, we will still need a fuel, and the amount of energy we choose to accumulate is a strategic question. Fortunately, we have options beyond fossil fuels, and even better, these options are renewable. Still it would be prudent to retain supplies of both fossil and nuclear fuels for large catastrophic events (i.e., a major volcanic eruption) that impede renewable production for months if not years. So, let's briefly compare and contrast the properties of hydrogen and natural gas.

Way back in chapter 2 we studied the heat content for many of our fuel sources and learned that hydrogen has just over three times the heat content (BTUs/kg) of natural gas. This lends confidence that hydrogen can substitute for natural gas in industrial heat generation processes. There are combustion differences between the two gases that render appliances incompatible, but these difference are easily overcome with hydrogen-specific burner head designs. Hydrogen at standard conditions is the lightest of all gases, with a density about 10% that of natural gas. Consequently, a cubic meter of natural gas has nearly three times the heat content of a cubic meter of hydrogen, so volumetrically more fuel will need to be distributed. Thankfully, hydrogen is nontoxic, nonpoisonous, and noncorrosive. The fact that hydrogen is colorless and odorless and its pale blue flame is nearly invisible presents safety challenges that will require development. For natural gas, for instance, suppliers add a sulfur-containing odorant to alert the consumer of its presence. However, there are no known odorants identified that "travel" with hydrogen, and even so they would need to be compatible with fuel cells. So, research and development are working on hydrogen-specific approaches to ensure safety. Finally, utilizing existing natural gas pipelines to distribute hydrogen is possible depending on the pipelines' construction and operating pressure. High-pressure transmission pipelines are not likely appropriate for pure hydrogen. So, only portions of our natural gas infrastructure can be leveraged, which means investment.

In conclusion, the combustion properties of hydrogen and natural gas are similar, and the differences are manageable. There are already locations that distribute a mix of natural gas and hydrogen to lower their carbon footprint. Replacing fossil fuels for heat generation applications is difficult, but the heir apparent—hydrogen—is materializing.

Centralized CHP: District Heating

District heating has been around for a long time, nearly as long as we've had inefficient energy conversions at power plants. In district heating, energy losses from centralized power generation are distributed for heating applications via a thermal grid. Most often the source of thermal energy is a large plant nearby a large demand, whether industrial or residential. District heating is not an energy source per se but, rather, a method to utilize otherwise wasted energy. There are practical limits to the applicability of a thermal grid; thermal energy just doesn't travel as efficiently as electricity. Still, wherever there is a large source of wasted thermal energy "close" to a large demand, district heating can be an option. When you consider the future with renewable sources, geothermal seems an ideal candidate for district heating. Rather than convert thermal energy into electricity, heating applications can more efficiently use the source directly. District heating will be useful where favorable conditions exist, but on a global scale it will play a minor role as micro-CHP develops.

Solar Solutions

There are a couple different types of products applied to the collection of thermal energy, and most of the capacity thus far is used to heat water. Space heating applications are beginning to occur; as an example, Australia is installing combination systems that include hot water. There are two types of thermal panels I will introduce: the flat plate collector and the evacuated tube collector. And believe it or not, these same panels can be applied in space cooling. Currently they number in the thousands and are predominantly located near the Mediterranean.

Not every panel you see on rooftops is solar PV; some of these are used to deliver thermal energy. A flat plate collector panel is made up of an absorber, a top surface glazing, heat-transfer fluid piping, and an insulated backing. The glazing reduces heat loss and keeps the inside surfaces clean to maximize solar absorption. The absorber plate is the crux of the system, designed to maximize the conversion of solar radiation into thermal energy and effectively deliver this to a transfer fluid. Circulating fluid flowing through piping attached to the absorber plate transfers the thermal energy from the panel to the end application. The fluid can be water or a water/antifreeze mixture to support lower operating temperatures.

Nature abhors a vacuum, but engineers love it. Evacuated tube collector panels utilize a very old invention to capture and transfer thermal energy: the heat pipe. Under standard conditions, water boils at 212°F (100°C); in an evacuated tube (i.e., 0.042 atm) the water boils at 86°F. So, as the system absorbs solar radiation it causes vaporization inside the heat pipe. The vapor rises and delivers its thermal bounty inside a manifold. The lower temperature of a circulating fluid causes the vapor to condense and release the water's latent heat of vaporization. The condensate falls like rain within the heat pipe, where it can begin the cycle again. To make the system even more efficient, the heat pipes are encased in vacuum-sealed glass tubes that reduce the loss of thermal energy. These capabilities allow an evacuated tube collector system to operate across a broad range of operating temperatures.

China is far and away the leader in the development and adoption of thermal solar systems. Millions of installations across the country deliver hot water to residences, and some of the systems are core elements of a fully integrated central heating and cooling application.

Ground- or Water-Assisted Heat Pumps

Closely related but not to be confused with accessing pure geothermal energy is geo-exchange, where heat pumps use ground- or water-source thermal energy to boost the efficiency of heating and cooling systems. Depending on location they can dramatically improve heat pump efficiencies by 30% to 60%. They dump thermal energy when cooling and collect thermal energy when heating. Although it is often referred to as geothermal, the energy is primarily imparted by the ground's absorption of solar energy and consequently would be better named geo-solar. With geo-solar we are still using an indirect energy source, electricity, but much less.

An Implausible Application: Cooling

At first blush it seems beyond unlikely that solar thermal energy could be useful in cooling one's residence or office space. Well, it works. More as a celebration of out-of-the-way thinking than as a world-turning advancement, we look at the basic science of vapor absorption refrigeration systems (VARS). When I was working a summer construction job between semesters at the University of Arizona, we would fill a canvas bag, a "Desert Bag," with water at the start of the day. The bag kept the water cool all day long in the sun. The evaporating water confiscated the latent heat of vaporization as it escaped, thereby cooling the bag. Tasting cool water all day was infinitely satisfying. Today, it is interesting to note once again that basic principles can be applied in lieu of more complicated technology. On hot days, air conditioners and their electrically powered compressors challenge power grids the world over. Solar cooling has the potential to reduce peak loads in the near term while slightly shifting our energy mix to renewables.

A couple of solar technologies have been applied to the application of cooling: absorption refrigeration and desiccant types of systems. Absorption cooling isn't a new idea; the French scientist Ferdinand Carré first invented it in 1858. Talk about thinking outside the box—a scheme to use a heat source to drive a cooling process that actually works is the essence of genius. Here in the twenty-first century, the most common VARS refrigerants are lithium-bromide (LiBr)–water and ammonia, both of which are environmentally benign. The second method employs solar heat to regenerate desiccants (i.e., silica gel), which are used to dry the air. The removal of moisture from the air in humid environments provides a cooling effect.

Most of today's refrigerators/air conditioners use mechanical compressors to elevate the refrigerant's temperature above ambient, and this allows the system to "dump" heat through the condenser. Alternately, the absorption system uses heat instead of a compressor to increase the refrigerant's temperature. VARS operate quietly except for the pumps that move the system fluids around, and the electricity required to operate pumps is less than that required for compressors. Figure 13.2 helps explain the principle of operation. When LiBr is added to water it reduces the vapor pressure, and this pressure differential is what causes the circulation of water, just like atmospheric pressure differentials cause the movement of air (wind). There are two reservoirs of LiBr and water, one maintained with a strong and the other with a weak concentration. The weak concentration reservoir is where heat is removed, while the strong concentration reservoir facilitates the "dumping" of heat. Amazingly, the net effect is space cooling below ambient temperatures.

Incoming air or water from the space being cooled (1) passes through the evaporator (3) and returns chilled (heat removed). The pure water container (2) cools as water evaporates and escapes to the LiBr solution, somewhat akin to the "Desert Bag." This allows the evaporator to extract heat and provide a pleasant cooling effect. The water flows from the evaporator to the lower LiBr solution, and this reduces the concentration of LiBr. To keep the refrigeration cycle operating, though, the concentration of LiBr needs to be maintained and the evaporator's water needs to be replenished. The first step was the removal of heat; now the system needs to exorcise the heat. The solar collectors (4) allow this to occur. When heat is applied to the top vessel the vapor pressure increases, causing the release of water vapor, and this in turn causes the concentration of LiBr in the top reservoir to increase. The water vapor exits to the condenser (5), where it "dumps" its heat, and returns to the evaporator to start the process again. There is an exchange between the strong and weak LiBr solutions, and this maintains the vapor pressure differentials that power the refrigeration cycle. These are the basic principles of operation behind today's commercialized VARS. Interestingly, Albert Einstein and Leo Szilard devised a refrigerator so simple it didn't even require the use of pumps. A century later, efforts are still under way

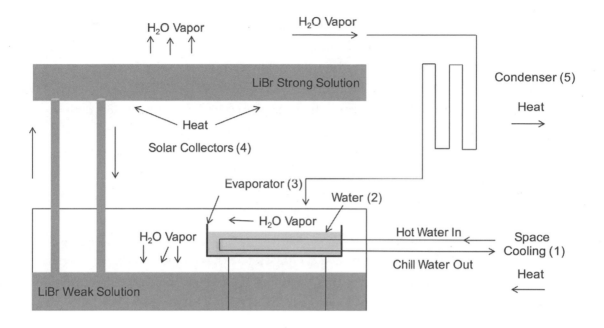

Figure 13.2. Lithium-bromide refrigeration cycle.

to help realize the concept, which could help remote locations lacking access to electricity.

Along with desiccants and absorption, there is a third method of solar cooling. Solar-assisted air conditioning reduces the workload of the conventional heat pump's compressor. Solar thermal panels can preheat the refrigerant and reduce the work of the compressor.

Today, 81% of solar cooling installations are in Europe, typically in sunny and dry climates. Adoption is starting to increase elsewhere, in Australia, India, and the Middle East.

Heating and Cooling Applications Too Big to Ignore

The key challenge for solar heat is the need for supplemental systems. In most cases you cannot or will not forgo your conventional water heater, so solar heating is supplemental. A 2014 report from the European Technology Platform on Renewable Heating and Cooling, cofunded by the European Union, shares a technology road map that it hopes will spark adoption of solar heating technologies.[2] The study sites the flaws of the current renewable energy marketplace, where consumers must gather pieces like they are putting together a puzzle to assemble a working system, as the main obstacle to adoption and reliability. In contrast to that multistep piecemeal process, the report describes a Solar Compact Hybrid System, a new type of home appliance including a hot water tank and a backup water heater all integrated with a solar heat collection system. No doubt this will be a welcome integration and stimulate placements. It is easy to envision that an enterprising organization that agrees with the report's assessment will develop the new-age appliance in the foreseeable future.

According to REN21's *Renewables 2014 Global Status Report*, slightly more than 1,000 solar cooling systems have been installed worldwide—small numbers at this point in time. These systems are more complex than heating-only systems. The durability, maintenance, and cost of the systems will need to continue to improve in order to see expanded adoption. Worldwide installation growth was 40% annually between 2004 and 2012. Smaller systems (<20 kW) are opening new market opportunities for this technology.

Notes

1. Paul E. Dodds, Iain Staffell, Adam D. Hawkes, Francis Li, Philipp Grünewald, Will McDowall, and Paul Ekins, "Hydrogen and Fuel Cell Technologies for Heating: A Review," *International Journal of Hydrogen Energy* 40, no. 5 (2015): 2065–2083.
2. Aleksandor Ivanic, Daniel Mugnier, Gerhard Stryi-Hipp, and Werner Weiss, "Solar Heating and Cooling Technology Roadmap," European Technology Platform on Renewable Heating and Cooling, June 2014, http://www.rhc-platform.org /fileadmin/user__upload/Structure/Solar__Thermal/Download/Solar__Thermal__ Roadmap.pdf.

14

Technological Readiness

God gives the nuts, but he does not crack them.
—Franz Kafka

We've looked at how worldwide energy demand is forecasted to grow over the next several decades. We've explored noncarbon energy sources and their raw potentials. Finally, we've looked at technologies to understand their realistic ability to harness this energy. Now let's integrate these pieces to get a sense of our transition readiness and understand the impediments to replacing fossil fuels. We are starting to answer some of the big questions as we next embark on a glance at selective country energy plans.

The hard work of energy planning is choosing the toll we accept for the sources we deploy and the degree to which energy conservation can reduce our footprint. So, technology preferences will be subject to value systems, and those preferences vary nation by nation. We will understand why a single formula for a noncarbon energy mix is not practical on a local level when we examine individual countries. Some countries may be less risk-averse to nuclear power, others may prefer centralized to decentralized solutions, and others may place a premium value on the use of land. A country's political environment can also influence its ability to make tough decisions. Still, it is not uncommon to find literature that stipulates a precise energy mix promulgated by models that trivialize the relationship between local values and choice. Do we build more

We must put in the hard work of energy planning to choose the toll we can accept for the sources we deploy and the degree to which conservation can reduce our footprint.

dams, construct nuclear plants, and continue to centralize? Do we continue to extend our surface occupancy by millions of acres to satiate our energy appetite? These are neither simple nor easy choices, and one scenario will not fit all governments or localities. Aggregating globally, however, we can forecast a formula: we should harness energy from the opportunistic renewables (geothermal, hydro, wind, and tidal) wherever possible, with the balance coming from a combination of solar and nuclear. We will always do well if we can curb our appetites, no matter the energy mix.

Renewables and nuclear energy generated about one-third of the world's electricity in 2012. As we move toward 2040, new capacity in wind, hydro, biomass and waste, and geothermal power will be hard-pressed to maintain share levels, as global electricity demand is set to increase 82% between 2015 and 2040. Looking back, worldwide electricity consumption increased 112% from 1987 to 2012. Although the projected percentage is slightly lower, the absolute growth is quite similar, and the implication is that our world is still trying to turn on the lights as we confront a major energy transition. Figure 14.1 shares the world's renewable power capacity along with the solar thermal capacity in place at the end of 2014. We crossed an important line in 2014; global nonhydro renewable capacity for the first time exceeded hydro capacity (when thermal is included).

Nuclear investment is critically important, but the fast nuclear reactor is a decade away, assuming successful performance testing. Fusion, an even longer-term bet, has yet to demonstrate laboratory feasibility. In the 1990s nuclear energy provided 17% of the world's electricity, while today that figure is 11%. Nuclear fission is in a holding pattern in the United States, but elsewhere nuclear power is expanding, and new technology is entering the market.

That leaves solar, the most potent renewable, to provide the balance of alternative energy. The rapid pace of change in solar technology fosters a tendency to delay adoption until the advances stabilize. There is no such thing as a mature technology at launch. Examples abound of the rapid rate of product evolution after commercialization, including cell phones, televisions, and computers. Moore's law is obvious across many

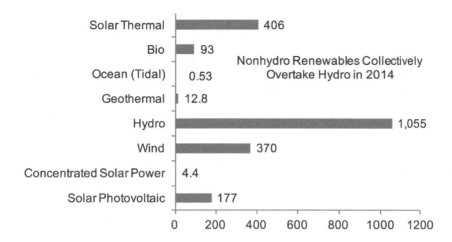

Figure 14.1. Worldwide renewable capacity (GW) in 2014. *Source Data:* REN21 Renewable Energy Policy Network for the 21st Century, *Renewables 2015 Global Status Report* (Paris: REN21 Secretariat, 2015), http://www .ren21.net/wp-content /uploads/2015/07 /REN12-GSR2015 _Onlinebook_low1.pdf.

products in the electronics industry, so much so that features of the initially launched product seem primitive and obsolete in just a few years. Once a product is launched, cost pressures and new features that attract the customer drive further evolution of the product. Launching a product actually speeds its evolution, though, so waiting for the ideal product is counterproductive. Many governments should be given credit for engaging policies that stimulate initial demand.

Lessons from early movers inform us of the critical role underpinning technologies play in shifting the energy mix toward renewables. The most common manifestation of this is seen in curtailment, where the grid is swamped with oversupply and renewable supplies are dropped (lost). Without operational changes, curtailment of wind and solar power can range above 10%. Eliminating curtailment altogether will require smart grids and storage technologies.

A full-scale plan to establish a hydrogen system for the accumulation, distribution, storage, and retrieval of energy is lacking. This applies to hydrogen used for heat generation, transportation, and the power segment. We can do without this new-age fuel system during a transition period when we continue to use fossil fuels, but we will never make the final leap from fossil fuels without these support technologies in place.

Table 14.1 summarizes the status of support technologies. The "Time Frame" column estimates the amount of time it will take with sustained

Table 14.1. Supportive Technology Readiness.

Support Technology	Technology Readiness	Time Frame (Years)
Hydrogen fuel		
Production	☑	50
Transport and storage	☐	50
Fuel cells	☑	50
Storage (battery)	☑	30
Smart grid	☑	30
Transportation stations		
Electric recharging	☑	20
Hydrogen refueling	☑	50

There isn't a technical obstacle to transitioning; what we need is decades-long steady investment behind infrastructure to pave the way for higher levels of alternatives.

funding to build the necessary infrastructure in order to completely replace fossil fuels. The technology necessary to harness large amounts of renewable energy is ready, with the exception of choosing a method for the transport and long-term storage of hydrogen. There isn't a technical obstacle to transitioning; what we need is decades-long steady investment behind infrastructure in order to pave the way for higher and higher levels of alternatives. Where conservation and efficiency are priorities, smart technologies will bring the consumer the practical tools to act.

Regarding our energy sourcing technologies, Table 14.2 summarizes the status as discussed in this section. With the exception of nuclear technologies, the renewable technologies are ready. Deploying fission, assuming the fast reactor, is zero (Russia) to 20+ (U.S.) years away, while fusion is at least 20 years away and likely much more. Land use is straightforward and is included because it will be a factor in the selection of renewable technologies given such low power densities. The multiuse designation indicates whether harnessing energy can be accomplished without dedicated use of the land. For instance, rooftop/roadway solar panels and wind farms leverage surface area already applied to other purposes. In addition to land use, the cost of energy from temporal sources such as wind should be discussed while considering the round-trip efficiency of storage. The value of fluctuating renewables will depend on a combination of the two costs. What we can assume is that the cost curves for solar PV across segments will cross those of fossil fuels no later than 2040. The cost of wind is already competitive, so its box in the "Cost" column has been checked, and nuclear fission is checked on the basis that, if it is proved, we have the technology to generate electricity effectively. Still, the story of energy costs is a little more complicated, so chapter 28 will discuss this in further detail.

These tables outlining time frames for infrastructure development and sourcing technology readiness imply that we must focus on both solar PV and the fast reactor. Without a sizable contribution from these two sources, the lights will be dimmer in the future. The early phase of the transition was easy; now we need to make complementary investments in energy management systems and smarter grids to further advance our changeover.

Going beyond Depletion to Sustainable Energy Collection

Long before the term *sustainability* became almost ubiquitous, the Iroquois Nation in North America governed in accordance with its "seventh

Table 14.2. Alternative Energy Technology Readiness.

Energy Source	Technology Readiness	Time Frame (Years)	Land Use	Cost of Energy
Wind	☑	Ready	Multi	☑
Solar photovoltaic	☑	Ready	Multi	☐
Concentrated solar	☑	Ready	Single	☐
Nuclear fission	☐	0–20+	Single	☑
Nuclear fusion	☐	20+	Single	TBD

generation principal." Every decision must consider how descendants seven generations into the future will be affected. As the Iroquois Great Law of Peace advised, making decisions with this time window often required "skin as thick as the bark of pine." Today we would be wise to share this longer-term view in making decisions and taking actions.

We are armed with a good understanding of the world's energy appetite, the fundamental science of alternative sources of energy, their potential, and the readiness of technologies. Now we glance around the world to see what countries are doing, how they are trying to drive supportive behavior, how they apply different value choices, and what we can learn from each other.

Where conservation and efficiency is a priority, supportive technologies will bring the consumer the practical tools to act.

Vignette: The Arc of Discovery

Our knowledge seems to increase in fits and starts, and some momentous discoveries find immediate application, while others create the spark that opens doors to entirely new fields. Alessandro Volta's creation of the "modern battery" in 1799 is one spark. Upon learning of Volta's battery, researchers around the world built their own versions and began experimenting. Almost immediately William Nicholson in 1800 observed that when leads from a battery were placed in water, hydrogen and oxygen formed separately. This was the world's first electrically induced chemical reaction and uncovered the production part of the hydrogen cycle, electrolysis. Later in 1839, Sir William Robert Grove invented a "gas battery," the precursor to today's hydrogen fuel cell. Hydrogen ionized in the presence of a catalyst, allowing the generation of electricity. Water was reformed when the hydrogen ion, the electron, and oxygen combined. The complete and reversible hydrogen cycle was understood, and today it promises to be the means of accumulating energy in the future. As we already know, in the same year of 1839, a 19-year-old Edmond Becquerel observed the "photovoltaic effect." As we think about an energy future absent fossil fuels, hydrogen fuel cell and photovoltaic technologies will be considered some of the most vital. The year 1839 should be remembered as a pivotal year of discovery for alternative energy technology critical for the twenty-first century and beyond.

The second half of the nineteenth century was equally remarkable, recording the extraordinary genius of Nikola Tesla and his powerful portfolio of inventions, which included patents for the generation and transmission of alternating current. As the nineteenth century was drawing to a close, discoveries were revealing yet another field critical to the future: nuclear. Einstein's 1905 simple equation relating energy and mass with the speed of light placed the foundation for understanding the power behind the stars.

Two Major Technology Arms of Alternative Energy

Renewables

Third Century B.C.	Water Mill—Greek Engineer Philo of Byzantium
First Century A.D.	Parthian Battery—Mesopotamia
Sixth Century	Windmill—Persia
1799	Modern Battery—Italy, Alessandro Volta
1800	Electrolysis—United Kingdom, William Nicholson
1839	Fuel Cell "Gas Battery"—United Kingdom, Sir William Robert Grove
1839	Photovoltaic Effect—France, Edmond Becquerel
1858	Absorption Refrigerator—France, Ferdinand Carré
1887	Alternating Current Generation and Transmission—Serbia, Nikola Tesla

Nuclear

1896	Radioactivity of Uranium—France, Antoine Henri Becquerel
1905	$E = mc^2$—Germany/Switzerland, Albert Einstein
1920	Principal of Fusion—United Kingdom, F. W. Aston

1932   Discovery of the Neutron—United Kingdom,
            James Chadwick
1932   Helium-3 and Tritium Fusion—Australia,
            Mark Oliphant
1938   Uranium Atom Split with Neutrons—
            Germany, Otto Hahn and Friedrich
            Wilhelm Strassmann
1939   Nuclear Chain Reaction—Italy and Hungary,
            Enrico Fermi and Leo Szilard

PART III

A Scan of Country Energy Plans

The Globe We Share

15

Fossil Fuel Wealth and Energy Consumption Framework

There are no rules for good photographs,
there are only good photographs.
—Ansel Adams

Thus far the discussion of climate change in this book has been minimal in order to give other influences on energy sourcing a fair share of the dialogue. Like being in a conversation where you can't get your voice in edgewise, examining the topic of energy without climate change taking over the discussion is difficult. There are many forces acting against the continued use of fossil fuels as an energy source, and climate change is a subcategory in just one of these forces. When we discuss energy policy only from the perspective of climate change we create unnecessary divisiveness and we allow ourselves to be distracted from the bigger picture—the discontinuance of fossil fuels as an energy source. However, because climate change is currently the centerpiece for global coordination, it warrants a brief overview and a look into how it vies with other energy priorities.

Climate change is often framed in the context of the world's use of fossil fuels. Today in places the lines are blurring between climate policy and energy policy. Just as climate policy shouldn't be limited to the energy sector (for example, agricultural patterns and deforestation have major climate change impacts), energy policy shouldn't be narrowly defined by GHG emissions (for example, independence and access priorities may conflict with those of GHGs). Where climate change and energy priorities meet, ideally both will be served.

Furthermore, energy policy that sets greenhouse emission reductions as a top priority and omits other strategies to convert fully to alternative sources and lower energy intensity (conservation and efficiency) is like climbing partway up the mountain to base camp only to learn that the path has no upward access to the summit. If we myopically pursue the goal of reducing greenhouse gases, we end up with lower emissions in the 20% to 30% range. But we still will be an economy and a society powered by fossil fuels that will become unruly in the years ahead. We will not have eliminated the carbon problem, and we will be as unprepared as ever for life after fossil fuels. We will have invested effort and money into dead ends and robbed time from more permanent solutions.

As long as the illusion of limitless supplies of fossil fuels remains and the world's economy falsely assigns great wealth to reserves, many will see little reason to change.

As an example, we can significantly lower GHG emissions by switching from coal to gas, providing flaring and methane emissions are addressed, but we are still switching from one economically finite source to another. Retrofitting coal plants to natural gas and making cars and jet engines more efficient are valuable but no substitute for building noncarbon capacity and halting the use of fossil fuels. Carbon reduction, while urgent, doesn't fix the looming problem that fossil fuel supplies have limits for all the reasons described, and lacking alternatives, we place global stability at risk. Large investments in the marginal reduction of carbon footprints shouldn't distract us from the full pursuit of building fossil fuel alternatives. Conversely, if our goal is to address the limits of our carbon energy sources by building significant alternative capacity, we solve both the finite supply problem and climate concern. The right goal will lead to the summit—a smooth transition to alternative energy supplies that are both plentiful and for the most part carbon-free.

The United States is admittedly one of the globe's leading consumers of fossil fuels, so it's instructive to use U.S. public opinion to illustrate again the disadvantages of just focusing on climate change. A 2010 Pew Research survey showed that while 40% of the U.S. population denies human impact on climate change, 75% accept that fossil fuels will one day run out.[1] A more recent 2014 Pew Research study reported that 61% of Americans believed that there is solid evidence that the earth is warming but only half considered the warming a significant threat.[2] So, in the United States, seven of 10 people don't think that climate change is a priority. Therefore, when the United States discusses shifting energy sourcing because of climate change, 70% of the population is disengaged. Not surprisingly, the opinions were sharply divided by political affiliation. Rather than remaining trapped in debate over climate change, almost everyone can agree with the idea that a transition from fossil fuels to alternate sources will need to occur someday. If the lead time to transition to alternative energies is similar to the depletion window for affordable fossil fuels, most of us can feel a responsibility and an urgency to take up the task.

The concern with greenhouse gases, global warming, and other climate changes is vital, though. It adds incentive for faster action within a broader energy plan to replace fossil fuels altogether. Investing in carbon capture and sequestration technology and power plant retrofits should be in addition to, not in lieu of, alternatives. Included at the end of this chapter is a brief examination into the facts and fiction of mitigating CO_2 emissions while continuing the use of fossil fuels. As long as the illusion of limitless supplies of fossil fuels remains and the world's economy falsely assigns great wealth to reserves, there will be less impetus to change.

A Glance around the World

Critically judging a country's energy policy from the outside is neither fair nor productive, since every country faces a unique set of conditions. What is needed is understanding and empathy; there is no magic wand for transitioning from fossil fuels. Fully developed countries with abundant reserves of fossil fuels will likely take a measured position on the issue of if, when, and how to convert, while developed countries that are poor in reserves will build domestic capacity sooner rather than later. These import-dependent countries are further incentivized to harvest domestic sources of energy in the face of clear economic signs forecasting price volatility as cheap supplies will absolutely give way to more expensive extraction. Countries with reserves will want to exploit them as an economic asset even while acknowledging consequential emissions.

For those countries without reserves, supply instability in the future is a direct threat to their long-term economic and national security.

Another predictor of behavior relates to a nation's energy consumption level. If you are one of the 22 countries already using 75% of the world's fossil fuels, you are far more likely to put carbon caps or reduced carbon emissions on the table. If you are a country trying to bring a large percentage of your citizenry "to the grid," such as India, the priority of achieving energy access is way ahead of capping carbon emissions, and who can blame you? Developed countries will need to counterbalance this if the world intends to decarbonize the atmosphere.

Fair judgment of any country's energy plan, then, requires an understanding of its history and current circumstances. Energy plans will always wrestle with balancing near-term responsibilities such as access and affordability with longer-term responsibilities of national security, environmental stewardship, and sustainability. With few exceptions, a country's energy plan reveals that its reserve wealth (its supply of fossil fuels) clouds its political, economic, and environmental stand on addressing the limits of its fossil fuels. A review of energy plans from around the world and a basic acknowledgment of human traits tells us that those countries with reserves in the ground will surely be tempted to get them out and use them up, either for their own nation's consumption, or for export, or for both.

The following framework allows us to anticipate a country's priorities and what will motivate change. The Fuel Wealth and Consumption framework calculates two factors: a nation's relative wealth in terms of its fossil fuel reserves per capita—Fossil Fuel Wealth Factor—and the amount of primary energy that country uses per capita relative to all other countries—Primary Energy Consumption Factor.

Fuel Wealth, denoted by the Fossil Fuel Wealth Factor (FFWF), is displayed along the horizontal axis in Figure 15.1. Think of the FFWF as simply a measure of a country's worldwide share of proven fossil fuel reserves. The higher this value, the wealthier the country. If a country's FFWF is less than 1 (left of vertical line), it has less than its "fair share," on a per capita basis. Conversely, large numbers convey that the country has a disproportionately high share of the planet's fossil fuel wealth.

Figure 15.1. Fossil Fuel Wealth Factor by Primary Energy Consumption Factor. *Note:* See Table 15.1 for an explanation of how the factors are calculated.

The Primary Energy Consumption Factor (PECF) is a measure of a country's relative energy appetite. The PECF is displayed along the vertical axis. A low value indicates that that country is using proportionately less energy than the world per capita average. Energy consumption reflects many contributing factors such as percentage of population with access to energy, societal behavior, policy influencing conservation, climate, and economic orientation such as heavy versus service industry.

These measures are helpful in predicting responses to the ultimate limits of fossil fuel, since no country has a blank sheet of paper on which to plot its path to the future. Infrastructure, industry, transportation systems, population demands, and natural resources frame their ability and desire to change their energy course. So unless they face immediate crisis,

If they fail to adapt, countries wealthy in fossil fuels may find themselves at a great disadvantage in the post–fossil fuel era to come later this century.

countries will likely plan long transitions rather than adopt quick shifts. Problematically, most country energy plans are concerned with near-term access and the economics of conventional energy. But where we find longer views or a low FFWF, alternative energy capacities are increasing.

Looking at where a country falls in Figure 15.1 can help us anticipate the basis from which it draws its energy plans. Is the country struggling to develop? Is the country energy self-sufficient? Or is the country poor in resources and import-dependent because of its high energy consumption? The chart plots 14 representative countries by their fossil wealth and energy consumption factors. The 45-degree line shows the point where a country's fossil wealth and energy consumption factors are equal—in other words, where a country is balanced in terms of reserves and consumption. We can also anticipate that those countries whose fuel wealth and energy consumption are matched (close to the 45-degree line) will place a priority on self-sufficiency (economic concerns) over conversion to alternatives. They won't feel supply pressures. The vertical line helps predict which countries would emphasize seeking alternative energies and mitigating the risks within their fossil fuel supply networks (those located to the upper left of the line). Finally, the horizontal line anticipates one of two environments: countries in the lower left are much more likely to emphasize economic development and energy access over issues of health and environment, while developed countries in the upper left will advance energy efficiency and conservation programs and the pursuit of domestic alternatives. This framework gives us good insight into whom to watch for what.

Looking at Figure 15.1, you can expect Russia and Saudi Arabia to be exporters of fossil fuels. You can surmise that China, Brazil, and India will emphasize economic development followed by the securement of stable energy supplies. Italy, Spain, Japan, and South Korea, poor in fossil fuel wealth but considered developed, will pursue alternatives and conservation measures. The United States, Canada, Germany, and Norway, then, should be expected to emphasize self-sufficiency first, with renewables as a secondary goal. Germany actually defies this prediction. Germany's FFWF is largely due to some of the world's dirtiest coal, and consequently its fossil fuel wealth is overstated as pressures to discontinue its lignite usage increase. When this is considered, Germany is similar to Japan, and the framework proves decidedly predictive. Canada's and Norway's significant hydropower capacity bumps them fully into the exporting category. Table 15.1 shares the means used in calculating the FFWF and PECF.

There are 146 countries with both factors below 1. Another 51 have an FFWF below 1 and a PECF above 1, meaning that they have relatively lower fossil fuel reserves but a relatively high consumption dependence. Together nearly 200 countries have limited domestic supplies of fossil fuels. Most of these same countries do, however, have more than enough domestic potential with renewables. The sun shines brightly in many places, so transitioning to renewables for these countries can unleash economic development, mitigate supply risk, and provide some measure of energy security.

There are only 27 countries with both factors above 1. These are the privileged few. As we've seen, fossil fuels were not evenly deposited across the planet. This uneven distribution created inherent advantages for some nations and continues to provide benefit in trade for the petroleum-rich. The disparity has already led to conflicts, and in the future it will become

Table 15.1. Fossil Fuel Wealth Factor (FFWF) and Primary Energy Consumption Factor (PECF) Calculation.

Variable	Meaning
Rw	= Worldwide reserves
Rc	= Country reserves
Pw	= Worldwide population
Pc	= Country population
Ew	= Worldwide energy consumption
Ec	= Country energy consumption
FFWF	$= (Rc/Pc) / (Rw/Pw)$
PECF	$= (Ec/Pc) / (Ew/Pw)$

Note: Reserves of oil, natural gas, and coal are converted to a common unit of BTUs using established heat values.

a source of increasing regional and global instability absent alternatives. For further reference, appendix A shares the FFWFs and PECFs for most of the world's countries.

Renewable energy potential is far more equitably distributed. The basis of competition will change, and countries blessed with fossil fuel reserves may be slow to transition. If they lag too far behind, however, the seesaw may fall the other way; fossil fuel–wealthy countries failing to adapt may find themselves at the greatest disadvantage in the post–fossil fuel era later this century.

In the meantime, countries whose energy costs are likely to rise within the depletion window will feel a certain kind of economic urgency. For example, China's low cost of labor and infrastructure investments have helped attract and establish a large manufacturing base. As China's wages rise, its manufacturing base will be at risk, and if China cannot continue to secure cost-competitive energy, this critical economic base will be a flight risk.

Examining country energy plans will reveal each nation's sense for the energy transition already under way. Are they leading, lagging, or simply hopeful for others to find solutions?

A Brief Diversion on the Subject of Nuclear Energy

Nuclear power warrants a brief discussion before launching into the energy plans of specific countries. Nuclear power doesn't emit CO_2 but has minimal greenhouse emissions related to the extraction of uranium and plant construction and operations. There are insufficient uranium fuel reserves, given projected consumption levels, to sustain nuclear power beyond the twenty-first century without successfully adopting new technologies, and the confidence is high that this will happen. The innovators are banking on the fast reactor.

On a worldwide basis, nuclear power is projected to grow 60%, from 392 GW in 2013 to 620 GW by 2040, according to the Institute for Energy Research (IER).[3] China accounts for 45% of the anticipated growth. The World Nuclear Association reports that globally 65 nuclear reactors are under construction and another 150 are in active planning

The US and France
Supply 50% of the World's
Nuclear-Generated Electricity

Figure 15.2. Countries practicing civil nuclear power and their percentage of worldwide electricity generation. *Source Data:* International Atomic Energy Agency, "Nuclear Share of Electricity Generation in 2015," accessed August 1, 2016, https://www.iaea.org /PRIS/WorldStatistics /NuclearShareofElectric ityGeneration.aspx.

stages,[4] aligning with the IER projection. There were 28 countries generating nuclear energy in 2012, while the number of countries planning nuclear power will increase this figure to 48 countries, indicating that for many the source is a viable alternative to fossil fuels.

Currently, just eight countries generate 80% of worldwide nuclear energy. Figure 15.2 displays the countries that have access to civil nuclear power along with the portion of worldwide nuclear electricity they generated in 2015. Japan's figures are below what can be expected, as it restarts plants, while Germany is higher than expected, as it surprisingly shuttles plants. Many of the top generators are also leaders in the export of nuclear technology.

Looking forward, the role of nuclear power may well increase beyond the IER projections. If the practical and acceptable deployment of renewables falls short of worldwide energy demand, nuclear is the other option. The hurdle to exercising the nuclear option has been discussed. Safety liabilities have held nuclear power at bay, but the moment these burdens are practically eliminated is when nuclear power can become a vital contributor in the retirement of fossil fuels.

The FFWF and PECF framework gives us a way to look at representative countries around the world, understand the motivations behind their energy plans, and assess their progress in developing renewable and/or nuclear capacity. There is no doubt that many other countries are working hard on progressive energy plans. The goal of this book is not to detail worldwide energy plans per se but, rather, to answer questions about alternative energies and whether these sources have the potential, the technologies, and the attractiveness to replace fossil fuels. Examining country energy plans helps answer these questions. The answer to who has the correct plan is a local matter. Every country will feel that it has the right plan for its situation. Objectives to reduce GHG emissions endorsed by the Intergovernmental Panel for Climate Change are noble, but they often clash with local priorities.

The countries we study represent about 50% of the world's population and provide a sample from the important categories described in the FFWF and PECF framework. We study Japan, South Korea, and Germany, considered vulnerable given their lack of domestic fossil fuel reserves and relatively high primary energy consumption. Representing the dually challenged category are China (vulnerable/dually challenged), Brazil, and India, with both of their factors relatively low. The rest of the countries we study fall into the fortunate category, including the United States, Canada, Russia, and Saudi Arabia. Although this group is characterized as fortunate for the present, they may be vulnerable to the resource curse, lingering too long at the well. Adapting their

Table 15.2. Baseline 2015—Global Renewable Energy Capacity Rankings by Total Capacity and Capacity Per Capita.

ENERGY SOURCE	RANKING				
	1	2	3	4	5
Total Capacity					
Renewable power (including hydro)	China	United States	Brazil	Germany	Canada
Renewable power (not including hydro)	China	United States	Germany	Spain/Italy	Japan/India
Biopower generation	United States	Germany	China	Brazil	Japan
Geothermal power	United States	Philippines	Indonesia	Mexico	New Zealand
Hydropower generation	China	Brazil	Canada	United States	Russia
Concentrated solar thermal power	Spain	United States	India	United Arab Emirates	Algeria
Solar photovoltaic	Germany	China	Japan	Italy	United States
Wind power	China	United States	Germany	Spain	India
Capacity Per Capita					
Renewable power (not including hydro)	Denmark	Germany	Sweden	Spain	Portugal
Solar photovoltaic	Germany	Italy	Belgium	Greece	Czech Republic
Wind power	Denmark	Sweden	Germany	Spain	Ireland

Source Data: REN21 Renewable Energy Policy Network for the 21st Century, *Renewables 2015 Global Status Report* (Paris: REN21 Secretariat, 2015), http://www.ren21.net/wp-content/uploads/2015/07/REN12-GSR2015__Onlinebook__low1.pdf.

economies may prove to be a more difficult task than the establishment of domestic energy for the rest of the world.

Table 15.2 provides country rankings for renewable capacity according to a Renewable Energy Policy Network for the 21st Century report. The bottom section of Table 15.2 provides rankings on a per capita basis,

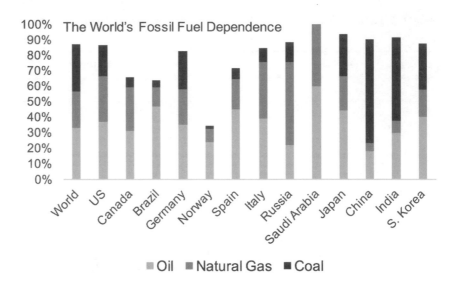

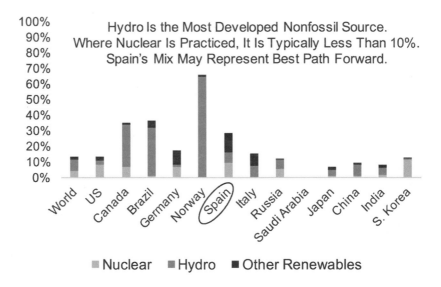

Figure 15.3. Study group countries' primary energy mix, 2013. *Source Data:* BP, *BP Statistical Review of World Energy June 2014*, June 2014, accessed April 9, 2015, http:// www.bp.com/content /dam/bp-country/de _de/PDFs/brochures /BP-statistical-review- of-world-energy-2014 -full-report.pdf.

and this helps reveal other countries that are building substantial renewable capacity. Western Europe dominates the per capita top five across the board. What is it about Europe that stimulates this production? A lack of fossil fuel reserves and an energy-dependent developed economy! In the chapters ahead we examine the energy plans for many of these top-ranking countries.

Figure 15.3 displays the primary energy mix for the group of countries we are about to examine. The 2013 primary energy consumption data are from a BP report. It will be interesting to update these graphics every decade and observe the changes in the energy mix. As we embark on country examinations, these graphics will help us compare and contrast the environments within the study group. In the upper graph we can see that a mix of fossil fuels support the vast majority of the group's primary energy. We also notice that the challenge for China and India is to reduce their relatively high dependence on coal. Russia, home to some of the world's highest natural gas reserves, readily relies on this particular fossil fuel for most of its primary energy. The chart on the bottom of Figure 15.3 breaks out the alternative contributions to primary energy. One can anticipate that the fossil fuels will furnish less and less primary energy as the alternative sources emerge. For today, though, only hydropower has established a strong position in primary energy, most notably in Norway, Brazil, and Canada. Collectively, nuclear and renewables are making significant contributions in Spain, and this mix may represent the best way forward. The rest of the countries in the study group roughly mirror the worldwide averages in the use of fossil fuels. The exception would be Saudi Arabia, which is solely fossil-fueled.

Let's take a closer look at specific country energy plans and get a sense of our global position as we progress through the twenty-first century.

Comment: Climate Change Mitigation with Fossil Fuels: Fact or Fiction?

There are many forces striking back at the continued use of fossil fuels, not least of which is climate change mitigation. Is it conceivable that we can still enjoy the energetic benefits of carbon-based fuels while avoiding emissions of GHGs? Is migration to noncarbon energy sources the only way forward? If carbon capture and sequestration (CCS) could remove better than 90% of a power plant's CO_2 emissions, it would help relieve the climate change pressures that are part of a broader health and environmental category. Later, in chapter 30, we discuss four forces that are applying constant repressive pressure on the use of fossil fuels. If the strength of any one of these forces can be lessened, it buys the energy transition time, and this could be valuable. The following discussion is by no means an exhaustive technology review of carbon capture development but, rather, an overall assessment of the potential to mitigate GHG emissions with the continued use of fossil fuels.

Most but not all CCS development is intended for the power sector and particularly for coal power plants that have higher relative CO_2 emissions per heat unit as compared with natural gas. To a big extent inventive efforts are trying to find solutions that provide fossil fuels a pathway to continued relevance in the age of reduced carbon emission commitments. The stringency of the CO_2-reduction targets that define the implementation of CCS will be telling. CCS employed in the 30% reduction range would reflect goals to lessen carbon emissions without substantially altering the economics of fossil fuels in the power sector. On the other hand, CCS employed above 90% would confront the need to decarbonize the atmosphere, which then creates a fair playing field for noncarbon alternative sources. It will be interesting not only to see the evolution of technologies but also to follow GHG emission reduction commitments in implementation.

Broadly, there are two research and development fronts to CO_2 capture for the power sector: pre- and postcombustion isolation. Postcombustion, as expected, captures CO_2 from a plant's exhaust flue. Today most fossil fuel power plants use atmospheric air to burn the fuel. One method for reducing the cost of capture is to burn the fossil fuel with pure oxygen (i.e., oxyfuel combustion), creating a relatively pure stream of CO_2 that lowers the capture costs. A second postcombustion method utilizes sorbents to capture the CO_2. Subsequently the sorbents release the CO_2 in a pure stream, and this also helps reduce the cost of implementation. A third class of postcombustion capture takes advantage of fuel cell technology. Specifically, molten carbonate fuel cells can simultaneously capture and purify postcombustion CO_2 streams from a coal plant flue while also producing electricity. In this scenario, coal power plants would operate alongside a plant integrated with a natural gas reforming process that produces the hydrogen used in fuel cells. This last approach may have a leg up on the other methods because it generates surplus electricity while lowering capture costs.

The technology is not fictional; it is stymied by the hierarchy of valuing low-cost energy above reducing greenhouse gas emissions.

Precombustion takes an altogether different tack to the problem. In the case of an integrated gasification combined cycle coal power plant, a multistep gasification process produces a relatively pure stream of hydrogen gas used for electricity generation, while the CO_2 stream can be separated and sequestered. A similar approach can be taken with natural gas that is reformed to produce a pure stream of hydrogen and CO_2. Finally, methane cracking, a chemical process that separates CH_4 into hydrogen gas and carbon, is yet another potential solution to accessing the energetic properties of a carbon-based fuel without the detrimental effects of GHG emissions. In the case of cracking, carbon is isolated as opposed to CO_2, which is a much more practical approach to capture.

Today, the pre- and postcapture approaches are considered too expensive for broad commercialization. The technologies are estimated to increase the costs of energy by as much as 80% and result in operations that are 20% to 30% less efficient. That is why CCS has not been implemented on a mass scale. The U.S. Department of Energy (DOE) is targeting the mid-2020s for carbon capture at $40/ton of CO_2. At this level the DOE believes that it could be viable for broad commercial adoption.

Excluding methane cracking, even if carbon capture becomes viable, no universal solution for CO_2 sequestration exists. The current plan is the storage of carbon dioxide in massive underground geologic formations. The DOE estimates that 40% of coal power plants are situated above suitable geologic formations. The others will need to transport CO_2 to appropriate sites. Power plants around the world produce billions of tons of CO_2 per year, which would need to be transported and stashed. This would be hoarding on a massive scale.

To complete this discussion, it is worth a brief mention of geo-engineering. Geo-engineering is the application of large-scale interventions that reduce the impact of GHGs on climate change. The sense of applying geo-engineering measures assumes that we fix the underlying problem—the increasing atmospheric concentrations of GHGs from human activity. One approach seeks to temporarily increase the reflection of incoming solar radiation, a sunscreen applied to our atmosphere. This would hold temperatures until the concentrations of GHGs are reduced sufficiently to halt the rise of temperature. Another approach draws inspiration from the natural removal of CO_2 with the growth of biological matter. Ideas include methods to stimulate the growth of plankton, reforestation, or the increased use of biomass in power production. Geo-engineering would likely require a doomsday scenario to gain global support.

The technology isn't fictional; it is held back by the hierarchy of valuing low-cost energy above reducing greenhouse gas emissions. Instituting a carbon tax would usher change and vet some of the concepts described. Without the proper economic relevance, the advancement of solutions is hampered. The marketplace won't adapt to rising concerns with greenhouse gases until the economics reflect the negative impact. Fix the economic formula, and the marketplace will respond with innovations and solutions.

Notes

1. Russell Heimlich, *Public Sees a Future Full of Promise and Peril*, Pew Research Center for the People and the Press RSS, June 22, 2010, accessed April 16, 2015, http://www.people-press.org/2010/06/22/public-sees-a-future-full-of-promise-and-peril/.
2. Seth Motel, "Polls Show Most Americans Believe in Climate Change, but Give It Low Priority," Pew Research Center RSS, September 23, 2014, accessed April 14, 2015, http://www.pewresearch.org/fact-tank/2014/09/23/most-americans-believe-in-climate-change-but-give-it-low-priority/.
3. Institute for Energy Research, "IEA's World Energy Outlook 2014," November 21, 2014, accessed March 26, 2015, http://instituteforenergyresearch.org/analysis/ieas-world-energy-outlook-2014/.
4. World Nuclear Association, "Nuclear Power in the World Today," January 2016, accessed July 18, 2016, http://www.world-nuclear.org/information-library/current-and-future-generation/nuclear-power-in-the-world-today.aspx.

16

The United States

You cannot escape the responsibility of tomorrow
by evading it today.
—Abraham Lincoln

Despite ongoing climate discourse, thoughtful energy planning has not been at the national forefront in the United States for many decades. Particularly since 9/11, presidential and party platforms have emphasized immediate concerns of national security and the economy over the longer-term need to transition to alternative energy sources. A second complicating factor for the United States is the strikingly different positions that characterize Democratic and Republican party platforms regarding energy. Energy planning everywhere would benefit from a steady hand at the helm, but as party influence sways, so swings the national energy direction.[1] In general, the Republican Party favors open access for fossil fuel development, nuclear power, and lowering regulatory hurdles for domestic petroleum production. The Democratic Party supports strong regulation (i.e., lowering CO_2 emissions), establishing renewable energy capacity, and the protection of environmentally sensitive areas while opposing nuclear power. Overall, it is illustrative that while many countries including the United States have security as a top energy priority, there is little long-term planning for the chaos that will occur if cheap and easy fossil fuels are depleted before we can replace them.

Because U.S. energy priorities and strategy are viewed through a lens adjusted to national security and economics, the top drivers influencing policy are self-sufficiency and low prices at the pumps and meters. Although human-influenced climate change continues to be debated and denied, it is stumbling its way into national energy policy via the Clean Power Plan.

In 2013, the United States was well on its way to achieving its pursuit of energy independence, producing 65% of its petroleum requirements. Domestic production of natural gas and coal was even higher, reaching 93% and 114% of domestic consumption, respectively. Interestingly, despite a "Shale Boom," the United States is still a net importer of oil and natural gas. But the twin strategy objectives of independence and cheap energy are likely to be in conflict, as demonstrated by the 2014–2016 drop in worldwide oil and gas prices. A safe bet is that cheap energy from imports will always supersede higher-cost domestic supply. Indeed, by 2016 imports of crude oil had increased 20% above May 2015 levels.[2] This reinforces the point that sound and strategic energy planning must have a steady hand to be effective. The United States has a whipsaw energy plan of sequentially hunting domestic fossil fuels and not.

U.S. natural gas reserves have increased 62% since 2000 courtesy of horizontal drilling and hydraulic fracturing technologies that are able to unlock resources previously deemed uneconomical. However, the 2014–2016 worldwide decline in oil and gas prices has altered what is economical, highlighting that economics, not technology, will determine tomorrow's energy sources. Building alternative capacity to replace fossil fuels seems incompatible while experiencing a "Shale Boom" and a worldwide oversupply of petroleum. Policies that disproportionately prioritize short-term domestic fossil fuel development and position shifts among fuels as an acceptable means of addressing climate change are operating on too short of a planning horizon. The meaningful establishment of alternative energy sources is the heavy lifting that needs to occur this century, and the longer we delay, the more exposed we become to volatile supplies and pricing—and the further the United States falls behind other countries innovating with far more urgency and progress.

Through the end of 2014 the United States ranked second in the world in renewable power capacity; that aligns with being the world's

The role of energy in global and national security is grossly underestimated.

second-largest consumer of primary energy. The problem is that renewables supplied just 5.3% of U.S. primary energy in 2013.

What We Can Anticipate (PECF = 4.2 × FFWF = 3.6)

The framework shared at the beginning of this section was developed as a tool to allow us to anticipate how a country will develop its energy plan and where the adoption of alternative energies will be a top priority. The United States is home to 4.5% of the world's population. It has a Fossil Fuel Wealth Factor of 3.6 and a Primary Energy Consumption Factor of 4.2, suggesting that it will adopt a strategy of fully leveraging its carbon-based reserves while dabbling in alternative sources. Climate change concerns will be in conflict with this measured approach, since fossil fuels remain in this scenario for the near term. The "Shale Boom" will cause the United States to focus efforts on the development of its domestic resources, and given such a large share of the world's coal reserves (26%), one would also expect research on methods to establish clean coal operations.

The model suggests that the top U.S. priority would be to harness the advantage of its fossil fuel reserves in the interest of self-sufficiency. So addressing climate change in the United States means reducing fossil fuel CO_2 emissions, not abandoning their use. Commitments to reduce greenhouse emissions will then correlate primarily with shifts to natural gas along with conservation and efficiency activities. Developing and deploying renewables and even carbon capture will occur on their economic merits. The United States prioritizes national and global security, but in practice economics play the stronger hand in energy decisions. Ironically, building renewable capacity would both enhance national and global security and reduce health and environmental consequences. Once built, renewables are the one energy source available for eternity, the great equalizers in the world's access to energy.

Figure 16.1 informs us on how the energy sources on the left serve the various energy sectors on the right. The graph represents the basis from which the United States confronts its energy transition; this is its "blank sheet of paper." In 30 years, will the transportation sector that

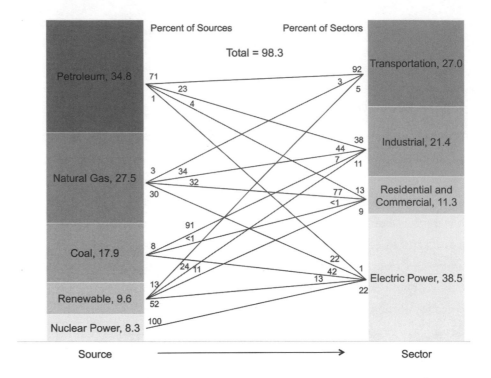

Figure 16.1. U.S. primary energy consumption (quadrillion BTUs) by source and sector, 2014. *Source Data:* U.S. Energy Information Administration, "Primary energy consumption by source and sector, 2014," accessed March 14, 2016, http://www.eia.gov/energy_in_brief/article/major_energy_sources_and_users.cfm.

today is 92% dependent on petroleum significantly shift to alternatively powered vehicles? What is the role of nuclear power in the future's power mix? Today, coal, natural gas, and nuclear energy provide most of U.S. electrical power, the largest sector of energy. Renewables will first find application in the power sector and indirectly affect the transportation sector with alternative vehicles. In time all the energy sectors in Figure 16.1 will completely transition to alternatives; it is just a question of when.

Three overarching influences will drive changes in the U.S. energy mix over the next few decades: national goals, climate regulation, and state renewable portfolio standards. In addition, the U.S. Department of Energy has supportive initiatives related to technology development.

National Goals: Measured Change

At a national level, in the context of climate policy, President Obama in November 2014 committed the United States to reducing greenhouse emissions by 26%–28% below 2005 levels by 2025. The president also set a goal to double renewable generation from wind, solar, and geothermal sources from 2012 levels by 2020. For reference, nonhydro renewables generated just 5.7% of electricity in 2012, with leading contributions from wind followed by biomass and waste, making this an achievable but underwhelming goal. To support this goal, the United States has targeted licensing 10 GW (~6% of target) of renewable capacity on federal lands.

Like most other country plans, the U.S. plan does not acknowledge a time when fossil fuels will no longer be used as an energy source. It seems out of character that the United States is applying its considerable technical prowess to extracting the last drop of oil and puff of gas rather than directing spirit and drive to establish a leadership position in the regeneration of the global energy sector. A century and a half ago, as we were quick to enter the fossil fuel era, domestic supplies were an advantage. Now as we enter the twilight of the fossil fuel era, domestic supplies can be a weight holding us back.

Climate Regulation: The Clean Power Plan

Before we discuss the Clean Power Plan (CPP), we should take a look back at the Environmental Protection Agency's history with regulating air quality. The Clean Air Act of 1970 authorized the EPA to establish National Ambient Air Quality Standards to protect public health and welfare through the regulation of hazardous air pollutants from stationary and mobile sources. The vehicle mass emission standards were part of the broader Clean Air Act. Absent from the list, though, was CO_2, which was not classified as a public health pollutant in 1970. Since then it has become the chief concern and the identified cause of human-induced climate change. The Clean Air Act and EPA have been immensely effective under the 1970 authorization, and now the CPP is extending regulatory reach to include CO_2 among the regulated substances.

The United States has chosen to apply regulation as a policy tool to reduce the emissions of CO_2 from the power sector. While atmospheric levels of GHGs are the underlying cause of human-induced climate change, regulating CO_2 emissions may not be the best policy option, because it doesn't solve the real problem—the need to transition entirely away from fossil fuels. Halting the progression of GHG concentrations in our atmosphere can only really occur by decarbonizing our energy sources, and this needs to be our metric. Beyond being the wrong metric, the CPP takes a big leap from health to climate regulation. To the extent that there is ambiguity surrounding the cause and effect of fossil fuel GHG emissions and climate change, legal challenges to the regulation will occur. As an example, in February 2016 the Supreme Court, by a five–four decision, hit the pause button on meeting the obligations in the CPP regulation. At least temporarily, the ruling unraveled the ability of the United States to deliver firm GHG reduction commitments by April 2016 as part of ratifying the U.N. climate accord in Paris. A contentious CO_2 regulation unfortunately saps the spirit and leadership from what should otherwise be a positive endeavor, transitioning our energy sources.

Specifically, the CPP establishes emission performance rates for Electricity-Generating Units (EGUs) in the lower 48 states, which will collectively deliver a 32% reduction from 2005 levels by 2030, per Table 16.1. This regulation constitutes what can be done with the continued use of fossil fuels, as opposed to what needs to be done.

There are two classes of EGUs: fossil fuel–fired steam-generating units and stationary combustion turbine units, which have emission rates set at 1,305 lbs CO_2/MWh and 771 lbs CO_2/MWh, respectively. The EPA refers to the "best system of emission reduction" that EGUs can employ to meet these reduction targets. Three of the four methods allow the continued use of fossil fuel. EGUs can apply best practice in existing fossil fuel plants, convert to less CO_2-intensive natural gas, and implement demand-side reductions related to efficiency and conservation. A fourth action step, and arguably the most impactful, would be to build clean energy sources (i.e., renewables and nuclear), but the CPP itself offers no preferential incentives to support that direction.

The EPA's CO_2 emission reduction goals in Table 16.1 display the national baseline ("do nothing") and the CPP targets. Note, CO_2 emissions are declining in the base case, so the regulation is pushing along an existing trend. The CPP will also foster reductions of other pollutants regulated under the Clean Air Act authorization (i.e., carbon monoxide

Table 16.1. Clean Power Plan CO_2 Reduction Goals.

	CO$_2$ EMISSION REDUCTIONS BELOW 2005 LEVELS (% REDUCTION)		
	2020	2025	2030
Base case	11%	8%	7%
Final rule	22% to 23%	28% to 29%	32%

Source: U.S. Environmental Protection Agency, Federal Register (Vol. 80, no 205, p. 64679, Friday, October 23, 2015 , Rules and Regulations).

and nitrous oxides). It opens the door to emission trading programs as a means for complying with the regulation; this interjects flexibility but also dilutes the results. Time will tell, but the bar has been set too low to drive meaningful alternative capacities or halt the progression of atmospheric CO_2 concentrations. The path of least resistance is marginally lower CO_2 emissions with subtle shifts in energy mix, which is what the CPP embodies. In contrast, higher targets would encourage the establishment of meaningful noncarbon capacity. Endorsement of the CPP, then, depends on whether you believe we need to reduce CO_2 emissions with the continued use of fossil fuels or hold the position that we need to institute more stringent targets to spur the development of noncarbon energy sources. The CPP targets undoubtedly represent what was politically expedient, not what needs to be done. Legal flip-flopping layered on top of political shortsightedness plays into the hands of inertia, while the problems accumulate in the atmosphere.

Of the fossil fuels, natural gas has the lowest emissions of CO_2, so where access to economical supplies permits (i.e., the United States), the power sector is retiring coal plants. Over the course of just a few decades, petroleum for the transportation sector and coal for the power sector have lost attractiveness, and in many countries they are coming off the menu. If the behavior that these regulations spawn shifts our use from coal to gas, we will finish one race and immediately need to enter another, albeit late to the starting block for having come to a dead end. Moving to a cleaner finite fossil fuel is kicking the fuel can down the road—swapping one risk for another. Do we defer to the next generation the challenge of burning coal cleanly because we've depleted "clean" natural gas reserves?

Comment: Regulation Is the Wrong Tool to Combat Climate Change

The U.S. EPA's Clean Power Plan seems like an old solution to an old problem, a policy that accepts the use of fossil fuels and seeks to limit CO_2 emissions as opposed to stimulating noncarbon energy capacity. It is solving the problem with the same brain that created it. After decades the vehicle emission regulations didn't stimulate carbon-free vehicle choices; they forced lower emissions and cleaner fuels from refineries. Similarly, the CPP will modestly reduce emissions with a sustained use of fossil fuels. This plan is said to be flexible when in fact that flexibility encourages wasted effort. If one believes that global fossil fuel production and supply will become disruptive this century, then efforts exerted to convert from one fossil fuel to another are shortsighted. The United States is spellbound by its underground fossil fuel resources. Technological prowess is directed toward the extraction of fossil fuel supplies with decades of potential instead of advancing alternative solutions for a sustainable energy supply.

Only slightly opening Pandora's box, there is another argument that favors aggressive non-carbon-based targets rather than the Clean Power Plan's shell game of switching from one carbon fuel source to another. The CPP regulation will have marginal impact when you consider methane flaring and leakage from oil and gas production, the cost of converting from coal to gas, and time spent that strands capital assets versus converting them to noncarbon sources. The CPP is a dilutive climate policy that pales in comparison to what could be achieved if we set aggressive standards for noncarbon capacities.

Frustratingly, utilities whose compliance relies on the conversion of coal to natural gas operations may not reduce greenhouse gas emissions' impact on global warming as much as we hope or think. At first blush, measuring emissions from the smokestack of a power plant that converted from coal to natural gas would report significantly lower levels of CO_2, we would reasonably conclude that global warming was slowed, and the utility would achieve regulatory compliance. However, applying the measurement of GHGs from a power plant smokestack is misleading and overstates the regulation's impact. If the regulation also comprehended GHG emissions from oil and gas extraction, production, and delivery and included another dangerous greenhouse gas, methane, that equation would better measure the merits of converting from coal to natural gas in the power segment.

Methane, the main constituent of natural gas, is a powerful greenhouse gas, estimated by the U.S. EPA to be 25 times more potent pound for pound than CO_2 over a 100-year period.[3] That means it traps 25 times more heat than CO_2 in the atmosphere. Natural gas can be lost at all phases, from production, to transportation, to storage and distribution. Flaring, the process of burning natural gas, is quite common in the production phase of oil and gas. Satellite images of oil and gas production flaring can be found online (search skytruth methane). In an act that is the lesser of two evils but still highly wasteful, extraction companies choose to burn natural gas, thereby producing CO_2 and water, rather than allowing it to escape noncombusted into the atmosphere. The World Bank has launched an initiative to zero natural gas flaring from petroleum production by 2030.[4] It further estimates that 140 billion m^3 of natural gas are flared, which represents 4% of the world's 2013 production. That's a lot of emissions before the natural gas even arrives at the power plant. Natural gas is touted for its lower CO_2 combustion emissions, but when the production and development emissions of GHGs are considered, the case that it is a cleaner fuel is more precarious.

In addition to the release of CO_2 from flaring, methane leakage directly into the atmosphere is not comprehended in the regulation, and this further distorts the advantages of natural gas. According to an April 14, 2014, report in the *Washington Post*, the natural gas contribution to global warming is only less than coal's if leakage is less than 3.2%. But recent estimates reveal that anywhere from 2.3% to 17.3% of the methane escapes.[5] The U.N. Framework Convention on Climate Change more specifically reported that 3% of natural gas production escapes as methane emissions.[6] Given the combined estimates for flaring and direct methane emissions from the production of natural gas, coal is a victim of the fossil fuel shell game because the net climate impact in converting to natural gas is neutral at best. It seems that achieving compliance under the CPP from fossil fuel conversion is postponing the inevitable and far more desirable goal, a decarbonization of our energy sources.

If human-induced climate change is your concern, then the replacement of fossil fuels with noncarbon alternatives is the most powerful long-term solution, along with conservation and efficiency programs.

Many carbon capture and sequestration technologies exist, but the costs of applying solutions trump climate change concerns, and there is inadequate regulation to counteract the lack of economic pressure.

Delivering Change: The Renewable Portfolio Standards

While at the federal level the CPP attempts to clamp down on the CO_2 emissions from fossil fuels, the renewable portfolio standards (RPSs) are state policies that mandate the establishment of renewable energy sources. State RPS programs set the minimum portion of an Electricity-Generating Unit's supply that must be sourced from renewables. States choose from a broad menu of renewables, fostering healthy competition among renewable alternatives. A majority of states in the United States have adopted renewable portfolio standards during the past 10 years, while others have adopted voluntary goals. Some states are more prescriptive and set targets for specific renewable technologies (i.e., solar PV). Within the state RPS programs, alternative compliance penalties incentivize utilities to act or pay a fine. The supplier pays a fee for every MWh of electricity in its shortfall. Revenues from penalty fees are placed into a fund to support renewable technologies.

Figure 16.2 shares a sampling of state RPS goals as published by the DOE's Database of State Incentives for Renewables and Efficiency. The percentages in the figure refer to the target for the renewable portion of electricity. Hawaii has targeted 100% of electricity to be generated from renewables by 2045 and will tap its large potential of geothermal and wind to retire the use of fossil fuels in the power sector. The advantages of RPS policies are that once an initial milestone level is achieved and successfully integrated, new targets can be established: the path is open. Renewable portfolio standards only apply to the power sector, while the other energy sectors consume 60% of U.S. primary energy;

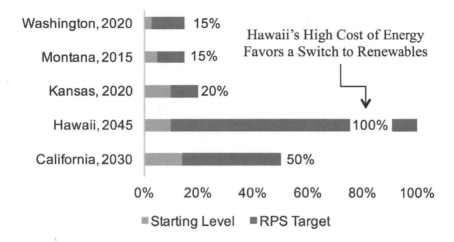

Washington, 2020 — 15%

Montana, 2015 — 15%

Kansas, 2020 — 20%

Hawaii, 2045 — 100%

California, 2030 — 50%

Hawaii's High Cost of Energy Favors a Switch to Renewables

0% 20% 40% 60% 80% 100%

■ Starting Level ■ RPS Target

so although Hawaii's commitment is impressive, we need to remember that it is 100% of 40%.

U.S. use of fossil fuels for primary energy is in line with worldwide levels, while nuclear power and renewables are above and below the global averages, respectively. Through 2012, solar was a mere 0.11% of electricity generation, while wind had made more impressive inroads at 3.48%. RPS programs will deliver the limited national goal of doubling renewables by 2020. One could argue that the national goals are just a summation of what the states are undertaking, as opposed to providing the driving force.

Figure 16.2. U.S. state-level renewable portfolio standard (RPS) targets. *Source Data:* N.C. Clean Energy Technology Center, Database of State Incentives for Renewables and Efficiency, "Renewable Portfolio Standards," Department of Energy 2013 Archive, accessed February 25, 2015, http://www.dsireusa .org.

Federal Help: The Department of Energy

Supporting federal and state programs are the DOE's technology objectives. SunShot is an example of a DOE initiative to help realize cost-competitive solar photovoltaic technology.[7] There are two approaches: transferring high-efficiency solar cell designs from the laboratories to commercial product and reducing the total installed cost per watt across the utility, commercial, and residential segments. The DOE's specific goal is to deliver $1.00/W, $1.25/W, and $1.50/W into the utility, commercial, and residential markets, respectively, by 2020. In 2010 the cost to deploy solar PV in the residential market was $6.00/W, and three years later it was still at \$4.90/W, so it remains to be seen whether the DOE can deliver its cost goal. A large portion of the cost per watt is now installation labor, a cost over which the DOE has little influence. In contrast, DOE translational funding to help bring higher-efficiency panels from the lab to the marketplace is something it could influence. Although the world record for solar cell efficiency exceeds 43%, the efficiency of residential solar panels remains less than 25%.

If we can see the need to continually upgrade military platforms, why can't we see the inevitable end of the fossil fuel era? Our failure to comprehend the role of energy affects our national security and the rest of the world.

Another of DOE's technology programs relates to CO_2. The DOE has targeted the mid-2020s for the delivery of commercially viable technologies for carbon capture and sequestration. Although many of the technologies already exist, the "prohibitive" costs of applying the solutions trump climate change concerns, and there is insufficient regulation to counter the lack of economic pressure. This is reminiscent of the DOE's goal to deliver safe and permanent storage for spent nuclear fuel by 1998. Nearly 20 years beyond that deadline, nuclear power plants are storing radioactive waste on-site, lacking a national solution. The Yucca Mountain Nuclear Waste Repository located northwest of Las Vegas, Nevada, was the plan, but funding for the project was canceled in 2011 for political—not technical or safety—reasons. Now we are left with a hole in the ground, accumulating waste at nuclear power facilities around the country, and a reminder that "no solution" is not a valid answer. The annual reports of utility companies note the national failure to meet its obligation.[8] As a result, the federal government is paying utility companies for the costs they incur to store and manage the spent fuel, which is hundreds of millions of dollars. This deadline is decades past, still unresolved with no plan in sight. The concern is that the CO_2 capture and sequestration obligation will mimic the elusive resolution of spent nuclear fuel, that political inaction, not technical capability, will block the achievement.

As spent fuel disposal is to nuclear power, so CCS is to fossil fuel power. Furthermore, the DOE's 2020s time line for commercially viable CCS seems out of sync with the need in the market. It seems well past the time to outfit fossil fuel plants to reduce greenhouse emissions. The horse has left the barn, and yet we remain 10 years away from a potential solution. If we were honest, we would admit that the United States is masking an economic barrier as a technology hurdle.

The Unspoken Alternative: Nuclear Power

Although there are only a few reactors under construction, the United States is investing nuclear R&D dollars on several fronts. About $5 billion is directed toward the International Thermonuclear Experimental Reactor project, which is testing the viability of nuclear fusion. There is also funding for next-generation nuclear fission plants and a demonstration project for hydrogen production from existing nuclear power plants. Producing hydrogen from steam is more efficient than from liquid-phase water, and that is why combining this capability with thermal plants makes a lot of sense. The DOE also refers to accelerating the time lines for the commercialization of small modular reactors through cost-shared agreements with industry partners.

The low-priority funding for fast reactors and fuel cycles seems out of step with the advantages. Despite successful work on the fast reactor decades ago, the United States has chosen to de-emphasize advancing Generation IV fission reactors, shutting down investments that are now advancing in Russia, South Korea, China, and elsewhere. Generation IV may be the best solution for managing spent fuel, but it seems it will be deployed elsewhere. The downside of nuclear regulation in the United Sates is that it has hindered the emergence of superior solutions. The regulatory burden of change is set so high that advances that constructively address risk factors cannot economically make their way to the market. There needs to be an option to bring superior technology forward. The Food and Drug Administration provides avenues for new drug therapies in response to public health concerns. It seems that a similar option should be employed to address the public health and diversion risks of spent fuel stockpiles. Regulation is crucial but shouldn't hold us hostage to the risks of current practices.

A Measured Approach Excludes Global Leadership

Opposing political parties in the United States have staked out positions that are mutually exclusive and seem determined to hold firm while the country loses decades in the energy transition. One party seeks energy independence (i.e., security) through increased access and reduced regulations for the petroleum extraction industry, while the other party wants to address climate change and safeguard the environment through regulations. It's like being for or against capital punishment, where the subject either lives or dies; a middle ground is hard to craft with such stark and entrenched differences. A summary of the Democratic and Republican Party platforms since 1992 included in appendix B is illustrative.

And another thing that hurts thoughtful energy planning is the distortion of influence from status quo incumbents. Legislative work on climate policy way back in 2009 was associated with lobbying expenditures of $175 million from oil and gas industries, against just $22 million from pro-environmental groups, demonstrating the stark differences in financial influence.[9] On the other hand, according to a 2014 Gallup survey, 42%, 31%, and 25% of U.S. citizens identified themselves as Independent, Democratic, and Republican, respectively. So, although the citizenry may be equally divided along party lines, the financial amplification from the petroleum industry tilts the legislation. It would be a rare political figure who takes a position on energy that could only incur costs during his or her tenure and who will be long gone before any benefits are realized.

On the other hand, we have only to look at military spending to see that long-term strategic planning is indeed possible and practiced in the United States. When it comes to national security, the United States outspent China by almost three to one in 2014. It is building new-generation weapons systems and support technology with time horizons well beyond those we apply to energy. If we can see the obsolescence of military platforms, why can't we see the inevitable end of the fossil fuel era? We fail to comprehend the role of energy not only in our national security but also for countries around the world. While the United States spent $610 billion on the military in 2014, it allocated only $16 billion across 10 years in the U.S. Energy Policy Act of 2005 for renewable and nuclear energy.[10] The role of energy in national security based on this funding is grossly underestimated. Many of the world's instances of geopolitical tension from the Middle East, to the Falkland Islands, to the South China Sea are in part amplified by significant quantities of recoverable oil and gas. In contrast, renewable energy has the potential to provide most countries around the world with long-term national energy security through perpetual domestic resources.

A way to move forward on common ground (although nothing will be easy in the current political environment) would be to prioritize energy in the context of national and global security. National security funding would become a mix of energy and military programs. If we begin to consider alternative capacity in the context of advancing national and global security, we may overcome flawed short-horizon economic models that sustain the disproportionate investment in fossil fuels. Funding the short-term benefits of petroleum would be rationalized against long-term strategic investments in alternatives. Further, regulation should be kept in place to guard against extraction methods that pose environmental risk. The special interest oil and gas lobby will press Congress for increased access, so regulation should be in place to protect sensitive areas and counter the unhealthy forces of inertia.

Despite state renewable targets, the U.S. national energy plan at least through 2020 is fossil fuel—centric; nearly 60% of the 2005 Energy Policy Act's funding through 2016 supports fossil fuels. Clearly, the United States remains intent on establishing domestic production designed to improve energy self-sufficiency at least for the near term. Greenhouse emission reductions from 2005 levels will be achieved largely through higher use of natural gas, efficiency, and conservation programs and less through substituting nonfossil energy sources.

The United States is following the path predicted by the FFWF and PECF framework: harvest its fossil fuel resources and institute a measured approach to renewable development with a keen focus on globally competitive energy costs. The change in energy mix appears to be appeasing climate change concerns as opposed to securing large-scale sustainable sources of noncarbon energies. Optimistically, 2020 will deliver the first phase of many state RPS successes, and the simultaneous infrastructure upgrades will pave the way for even higher levels.

Notes

1. Appendix B includes a table that synthesizes energy platform positions for the Democratic and Republican parties since 1992.

2. Christian Berthelsen and Lynn Cook, "Despite Shale Glut, U.S. Imports More Foreign Oil," May 4, 2016, accessed July 18, 2016, http://www.msn.com /en-us/money/markets/despite-shale-glut-us-imports-more-foreign-oil /ar-BBsCPeJ?li=BBnbfcN.

3. U.S. Environmental Protection Agency, "Overview of Greenhouse Gases: Methane Emissions," accessed April 16, 2016, https://www3.epa.gov/climatechange /ghgemissions/gases/ch4.html.

4. World Bank, "Zero Routine Flaring by 2030," May 22, 2015, accessed September 16, 2015, http://www.worldbank.org/en/programs/zero-routine-flaring-by-2030.

5. Patterson Clark, "Unexpected Loose Gas from Fracking," *Washington Post*, April 14, 2014, accessed April 16, 2016, https://www.washingtonpost.com/apps/g/page /national/unexpected-loose-gas-from-fracking/950/.

6. U.N. Framework Convention on Climate Change, "Transformative Action on Methane toward COP 21: Join the CCAC Oil and Gas Methane Partnership," August 19, 2015, accessed April 16, 2016, http://newsroom.unfccc.int/unfccc-newsroom /transformational-initiative-ccac-oil-gas-methane-partnership/.

7. U.S. Department of Energy, *Strategic Plan 2014–2018*, report no. DOE/CF-0067 (Washington, D.C.: U.S. Department of Energy, April 2014), http://www.energy .gov/sites/prod/files/2014/04/f14/2014__dept__energy__strategic__plan.pdf.

8. NextEra Energy, Investor Relations, *Annual Report 2013. Innovate. Grow. Invest.* March 21, 2014, http://www.investor.nexteraenergy.com/phoenix .zhtml?c=88486&p=irol-reportsCorporate.

9. Evan Mackinder, "Pro-Environment Groups Outmatched, Outspent in Battle over Climate Change Legislation," *OpenSecrets* RSS, August 23, 2010, accessed April 24, 2015, http://www.opensecrets.org/news/2010/08 /pro-environment-groups-were-outmatc/.

10. Stockholm International Peace Research Institute, "SIPRI Military Expenditure Database," accessed April 24, 2015, http://www.sipri.org/research/armaments /milex/milex__database.

17

Canada

If only we could travel in two directions at once.

Canada is a study in contradiction: nobly striving to build domestic capacities of alternatives while concurrently enjoying a vibrant fossil fuel export business. As evidence, 75% of the electricity that Canadians use is generated from noncarbon energy sources, while 40%, 38%, and 37% of oil, gas, and coal production, respectively, is exported.

Blessed with a diversity of natural resources, Canada holds some of the world's largest reserves of oil, has massive hydroelectric dams, and is replete with wide-open spaces from which to harvest wind power. Behind only Russia, Canada has abundant locations ideal for capturing onshore wind energy. Canada's Fossil Fuel Wealth Factor of 7.1 is one of the highest in the world. Combine this with its impressive and growing hydroelectric capability, and you can see that Canada is in a secure energy position. Like Russia, it is highlighting fossil fuel exports, but the two differ as they develop their bountiful natural resources.

For a country so rich in petroleum reserves, Canada uses 21% less fossil fuel for primary energy than the world average. Hydropower's contribution to power generation in Canada is similar to what fossil fuels contribute in the rest of the world, and its current nonfossil power mix (~79%) is what some countries will work hard to achieve by mid-century. For context, 33% of the world's electricity in 2012 was generated by nonfossil energy sources. That is less than one-half Canada's current mix!

With 35 million residents, Canada has 0.5% of the world's population, and it has 10.5% of the world's proven oil reserves. New methods of extraction and development promise to unlock even more resources. To support the flow of these products, Canada has built an extensive infrastructure of pipelines for both oil and gas, with more to come. There are 22,000 mi of oil pipeline and another 42,500 mi of gas pipeline.

Canada strives to build domestic capacities of alternatives while exporting fossil fuels.

Eight of the world's top 64 dams are located in Canada. The Robert-Bourassa Dam on the La Grande River and the Churchill Dam on the Churchill River annually produce 26.5 and 35 TWh of electricity, respectively. To put this in perspective, the largest dam in the United States, the Grand Coulee on the Columbia River, produces 20 TWh annually. Hydropower generation, ranked third in the world, provided 61% of Canada's electricity, with plans for further development. The balance comes from nuclear power with nearly 14%, nonhydro renewables just under 4%, and fossil fuels providing just 21%. This furnishes Canada with one of the world's cleanest power sectors.

What We Can Anticipate (PECF = 5.3 × FFWF = 7.1)

The framework predicts that Canada would exploit its fossil resources to create wealth, and its export figures support this conjecture. The model also expects a hesitation to adopt alternative energies unless the economics are favorable to fossil fuels. Additionally, the model anticipates that climate change would be addressed by conservation and efficiency, as opposed to turning away from the nation's fortune in fossil fuel reserves.

Canada's development of its hydro resource has allowed the electric power sector to avoid a large dependence on fossil fuels. Now, wind-generated electricity is cost-competitive with fossil fuels and poised for measurable contributions. Canada's Saskatchewan and Alberta provinces, where winds are particularly strong, hold promise for fertile development. Solar power, on the other hand, is lagging, as we see almost everywhere. The model predicts that Canada will pursue renewables on economic terms and approach reducing greenhouse emissions within a power energy mix that is already 75%+ non-carbon-based.

The Federal Government's Commitments

Canada is a federation of 10 provinces and three territories, which benefit from a relatively small population's access to disproportionate riches in both petroleum and natural gas sources. The federal energy plan is broad, providing autonomy for provinces and territories to pursue their own priorities. The provincial and territory plans provide specific actions within the federal framework, similar to state plans in the United States.

The federal government has identified the importance of promoting conservation, and the choice of the term *energy literacy* is interesting. The term epitomizes both the sixth and seventh signals from the book's first chapter, which suggests that no matter how we source our energy, there needs to be respectful use. Other priorities, directives that speak to Canada's dual pursuit of domestic clean energy sources and the exploitation of its fossil fuel assets, relate to the petroleum sector and include environmentally safe and secure networks for the transmission and transportation of domestic and export oil supplies. To support the production of petroleum, the federal government has committed to improving the timeliness and certainty of regulatory decision making while maintaining environmental interests. To coordinate the nation's petroleum trade, the federal government has taken the responsibility to formalize the representation of provinces and territories in international negotiations. Last but not least, the government is committed to facilitating the development of green and/or cleaner energy sources to meet both future demand and environmental goals.[1] Together these federal priorities portray a country unencumbered by obvious conflicts in its energy priorities.

Despite potent renewable energy resources, the federal plan is heavily tilted toward the continuation of a lucrative petroleum export franchise and less oriented toward a fundamental shift in the domestic primary energy mix.

British Columbia: A Representative Provincial Plan

British Columbia's energy policy has some decisive elements worth sharing.[2] All new power plants in British Columbia are regulated to have zero net greenhouse gas emissions. This plan will ensure that 90% of its provincial electrical generation continues to come from clean and renewable sources. New fossil fuel power will either include carbon mitigation or offset emissions by the removal of CO_2 elsewhere to comply. That will be hard to do. It seems to mean that new capacity will be renewables. British Columbia has plans to increase its hydropower capability with both large- and small-scale projects. It also identifies carbon-based biomass as another abundant alternative resource to help achieve the zero net goal, which implies the need for offsets elsewhere. As of 2016 all existing thermal generation power plants are to have zero-net greenhouse gas emissions. To accomplish this net zero, the utilities either retrofit plants with CO_2 capture technologies or lessen CO_2 from other sources through conservation and efficiency. British Columbia has a 90% stake in the ground, and its net zero smartly allows some latitude for B.C. power to address rising demands while also meeting the provincial goal.

Not located in British Columbia but worth mentioning, SaskPower in Saskatchewan has overseen a $1.24 billion project to retrofit its 110 MW coal plant, called Boundary Dam, with CCS technology designed to capture 90% of CO_2 and 93% of sulfur oxides.[3] This is the world's first full-scale deployment and perhaps a key milestone in retrofitting other coal-fired plants. SaskPower constructed a 66-km pipeline delivering a portion of the CO_2 to oil fields where it is used in enhanced oil recovery. The balance of the CO_2 is injected 3.4 km into an underground brine and sandstone formation. Both the CO_2 and sulfur oxides from Boundary Dam are sold, and that proved crucial in the economic viability of the project. Boundary Dam's CCS is a major achievement in the deployment of utility-scale technology to capture greenhouse gases, and its success exposes the real obstacle in the U.S. DOE's mid-2020s commitment date for carbon containment. The delay in implementing carbon containment isn't a technological problem; it is the result of placing economics above climate in the hierarchy of priorities.

British Columbia recognizes that its transportation sector is largely dependent on fossil fuels, representing 40% of its greenhouse emissions. It is in the early phase of bringing forward alternatives for consumers.

While most of the world sidesteps policies on taxing carbon emissions, British Columbia was the first location in North America to institute a carbon tax.

These include a demonstration hydrogen highway, renewable biofuels, and infrastructure for plug-in electric vehicles. Notably, and adding to the sense that the world is at a crossroads with nuclear power, British Columbia has chosen not to employ nuclear in its plans going forward, while elsewhere in Canada nuclear power is expanding. Nuclear power from 19 reactors, most of which are in Ontario, contributes nearly 15% of Canada's electricity. The provincial decision to discontinue further adoption of nuclear power undoubtedly was made easier by the other energy choices available in British Columbia, and its story exemplifies the advantages of local decision making.

While most of the world sidesteps policies on taxing carbon emissions, it is encouraging to learn that British Columbia was the first location in North America to institute a carbon tax. This is part of the province's overall Climate Action Plan. British Columbia's near-term goal is to reduce GHG emissions 33% by 2020. The tax rate on fossil fuel is based on a C\$30 per tonne of CO_2 emissions. The precise fuel tax rate is determined by the CO_2 emission factors for specific fuels. For instance, C\$30 per tonne translates into about C\$0.07 per liter of gasoline. Since the tax is applied to the use of fuels, conservation and efficiency behaviors allow consumers to minimize the impact by reducing consumption. Proper maintenance of vehicles including regular tune-ups and proper tire pressure can improve efficiency and help avoid the tax. The adoption of smart thermostats can also help conserve energy associated with heating and cooling. Interestingly, British Columbia promises to remain revenue-neutral, meaning that the carbon tax revenue is offset by the reduction of taxes elsewhere. The implementation of British Columbia's carbon tax was honed to the goal of reducing CO_2 emissions through conservation. Alternately, a tax that supports a wholesale energy transition would utilize the revenues to build the necessary infrastructure, and the tax would be levied where energy source decisions are taken, not necessarily where the fuel is consumed. Nevertheless, congratulations go to British Columbia for implementing a carbon tax; it places a toll on the atmospheric emissions of CO_2, and the comparative economic attractiveness of cleaner sources improves.

Petroleum-Rich and Conflicted

Few countries have the fossil fuel wealth of Canada, and even fewer have natural resources so conducive to the production of renewable energy. Many countries strive for some semblance of energy security, while Canada's biggest struggle may be choosing which of these resources to employ as an energy source. Canada's contradictory approach to energy is thrust upon it by virtue of its rich renewable and fossil fuel resources. Its energy policy is rational but nevertheless conflicted. It's almost as if Canada is saying, "Clean for us and fossil fuel for you." This would be fine, perhaps, if we only breathed Canadian air. If only we could travel in two directions at once.

Overall, Canada's plan has several novel elements. Canada is conserving and incrementally deploying renewables above already high levels. The Boundary Dam carbon capture project and the hydrogen demonstration highway are hopefully glimpses of the future. This is impressive, but before we place Canada too high on the renewable pedestal, we must acknowledge the irony of British Columbia's net-zero programs. If the notion of net-zero CO_2 emissions comprehended its fossil fuel exports, it would be quite positive. The earth's atmosphere does not recognize or respect borders. Canada and British Columbia specifically are acting as the framework anticipates.

Notes

1. Canada's Premiers, Council of the Federation, *A Shared Vision for Energy in Canada* (Ottawa: Council of the Federation, August 2007), http://canadaspremiers.ca /phocadownload/publications/energystrategy__en.pdf.
2. British Columbia, *The BC Energy Plan: A Vision for Clean Energy Leadership*, report no. 250.952.0241 (Victoria: Ministry of Energy, Mines and Petroleum Resources, 2009), accessed February 27, 2015, http://www2.gov.bc.ca/assets /gov/farming-natural-resources-and-industry/electricity-alternative-energy/bc __energy__plan__2007.pdf.
3. SaskPower CCS, "Boundary Dam Carbon Capture Project," February 20, 2015, accessed April 15, 2015, http://saskpowerccs.com/ccs-projects /boundary-dam-carbon-capture-project/.

18

Brazil

Brazil's sugarcane ethanol productivity of 6,800 L/ha is the
highest in the world.

In 2013 the population of Brazil surpassed 200 million, and it is projected to reach 240 million by 2030. Since 2000 its primary energy consumption has increased 42%, and it will increase further to keep pace
with rising population and economic development. Fortunately or unfortunately, depending on whether you are near- or farsighted, the world's
largest discovery of petroleum since 2000 lies off the coast of Brazil (i.e.,
Santos Basin).[1] The field is referred to as "pre-salt" because it is buried
5,000 m beneath a rock bed (including 2,000 m of salt) and 2,000 m
of water. The recovery of this resource will require advanced extraction
technologies and a high tolerance for environmental risk. Nevertheless,
Brazil expects liquid petroleum production to increase from 2.7 million
barrels a day in 2012 to 5.0 million barrels per day by 2021, allowing it to
join the ranks of petroleum exporters. Today, though, Brazil's net imports
are 10%, 43%, and 75% for oil, gas, and coal, respectively.

Historically, petroleum price instability in the 1970s had a profound
impact on Brazil. Its economy couldn't absorb for long the increased price
of petroleum, so it responded with big investments in domestic biofuel
production. Those early investments have established production levels
that rank second in the world at 28.4 billion L annually. Brazil's sugarcane ethanol productivity of 6,800 L/ha is the highest in the world, and

supporting this production requires the dedication of nearly 4.5 million ha of land. In 2009, Brazil used more liters of ethanol than gasoline.

Recently, Brazil has experienced major blackouts in the transmission of electricity, exposing the need for both capacity and infrastructure upgrades. Nearly 90 million people across Brazil were without electricity when the Itaipu hydropower plant shut down because of short circuits in the transmission lines. In response, Brazil has allocated 5% of its 2011–2021 energy investment for improvements in the nation's grid infrastructure. Access to electricity is a common pursuit, but once consumers make the connection it becomes a necessity, and disruptions are more than just mere inconveniences.

Brazil's use of fossil fuels for primary energy in 2013 was 23% lower than the worldwide average courtesy of disproportionately high contributions from both hydropower and biofuels. Hydroelectric generated an incredible 76% of total electricity, and according to internal estimates that represents only 30% of its potential. Quite amazingly, fossil fuels accounted for only 13% of Brazil's electricity, compared with a worldwide average of 67%. Brazil's proven reserves of petroleum will increase substantially as the pre-salt fields are evaluated, and it will be interesting to see whether this new resource is used domestically or exploited for exports. Will Brazil's impressive domestic production of biofuels decline as more petroleum is extracted? The sustained production and use of biofuels would render most of the projected increase in oil production available for export. Brazil may follow the way of Canada, simultaneously building nonfossil domestic resources while pursuing the benefits of exporting petroleum.

Brazil's sugarcane ethanol productivity is the highest in the world, and supporting this production requires the dedication of nearly 4.5 million hectares of land.

What We Can Anticipate (PECF = 0.8 × FFWF = 0.2)

The framework would consider Brazil dually challenged. Brazil is a major economy, ranking seventh in the world in gross domestic product (GDP), but one that is still developing and relatively poor in fossil fuel resources. In 2013, 1.4 million of its residents lacked access to electricity. Until the most recent discovery, Brazil had few fossil fuel resources and was emphasizing three renewable sources for new capacity: hydro, bioenergy, and

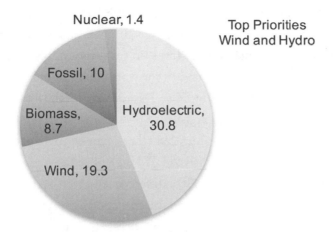

Nuclear, 1.4

Top Priorities
Wind and Hydro

Fossil, 10

Biomass,
8.7

Hydroelectric,
30.8

Wind, 19.3

wind. Now, though, the model predicts that fossil fuels would become a focus for development. Will Brazil take its foot off the pedal in building alternative capacity and shift efforts to harness its fossil fuel resources?

The Power and Transportation Segments Are Building Alternatives

Figure 18.1 shows where Brazil plans to establish new electrical capacity through 2021. A national plan initiated in 2004, and still bearing fruit, incentivized investments in small hydro projects, biomass, and wind. Further development of hydropower, however, will come with challenges. Large hydro projects, particularly in the Amazon region, contend with strong public resistance, and Brazil has recently proposed a novel approach for development within these sensitive regions modeled after offshore petroleum projects. There are no roads planned to the area, and workers are temporary residents, helicoptered in and out from the site. Theoretically at least, deforestation occurs to build the dam, but natural processes are allowed to return when construction is complete and the area is evacuated.

For the present, fossil fuel development is third, behind hydro and wind. To no one's surprise, early work in the Santos Basin field reports that the extraction technology will work. So production will begin to draw against this large resource. Oil and gas are receiving 68% of Brazil's

Figure 18.1. Projected growth in electrical capacity (GW) from 2011 to 2021 by source. *Source Data:* Cristiano Augusto Trein, "Jica Training and Dialogue Programs," Ministry of Mines and Energy, Brazil, Secretariat for Energy Planning and Development, Energy Development Division, June 2013, https://eneken.ieej.or.jp/data/5003.pdf.

2011–2021 energy investment budget. So, the framework is proving useful in anticipating behavior. Impressively Brazil ranks third in the world, behind China and the United States, in total renewable power capacity. Hydro capacity ranks second in the world, behind only China, with plans to add another 33 GW by 2021. Decades ago Brazil invested in biofuel production to limit its exposure to petroleum imports, and now it is home to the world's highest blended ethanol mandates of 20%. Across the country, 34,500 stations sell blended fuels, and flex-fuel vehicles represent 96% of new car sales. Fifty different varieties of sugarcane have been developed for various soil and weather conditions to maximize productivity. Finally, two nuclear reactors generate nearly 3% of the country's electricity, with a third reactor under construction. There are proposals for an additional four reactors to come on line in the 2020s.

The Lure of Petroleum Will Be Difficult to Ignore

Brazil was focusing on hydropower, bioenergy, and wind until it discovered the large pre-salt petroleum resource. This is exciting news for the petrochemical companies that will participate in the recovery of the resource. Rich in natural resources favorable to renewable energies and now fortunate with a petroleum resource, Brazil will undoubtedly adapt its energy planning. As in other parts of the world where large deposits of fossil fuels exist, the economic windfall is impossible to resist. Focus and priority will shift to the development and production of this newfound resource. Brazil now joins the club of energy-conflicted countries that simultaneously pursue clean domestic energy sources while determined to secure the benefits of their fossil fuel wealth through exports. Brazil realistically expects to achieve petroleum independence and soon after join the exclusive club of exporters.

Note

1. Government Offices of Sweden, *Energy Policy in Brazil: Perspectives for the Medium and Long Term* (Östersund, Sweden: Swedish Agency for Growth Policy Analysis, 2013), https://www.tillvaxtanalys.se/download/18.201965214d8715afd1 13b87/1432548740127/Energisystem%2Bbortom%2B2020%2BBrasilien.pdf.

19

Germany, Norway, Spain, and Italy

Well begun is half done.
—Aristotle

On a per capita basis, Western Europe leads the world in building renewable capacity across all categories. The prudent use of energy is a widespread practice, and policy is strongly influenced by its citizens. Generally speaking, leadership is more willing and better able to confront challenges when it has a mandate from the majority of its citizens. As with any transition, the move to renewables will involve initial costs and pain points prior to the realization of benefits. The environment in Western Europe thus seems particularly conducive to the task of steering a new course, and nowhere in Europe is this more true than in Germany, which refers to its plan as "Energiewende," or "an energy transition."

The power segment is worth highlighting because this is where the signs of the transition first appear. Nearly 76% of Germany's 2012 electricity was sourced from fossil fuels and nuclear energy, with plans to reduce this to 20% by 2050, a remarkable intention to shift its energy mix in just 35 years. Wind, biomass and waste, and solar already generate an impressive 24% of its electricity. Across all energy sectors, nearly 90% of Germany's primary energy in 2013 was sourced from fossil fuels and nuclear, both of which are now deemed unattractive and targeted for discontinuation.

In 2000, nuclear power was responsible for half of Germany's carbon-free electricity generation. On a world stage, Germany's nuclear

*Germany's
"Energiewende" is
a dual plan to stop
using fossil fuels
and nuclear power.*

power delivered 6.5% of the world's output. Despite the significant contribution nuclear power makes to its noncarbon energy mix, Germany has made the decision to close all its nuclear power plants by the year 2022. This comes on the heels of the Fukushima disaster and will create a significant energy vacuum. This implies that Germany either places safety first or simply believes that renewable technology is ready and capable. Aligning to the national shift, Siemens, a large German multinational firm, is discontinuing its nuclear business segment and instead entering the embryonic renewable field. Germany's energy transition, "Energiewende," is a dual plan, with goals to cease the use of fossil fuels while simultaneously terminating the use of nuclear power.

Because Germany is so poor in petroleum resources, its energy plan reflects a mighty determination to achieve energy independence and all that it conveys—national and economic security and environmental stewardship. The citizens largely support the bold adoption of renewables and are intellectually accepting of higher energy costs during the changeover. Almost 50% of renewable energy capacity is citizen-owned; they are invested.

What We Can Anticipate (PECF = 2.3 × FFWF = 2.3)

If we were to exclude coal reserves, Germany's Fossil Fuel Wealth Factor would be nearly 0. Its lignite has one of the lowest heat content factors in the world. Considering this, Germany would be classified as vulnerable. Consequently, the framework anticipates an urgent priority to develop domestic energy sources and measures to reduce consumption through conservation and efficiency. If health and environmental concerns escalate, that will only accelerate the country's move to alternatives. The framework predicts that the "vulnerable" will move first, and Germany is doing just that.

A Bold Move to Retire Both Nuclear and Fossil Fuels

Germany is a powerhouse economy by any measure, and its "Energiewende" sets inspiringly ambitious goals. Not surprisingly, the goals

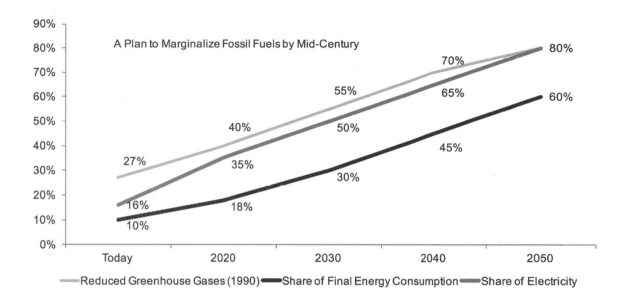

in Figure 19.1 are world-leading targets for the next 40 years. What makes them groundbreaking is how abruptly Germany is changing its energy direction. Renewables are starting from a baseline of 10% of final energy and 16% of electricity generation and moving to 60% and 80%, respectively. Other countries around the world likely will have higher portions of renewables by 2050, but none of those other countries are starting from such a low baseline. Alongside the renewable targets, Germany is planning an 80% reduction of greenhouse gases from a 1990 baseline by 2050.[1] Beyond these aggressive goals, perhaps the most difficult of the targets is related to conservation and efficiency: reducing overall energy consumption by 50% in 2050 from a 2008 baseline. Germany is already one of the developed world's most efficient consumers of energy, making this goal all the more impressive.

Figure 19.1. Goals of Germany's "Energiewende." *Source Data:* David Buchan, *The Energiewende– Germany's Gamble* (Report no. SP 26. Oxford: Oxford Institute for Energy Studies, June 2012), https://www. oxfordenergy.org /wpcms/wp-content /uploads/2012/06 /SP-261.pdf.

Beyond applying technologies to the task, Germany has pioneered innovative policies to help spur its transition. Feed-in tariffs are long-term (typically 20 years) guaranteed rates that are paid by the government for renewable energy. This attracts investors into the market. To fund the tariffs, a consumer surcharge is applied to electricity consumption. Tangentially it is also a policy that incentivizes energy conservation because prices are increased to cover the surcharges. With slightly more than 1% of the world's population, Germany ranks fourth globally in renewable power capacity. The feed-in tariff program was the spark plug that ignited its investments in renewables.

Although many countries are prioritizing other renewables ahead of solar, Germany ranks number one in solar PV capacity, and its Western European neighbors are also building large capacities of solar PV. What makes solar PV economical in these countries while elsewhere it is viewed as too expensive? Electricity costs are higher in Western Europe, and solar PV has hit the tipping point for cost-effective conversions.

Norway has accumulated an $860 billion post–fossil fuel fund (the equivalent of $170,000 per person) to support an employment base transition.

Already, in early 2014, 16% of Germany's businesses were self-sufficient for electricity.[2] This is demonstrating that investments in energy-efficient buildings and on-site renewable capacity can define a whole new energy model. At the same time this can be frightening to established utilities unprepared to adapt. Utilities in Germany are seeking to tax self-sufficiency to support grid infrastructure expenses. As mentioned earlier, this seems unfair but is actually quite rational. As self-sufficiency increases, utilities will be pushing less electricity through the line. As long as we connect to a grid, we should expect to finance its maintenance. Beyond businesses, big cities in Germany are establishing large capacities in renewables and setting even higher goals. The city of Munich, Germany's third-largest, impressively expects 100% of its electricity to be generated from renewables by 2025.

The World's Largest Hydrogen Production Facility

Here we get a glimpse of the future. As we move further into the energy transition, there will be more occasions when the supply of electricity exceeds demand, and we will need infrastructure to collect the energy. It is not surprising that those countries leading the world in renewable generation will be the first to confront this situation. Germany is building infrastructure to divert excess electricity to storage, and the elements of a new energy sector are forming.

Germany in July 2015 flipped the "on" switch for the world's largest dedicated hydrogen facility, able to power 6 MW of excess electricity to the production of hydrogen. Four nearby wind farm parks are welcoming the Energiepark Mainz plant with arms wide open. Avoiding curtailment now has an answer. Siemens has provided a highly dynamic polymer electrolyte membrane high-pressure electrolysis system, the cornerstone of the Mainz plant. It is anticipated that most of the hydrogen will be delivered to hydrogen refueling stations and support upward of 2,000 fuel cell vehicles. The building of Energiepark Mainz is also part of a plan to develop 1,000 hydrogen refueling stations. Any excess hydrogen will be fed into the natural gas grid for downstream consumer use.

Summarizing: A First Flight Example

Germany has set aggressive goals for the adoption of renewables and efficient use of energy. "Energiewende," though, has its share of detractors. Its big offshore wind farm projects have had setbacks, and critics are poised and ready to blast any missteps. Headlines have denounced increasing reliance on renewables for their temporal nature. It also doesn't help that high-value renewable manufacturing jobs are shifting to Asia. Finally, the reality of higher near-term energy prices because of surcharges is raising tensions in home budgets and multinational companies. Recently, citizen pushback on surcharges has occurred, with claims that households are bearing a disproportionate share of the costs. The pain point is arriving in Germany.

Although Norway is rich in petroleum, in early 2014 Norwegians purchased more electric vehicles than conventional automobiles, demonstrating the power of price.

Despite these battle scars, Germany seems determined to hold course. It is adopting renewables on several fronts: biofuels in transportation, and biomass, geothermal, wind, and solar in power. Germany's long-term energy plan is the world's boldest; it makes a complete break with the present and sets a course to move entirely to renewables. Germany is comprehensively and systematically building renewable capacity, to the point where fossil fuels will no longer be used for energetic purposes. When it makes mistakes, the rest of the world has an invaluable opportunity to watch and learn.

Astoundingly, Germany expects to reduce primary energy consumption by 50% by 2050. This is a remarkable target that is difficult to imagine even setting, let alone achieving. But by tying consumption behavior into its renewable capacity goals, Germany is comprehensively addressing its fossil fuel dependence. Home to 80 million residents, Germany is ranked in the top five on a per capita basis for many of the world's renewable categories in terms of capacity. Worldwide it is number one in solar PV (5% of Germany's electricity in 2013 was generated from solar PV), second in biopower generation, and third in wind capacity and overall nonhydro renewable capacity.

Germany is striving to transition to renewables without undermining its economy, a concern shared by nearly everyone. Unique among all the other countries studied, Germany has rejected the use of both fossil fuels and nuclear energy. If we believe that sooner or later we all need to transition to nonfossil energy sources, Germany is providing a first flight example.

Before we leave Europe, a quick review of a few other countries is warranted. Spain and Italy specifically are deploying high per capita amounts of renewable energy, while Norway is already producing almost all its electricity from renewables.

Very Briefly, Norway (PECF = 4.9 × FFWF = 4.2)

Some countries have it easier energetically, and Norway is one of them. Like Canada, Norway is blessed with both abundant fossil fuels and renewable resources. It has the largest reserves of oil and gas in all of Europe and delivers its exports of gas directly to the United Kingdom, Germany, France, and Belgium through a network of pipelines. Worldwide, Norway's exports of natural gas ranked third, behind Russia and Qatar, and overall oil and gas exports accounted for 23% of Norway's GDP.[3] Approximately 85% of Norway's exports go to its European neighbors.

Having it easier energetically doesn't mean that Norway is free of challenges, though. The "Curse of Oil and Gas" has created an unhealthy dependence on the highest oil and gas tax system in the world, with marginal rates as high as 78%, generating 30% of the country's revenues.[4] These revenues have artificially increased wages outside the petroleum sector and support large welfare spending on disability and sick pay. Perhaps as a reflection of the nation's fossil fuel reserve wealth, Norwegians have one of the shortest workweeks in the world at 33 hours. They do, however, see the inevitable end of fossil fuels and have accumulated an

$860 billion post—fossil fuel fund, the equivalent of $170,000 per person, to help support a different kind of transition: an employment base transition. The 2014–2016 price decline in fossil fuels has already forced Norway to tap into this fund, an ominous sign for countries so dependent on fossil fuel revenues. For Norway, oil and gas have been a boon and nearly a single-track part of its economy that is not sustainable. Although it has carefully prepared for it, its economy will face challenges in navigating a smooth transition when its export markets shrink. Replacing lost jobs in oil and gas with new types of employment will be difficult.

The FFWF and PECF model expects that Norway would exploit the economic opportunity afforded by its abundant reserves of fossil fuels. Indeed, Norway exports 88% and 95% of its production of oil and gas, respectively. Within power, hydro alone accounted for more than 97% of Norway's electricity, ranking seventh in the world in terms of capacity (29.3 GW). Overall, fossil fuels supplied 34% of the primary energy in 2013, well below the worldwide average of 86%. Norway has built sustainable domestic power capacity, primarily hydro, while investing in the development and production of fossil resources for exportation. Since 2000 the consumption of primary energy has declined 3%, which relieves some of the pressure in maintaining such a high contribution from renewables.

Remarkably given the country's petroleum wealth, in early 2014, Norwegians purchased more electric vehicles than conventional automobiles, demonstrating the power of price in consumer behavior. The policy of a high petroleum tax has driven a dramatic change in the transportation sector mix and illustrates the potential of a carbon tax in changing our primary energy mix. Also noteworthy is Norway's dabbling with heat pump renewables, opening a 14-MW hydrothermal plant using seawater for district heating.

Very Briefly, Spain (PECF = 1.8 × FFWF = 0.06)

Absent development of domestic energy sources, Spain is considered vulnerable by the framework, with imports of 97%, 100%, and 79% of its oil, gas, and coal, respectively, confirming the assessment. The use of fossil fuels for primary energy was 15% lower than worldwide averages in 2013, while renewables were a 9% higher portion of the country's electricity generation. Still, fossil fuels were used to produce nearly 50% of Spain's electricity, with nuclear and wind power contributing significant portions. Solar power is gaining investment momentum and expected to increase its contribution in the near term. Unlike Norway's, Spain's primary energy consumption has increased by 8.5% since 2000. So, Spain's plan includes demand-side conservation and efficiency programs along with supply-side shifts to alternative energy sources.

Similar to Germany, on a per capita basis Spain is a world leader in overall renewable power capacity, ranking fourth. Spain leads the world in concentrated solar power and ranks fourth in wind capacity. Looking forward, Spain expects geothermal energy to provide 5.8% of final energy supplies by 2020.

Very Briefly, Italy (PECF = 1.7 × FFWF = 0.02)

Italy, like Spain, is fossil fuel–poor and building measurable capacities of sustainable domestic energy sources. Italy imports 88%, 89%, and 100% of its oil, gas, and coal consumption, respectively, so energy independence is a faraway ideal. On the demand side, high domestic energy costs incentivize conservation and efficiency efforts, and this is always a helpful complement to building alternative capacity. Since 2000, primary energy consumption has declined by 5%, partially reflecting many factors, including the higher price signal. Overall, consumption of fossil fuels for primary energy in 2013 was in line with worldwide averages, with the balance contributed from renewables. As a measure of progress, domestic energy sources produced nearly one-third of electricity, with solar and wind building rapidly.

As with its Western European neighbors, Italy ranks high in the investment and establishment of renewable capacity. On a per capita basis, solar PV capacity, which generated 7.8% of its electricity in 2013, ranked second behind Germany. Through the end of 2014, Italy ranked fourth in the world in nonhydro renewable power capacity. Even higher domestic capacities will follow, reflecting Italy's impressive investments.

In 2013, renewable investments of $3.6 billion ranked 10th in the world. By leveraging its position along a border of the Eurasian Plate, Italy also ranked fifth in the world for geothermal capacity in 2013. Finally, beginning to address the transportation sector, Italy recently opened the world's largest advanced biofuel production facility.

A common thread is evident across Germany, Spain, and Italy. These countries are attacking reliance on imported fossil fuels from two sides. They are conserving energy with smaller energy footprints and saving energy through efficiency improvements. They are reducing energy demand from an already prudent level and highlighting the importance of behavior changes along with technology advances. All three of these countries have made renewable energy investments, ranking high in the world in per capita renewable capacity. The current pace of alternative energy development places Western Europe on track to replace fossil fuels by the end of the century if these countries stay the course.

Notes

1. Federal Ministry of Economics and Technology, *Germany's Energy Policy: Heading towards 2050 with Secure, Affordable and Environmentally Sound Energy* (Berlin: Federal Ministry of Economics and Technology, April 2012), http://www.bmwi.de/English/Redaktion/Pdf/germanys-new-energy-policy.
2. REN21 Renewable Energy Policy Network for the 21st Century, *Renewables 2015 Global Status Report.*
3. U.S. Energy Information Administration, "International Energy Statistics."
4. Balazs Koranyi, "Insight—End of Oil Boom Threatens Norway's Welfare Model," Reuters UK, May 8, 2014, accessed June 8, 2015,http://uk.reuters.com/article/uk-norway-economy-insight-idUKKBN0DO07520140508.

20

Russia

It's easier to resist at the beginning than at the end.
—Leonardo da Vinci

How do you win friends and influence neighbors? If you are fortunate to possess a large share of the world's extractable fossil fuel resources and you are inclined to export, that makes a good start. Even friendlier is sharing the resource at attractive pricing. Russia is afforded large shares of the world's proven reserves for natural gas and coal, with 25% and 18%, respectively. Proven reserves of petroleum of only 4% seem low in comparison, and yet Russia is consistently the world's number one or two exporter of petroleum along with Saudi Arabia. They are masters of extraction. There are two ways fossil fuel exporters can play the geopolitical implications of such enormous energy wealth: one that allows a country to exercise undue political pressure on its neighbors and another that allows it to serve as a stabilizing power. There have been several occasions when overnight political differences have resulted in supply disruptions. The energy supply card is simply too powerful to resist when tensions flare.

Russia is similar to many other fossil fuel–wealthy countries; its citizens and industry are able to access energy at artificially low prices. However, every unit of wasteful domestic consumption takes away from its export potential. Of the countries studied, Russia has the world's highest energy intensity by far. Energy intensity is the amount of BTUs consumed per unit of GDP, and disproportionately high isn't good. Russia and Saudi Arabia have energy intensities of 32,294 and 23,123, respectively. For reference, two other countries studied, Brazil and Germany, have energy intensity rates of 10,520 and 4,840, respectively. Russia's abundant supplies of fossil fuels have resulted in appetites much greater than necessary. This is both a big problem and a big opportunity for Russia. If Russia applied best practices and invested in energy conservation and efficiency, the results would be like discovering new fields of petroleum.

Just as the United States argued the construction of the Keystone pipeline, Russia's Black Sea natural gas pipeline was a subject of debate. In December 2014 Russia announced that it would cancel the Western European leg of the pipeline and redirect the pipeline to Turkey.[1] Pipelines make good theatrics for politicians everywhere. But much less often do we have intelligent discussions on changing our energy mix (i.e., Germany). Until a new energy mix is built that relies less on fossil fuels, pipelines, train tracks, and shipping lanes will be the paths by which these commodities flow.

A third of Europe's oil and gas is supplied by Russia, primarily via pipelines. Increasingly, however, Russia cannot help but shift its attention eastward to the growing appetites of Asian markets, and it is now developing crude oil processing centers in eastern Siberia to serve China. Exploiting its fossil fuel wealth will remain a top priority for this vast and fuel-wealthy nation for decades to come. Its business outlook for supply is positive, while economic uncertainty due to price volatility clouds the picture.

Russia's primary energy consumption has grown 1.45% annually since 2000. Natural gas provides more than half of Russia's primary energy, followed by oil and coal at 22% and 13%, respectively. Overall, Russia's fossil fuel mix was in line with worldwide averages, while its nuclear plants contributed higher than average levels in the power sector. Beyond hydroelectric, Russia has thus far deployed miniscule amounts of nonhydro renewable capacity. The electricity generation sector is nearly two-thirds fossil fuel, with the remaining source equally split between hydro and nuclear. The energy mix for the power generation sector has changed little since 2000 and will likely remain the same in the decades ahead.

Whereas Germany chose to close its fleet of nuclear plants following the Chernobyl accident, Russia followed it by investing in refining its nuclear technology.

What We Can Anticipate (PECF = 2.9 × FFWF = 7.8)

The framework considers countries with Russia's combination of factors fortunate. Abundant fossil fuel resources have afforded its citizens and industry a reliable and economical source of energy while also building a prosperous export economy. Almost uniquely, Russia is rich in all three fossil fuels. Exploiting its fossil fuel resources subordinates concerns with climate change. Indeed, the development of renewable energy sources amid such fossil fuel wealth is nearly impossible on purely economic terms. Furthermore, economically viable and distributed renewable technology runs counter to a centralized model; Russia will not lead the way in renewable deployment.

Beyond Fossil Fuels: Nuclear Power

There are ever-so-subtle changes in Russia's projected energy mix between now and 2030. The share of natural gas and coal in the primary energy mix will shift 6%, with natural gas going down and coal going up. Below the surface of energy mix charts, there are large investments in nuclear energy as a priority alternative. Less glamorous but perhaps more influential will be efforts to improve its poor use of energy as measured by the intensity factor.

Russia thus far has seemed disinterested in nonhydro renewable technology. It acknowledges the anthropogenic loads into the environment and the inevitable exhaustion of fossil fuels. Likewise, it admits concerns about health and quality of life with carbon energy sourcing, yet cessation will be determined on an economic basis. So much for that! Finally, Russia believes that renewables are not the only alternative to fossil fuels, which leads us to an exploration of Russia's view on nuclear power.

More so than any of the other countries studied, with the exception of China, Russia is articulating investments and plans for greater reliance on nuclear power in the future. A direct quote from Russia's *Energy Strategy for the Period Up to 2030* communicates its hopes for the technology: "The development of nuclear electric energy industry provides for elaboration of more advanced nuclear technologies enabling energy problems of mankind to be solved in the future."[2] Remember, the Ukraine was part of the Soviet Union when the Chernobyl accident occurred. Thirty years later, Russia is beginning to deploy a far more advanced nuclear technology. Where Germany to some extent chose to shuttle its fleet of nuclear plants following an accident that occurred oceans away in Japan, Russia followed the Chernobyl accident by investing in refining the technology. Russia is not only upgrading and improving its fission reactors; it is looking to produce hydrogen at its nuclear power plants as a method to accumulate energy. Russia is also a participant in the international project for nuclear fusion, while its fission strategy is a dual plan to develop closed fuel cycles and to colocate fast nuclear reactors alongside traditional thermal-neutron reactors.

The Energetic Inverse to Germany

Russia places a low priority on renewables, and decentralized energy sources may be antithetical to the Russian Federation. Russia's high energy intensity reflects how abundance can breed waste and how talk of climate change and the need to shift to renewable energy sources just doesn't resonate. It's like asking a gymnast to change her long-practiced routine midair. Exploiting and exporting fossil fuels has always served Russia well. The framework has proved predictive of Russia's energy priorities.

What is unique, given Russia's existing fossil fuel wealth, is its investment in nuclear technology. Russia went live in 2015 with its BN-800 fast reactor, believing that nuclear power will fill the gap in a reduced–fossil fuel era. The centralized aspect of fossil fuels has been a major factor allowing Russia to build and retain its federation. Moving to nuclear sources rather than distributed renewables not only is easier in an authoritarian-styled government but also continues to support the centralized paradigm upon which Russia's government is founded. We have seen two extreme positions from Germany and Russia on the role of nuclear energy. Germany has excluded nuclear as it moves to decentralized renewables, while Russia until now has dismissed a significant role for renewables and instead believes that nuclear power can fulfill its future energy demands. It seems that both positions are off the mark with their exclusions. The worldwide

Russia went live in 2015 with its BN-800 fast reactor, believing that nuclear power will fill the gap in a reduced–fossil fuel era.

sourcing of energy in the future will undoubtedly include both nuclear and renewables. To exclude one or the other undermines the pace at which we replace fossil fuels. Particularly given that nuclear fusion will some-day be commercialized, retaining a centralized nuclear tool set would be a prudent element in the world's long-term energy planning.

Notes

1. Stanley Reed and Sebnem Arsu, "Russia Presses Ahead with Plan for Gas Pipe-line to Turkey," *New York Times*, January 21, 2015, accessed April 15, 2015, http://www.nytimes.com/2015/01/22/business/international/russia-presses-ahead-with-plan-for-gas-pipeline-to-turkey.html?_r=0.
2. Government of the Russian Federation, *Energy Strategy of Russia for the Period Up to 2030* (Moscow: Ministry of Energy of the Russian Federation, 2010), http://www.energystrategy.ru/projects/docs/ES-2030_(Eng).pdf.

21

Saudi Arabia

When near-term pain and long-term benefits are pitted against each other, the short term typically wins.

Saudi Arabia holds 25% of the world's proven petroleum reserves, and these were first discovered less than 100 years ago. In 1933, King Abdulaziz bin Abdulrahman al Saud authorized Standard Oil of California (SoCal) to explore for oil. By 1939 SoCal had discovered large amounts of oil, and the Kingdom had begun exporting petroleum. The 1940s brought increased investment to the Saudi's oil fields when other American companies entered into a consortium with SoCal. This new Arab and American Oil Company took the name Aramco and by the 1970s had become the world's top exporter of oil. In 1980, the Saudi government assumed full ownership and named the entity Saudi Aramco, and it has been pumping world-leading quantities of petroleum ever since. The last major discovery of Saudi oil occurred in April 2005. A mere 83 years after the initial discovery, petroleum accounts for 45% of Saudi GDP and 80% of government revenues.

Great petroleum wealth doesn't immunize the Saudi government from challenges, though. Like many petroleum-wealthy countries, Saudi Arabia subsidizes domestic energy prices. Now higher domestic energy consumption is eating into exports, and increasing supplies elsewhere in the world are challenging its market share and pinching its domestic budget. As the Saudi government faces revenue shortfalls in today's environment of lower petroleum prices, its subsidy policy is coming under scrutiny.[1]

Although Saudi Arabia is still the world's largest producer and exporter of petroleum, its world market share has declined from 15% to 9.5% since 1980, when it took ownership of Saudi Aramco. Population growth, consumption rates, and the pace of production tell a "paradox of plenty" story. In 2013, the Saudi population was close to 29 million (including 8.5 million expatriates) and growing at just under 2% per year. Domestic consumption of electricity and petroleum has increased 115% and 85%, respectively, since 2000. So, although it produced 13% more petroleum in 2013 than it did in 1980, rising domestic consumption has led to an 11% drop in Saudi oil exports. Although Saudi Arabia is metaphorically running faster, it is falling behind; increasing production is no competition against rising consumption.

Saudi Arabia has signaled a plan to diversify energy sources, with 40% of its electricity generated from nonfossil fuels by 2040.

Since the 1933 decision to explore for oil, Saudi Arabia has seen one of the most remarkable transformations in the world. It has established top-quality education and health care systems for its citizens. In a few short decades its population has gone from meager consumers of energy to having world-class appetites. But well-intentioned subsidies, estimated by the United Nations to represent 10% of the country's GDP, may now be undermining Saudi economic health. About two-thirds of the subsidy supports transportation where the domestic price of gasoline is $0.40 per gallon; the balance of the subsidy supports lower electricity prices. Plans to improve conservation and efficiency will be thwarted by these subsidy programs.

What We Can Anticipate (PECF = 4.3 × FFWF = 12.7)

The framework predicts that Saudi Arabia's Fossil Fuel Wealth Factor will propel the country into a leading role in the petroleum marketplace by virtue of its abundant resources. And in actuality, Saudi Arabia's decisions on production strongly influence the global market. The production and supply of petroleum resources will preoccupy the Kingdom and make it difficult for Saudi Arabia to diversify its energy mix. For example, hydropower, the first major renewable to be harnessed globally, has no potential in the deserts of Saudi Arabia. Further, the Kingdom to date has not deployed nuclear power. Harnessing its fossil wealth was the easiest path forward, and it has allowed Saudi Arabia to finance and build a modern society seemingly overnight.

A long-term energy plan that excludes fossil fuels will be difficult to imagine and harder to consider as Saudi Arabia weathers the turbulence in the petroleum market. Wealth on this scale (FFWF = 12.7) can be both a blessing and a hex.

Realizing the Inevitable and Diversifying

No graph is required to convey Saudi Arabia's electricity generation by energy source. It is unanimously 100% fossil fuel. Similarly, gasoline-powered vehicles dominate its transportation sector; electric vehicles seem like an oxymoron. Across all the energy sectors, Saudi Arabia is singularly powered by fossil fuels. It will continue to build infrastructure to produce, refine, and export petroleum.

Recently, however, Saudi Arabia has signaled a plan to diversify energy sources that will have 40% of its electricity generated from nonfossil fuels by 2040. To meet this goal the Kingdom is contracting to build 17 nuclear reactors, representing a capacity of 17 GW.[2] In September 2013 GE Hitachi Nuclear Energy and Toshiba/Westinghouse signed agreements to provide the planned nuclear reactors. Taking advantage of its high solar insolation, Saudi Arabia is also building more than 40 GW of solar capacity, including both CSP and PV technologies.[3] Beyond diversifying the power sector's energy mix, the government has announced plans to develop 9 GW of wind capacity to support desalinization projects.

An Economy beyond Fossil Fuels

Saudi Arabia's well-intentioned goals for diversifying energy sources will struggle, however, with other funding priorities. Recent news from IHS, an industry consultant, lowers the forecast for solar PV products in Saudi Arabia.[4] This suggests that the volatility in its petroleum export market is forcing Saudi Arabia to make tough budget choices. Continuing to tap its proven reserves of fossil fuels is far easier than funding a long-term plan to diversify energy. When near-term pain and long-term benefits are pitted against each other, the short term typically wins.

The Kingdom's biggest challenge will be the adaptation of its fossil fuel–centric economy this century. What will be Saudi Arabia's contribution to world trade when its pumps are stilled? One could consider this challenge far more difficult than other countries building domestic energy sources. Without substantially addressing this looming eventuality, the Kingdom will confront social and economic upheaval.

Still, when Saudi Arabia sets goals to build substantial noncarbon electric power capacity by 2040, it's an acknowledgment of the inevitable limits of fossil fuels. Alternative energy sources are economically viable today, and fossil fuels will be subjected to lower and lower shares of the global energy market. The rest of us should take heed when the world's largest petroleum producer is building significant noncarbon capability. It's a signal that accessible fossil fuels are limited and under environmental attack and that the time scale for the change is the twenty-first century. The challenge for Saudi Arabia is to translate a plan into the actual construction of alternative capacity. The recent change in the market pricing of its exports will test the budget priority to diversify energy sources.

Notes

1. Saad ben Ali Alshahrani, "Saudi Arabia Reassesses Energy Subsidies," *Al-Monitor*, December 11, 2013, accessed April 15, 2015, http://www.al-monitor.com/pulse /business/2013/12/saudi-arabia-energy-subsidies-assessment.html.
2. World Nuclear Association, "Nuclear Power in Saudi Arabia," January 2016, accessed April 15, 2016, http://www.world-nuclear.org/information-library /country-profiles/countries-o-s/saudi-arabia.aspx.
3. Ian Clover, "Solar Power Key for Saudi Future, Says Energy Chief," Pv Magazine, October 28, 2014, accessed April 15, 2015, http://www.pv-magazine.com/news/details /beitrag/solar-power-key-for-saudi-future--says-energy-chief__100016969 /#axzz45ud6O7Ry.
4. Edgar Meza, "IHS Halves Five-Year Outlook for PV Installations in Saudi Arabia," *pv Magazine*, January 21, 2015, accessed July 18, 2016, http://www.pv-magazine.com /news/details/beitrag/ihs-halves-five-year-outlook-for-pv-installations-in-saudi -arabia__100017885/#axzz4ElT9n4I7.

22

Japan

What we think, we become.
—Buddha

Despite the major handicap of having few fossil fuel resources, Japan is one of the world's top economies. Until the disaster at the Tokyo Electric Power Company (TEPCO) Fukushima power plant, Japan was generating roughly 30% of its electricity from nuclear and 10% from renewable sources. The balance (60%) was produced via fossil fuels. Japan imported an enormous 97%, 96%, and 100% of its oil, gas, and coal supplies, respectively. Following the TEPCO disaster, the government temporarily decommissioned all of its 48 nuclear reactors in order to assess risk and implement safety upgrades. This obviously caused a huge gap in the supply of electricity, taken up by a substantial increase of imported fossil fuels. Japan increased its imports of crude oil and natural gas by 15% and 25%, respectively, between 2010 and 2012. As a consequence, the country experienced its first trade deficit in more than 30 years, reminding the world of the economic vulnerabilities associated with too much dependence on foreign sources for energy and, more fundamentally, the botched application of nuclear technology.

We can sense urgency in Japan's energy plan because of the TEPCO Fukushima nuclear disaster. Following the Fukushima incident, the Japanese government made the decision to close all its other nuclear plants to assess and implement safety standards, and this caused an overnight energy vacuum. Imagine the chaos of losing 30% of your electricity capacity and not only having to comb the world for fossil fuel suppliers but also having to urgently fire up plants to bring them on line. Countries like Japan, highly dependent on imported fossil fuels, are rightfully uneasy with the long-term risk associated with supplies from external partners. The world is unfortunately prone to regional and global tensions, and large exporters haven't hesitated to threaten supplies to exert political influence. Japan is confronting near-term economic challenges while acknowledging the long-term risks of its energy dependence.

Defining moments like the TEPCO disaster can marshal responses that can achieve remarkable advances. Can Japan seize the moment and invest in domestic energy sources, or will urgency in the present force a shortsighted plan that sustains its dependence on fossil fuel supplies? How will the Japanese weigh greenhouse gas emissions, nuclear safety, energy security, and economics in its energy policy?

Fossil fuels supplied 93% of Japan's primary energy in 2013, 7% higher than worldwide averages. No doubt the temporary loss of nuclear capacity has exacerbated the situation, but Japan's overwhelming reliance on imported supplies of fossil fuel is creating pressure to improve self-reliance with a combination of renewables and the restarting of its upgraded nuclear power plants.

The Fukushima incident is a call to upgrade safety standards, not a call to abandon nuclear technology.

What We Can Anticipate (PECF = 2.3 × FFWF = 0.01)

Having such a low FFWF and high relative PECF characterizes Japan as vulnerable. It is a fully developed economy with no significant domestic petroleum resources. Like Germany, Japan would be expected to prioritize improving its energy self-reliance. The framework anticipates a dual strategy to invest in domestic capacity (i.e., nuclear and renewables) and establish a stable fossil fuel supply network. Unlike Germany, Japan is including nuclear power in its plans to build alternative energy capacities.

A Desperate Desire for Energy Self-Sufficiency

Japan's own description of its situation and goals taken from its 2014 *Strategic Energy Plan* is worth sharing verbatim:

Japan is in a situation where its electricity supply-demand structure depends more on supply of overseas fossil fuels than at the time of the first oil shock, so it must be said that Japan's energy security is surrounded by a harsh environment.

    . . . Improvement of self-sufficiency by continuing mid to long-term efforts to strategically utilize renewable energy, nuclear power as quasi-domestic energy, and resources laying in Japan's exclusive economic zone, including methane hydrate and other offshore resources, as a domestically-produced energy.[1]

By 2030, Japan plans to double its self-sufficiency ratio, which means building renewable capacity, developing domestic oil and gas resources, and restarting its network of nuclear reactors. Directly aligned with the self-sufficiency target, Japan also expects to double its zero-emission power source ratio. In addition, there are goals to halve CO_2 emissions in the residential sector. This will follow a shift to cleaner power generation and zero-carbon transportation vehicles but also will encompass conservation and efficiency measures. Vital to Japan's industrial base are goals that improve energy cost efficiencies.

It is interesting to note that Fukushima Prefecture, the site of the TEPCO nuclear disaster, announced plans to generate 100% renewable electricity by 2040.[2] Cynically one could say that this is easy to do when the area has been evacuated within a 12.4 mi exclusion zone around the plant, but that would overlook an inspiring response. Literally from the ashes of a disaster a new energy structure is born.

Japan considers wind power economically attractive and assigns a top priority to building its capacity. To accelerate the establishment of wind capacity, Japan will streamline safety regulations and speed the process of environmental assessments. A large-scale offshore wind demonstration project utilizing floating platforms may prove successful and provide an environmentally attractive alternative to the deep-rooted fixed platforms we see most everywhere else.

Japan deems the current cost of solar energy too high for utility applications, requiring further optimization. Smartly, however, the Japanese are directing policy to advance distributed solar capacity where residential economics are favorable. Even despite this segmentation, Japan's solar PV capacity still ranks third in the world.

Geothermal, perhaps the most economical of all energy sources (from a variable cost perspective), is a different story. Japan, bordering the "Ring of Fire," already has the world's third-largest capacity for geothermal power in thermal applications. Geothermal potential is upward of 20,000 MW, which, if realized, would contribute nearly 20% of Japan's electricity. Today's geothermal power generates less than 1% of total electricity, so this represents a great opportunity. Previously, many ideal locations were off limits to development, and the initial capital investments are high. To support the establishment of this highly desirable domestic energy source, the government plans to introduce measures to encourage investment by reducing capital risk while also aligning regulatory processes.

Nuclear power, too, has many advantages to offer, providing that stringent safety standards can be implemented. Nuclear operations are greenhouse gas–free, they provide stable and economical energy supplies, and they are largely domestic. The Japanese government is committed to establishing long-term solutions for nuclear waste and establishing safety standards to provide its nuclear plants a path to restarting. The Fukushima incident has been a call for upgraded safety standards, as opposed to a call for the abandonment of the technology. In the longer term, Japan hopes to deploy commercial fast reactor technology sometime after 2050 on the assumption that it can meet the necessary safety criteria.

Where this island nation's energy plan is especially distinct and forward-thinking, however, is in its development of a means to accumulate, distribute, and retrieve energy. Japan envisions hydrogen fuel as a principal energy carrier that can be distributed and used much like natural gas. In literature, the term *carrier*, rather than *fuel*, is often used for hydrogen. This stems from the reversible hydrogen cycle, which is distinct from the depletion cycle of fossil fuels, and the fact that hydrogen, like electricity, will become a means of delivering energy. Hydrogen will be delivered directly to homes via the natural gas infrastructure to stand-alone units that convert hydrogen to electricity within residential and commercial buildings.

The ENE-FARM, a stationary hydrogen fuel cell appliance, was first launched in 2009 in Japan.[3] This joint development effort between Panasonic and Tokyo Gas is the world's first commercialized fuel cell system targeted for household electricity generation. The goal is to introduce 1.4 million units by 2020 and 5.3 million units by 2030. The unit draws hydrogen from the city's natural gas network and produces electricity on-site. The residual thermal energy is used to heat water. The unit is not only clean but quiet because there is no combustion, and the only by-product is water. Move over gas furnace and oil burner, the hydrogen fuel cell is taking your place!

Hydrogen will also be a new-age fuel for transportation. Japan's automobile industry is introducing impressive hybrid hydrogen fuel cell vehicles. Japan hopes to be a worldwide leader in hydrogen technologies and other advanced storage solutions.

Beyond Restarting Nuclear Power Plants: Investment in Alternatives Is Increasing

In the short term, Japan has been forced to import higher quantities of fossil fuels because its nuclear facilities were decommissioned for safety audits and upgrades. It is deeply concerned with import dependence for large portions of its energy requirements. Near term, to address this concern Japan is working to establish a diverse, flexible, and resilient supply chain to mitigate risk. Recent events have made it clear that despite the Fukushima event, Japan will restart its nuclear power plants as soon as possible to mitigate fossil fuel importation expenses and associated supply risks. The first reactor at the Sendai nuclear plant was restarted in August 2015, and many of the other reactors will follow as they satisfy new safety standards. This will allow Japan to regain a measure of energy self-sufficiency and stabilize its domestic energy costs.

Midterm, Japan will lower regulatory hurdles to hasten smooth development of new wind and geothermal facilities. Japan views these two renewables as technologically and economically ready to implement at the utility scale, while solar PV is deployed residentially. Less glamorous but still impressive, Japan's biopower generation ranks fifth in the world.

Longer term, reducing energy requirements through conservation, expected shifts to lower-intensity industries, and a declining population helps mitigate foreign dependence. Evidence of this is seen in Japan's primary energy consumption figures, which have declined 9% since 2000.

Japan is positioning itself for a leadership role in the embryonic hydrogen industry. This includes production, storage, fuel cell, and distribution technologies. Japan believes that hydrogen will prove to be the most efficient way to store and deliver energy as a secondary source. Its well-articulated plan to become a "Hydrogen Society" represents an eloquent and truly long-term solution, not seen in most other countries.

The Japanese have marveled the world with leading products in automotive and electronics industries. Can they do the same in renewables? In line with these aggressive goals, Japan increased its renewable investment by 80% in 2012, so it seems like a reasonable bet.

Notes

1. Japan's Ministry of Economy, Trade and Industry, Strategic Energy Plan, April 11, 2014, http://www.enecho.meti.go.jp/en/category/others/basic__plan/pdf/4th__strategic__energy__plan.pdf.
2. REN21 Renewable Energy Policy Network for the 21st Century, *Renewables 2015 Global Status Report*.
3. Tokyo Gas, "Development of the New Model of a Residential Fuel Cell, 'ENE-FARM,'" accessed April 15, 2015, http://www.tokyo-gas.co.jp/techno/english/menu3/2__index__detail.html.

23

China

For several decades to follow, China must continue to build new
capacity beyond the scale of many countries.

The energy headline for China is "Number One." China is the world's
largest energy consumer, the largest generator of electricity, and the largest producer, consumer, and importer of coal. It has the largest overall
renewable capacity, with leading positions in hydro and wind for the
power sector and leading positions in solar and geothermal for thermal
applications. China has the fastest-growing GDP on a net dollar basis,
and its energy demands won't peak for another 20 years. A population of
more than 1.3 billion people and a relatively high economic growth rate
help establish these rankings.

Headlines can be misleading though, particularly as they relate to
China's coal consumption. For instance, China consumed 49% and
extracted 46% of the world's coal in 2012. These statistics would suggest that China is too dependent on the dirtiest of the fossil fuels and
that its rampant use of coal is undermining worldwide efforts to combat poor air quality. Looking at the statistics for coal consumption and
extraction on a per capita basis, however, shines a slightly different light
on the story. China consumed and extracted 3.0 short tons of coal per
capita, almost exactly equivalent to the rates for the United States. The
United States consumed 2.8 and extracted 3.2 short tons per capita.
Among many conclusions we might draw, an obvious question is this: Is

China's use of coal "rampant," or is it merely a reflection of population? Morally, are we prepared to say that 2.8 short tons per capita is OK for a mature nation but 3.0 is too much for a developing country?

To further confuse matters, let's consider coal consumption and extraction on a per GDP basis. China consumes and extracts about eight-fold the amount the United States does, on a per GDP basis. Does this reflect the differences in the energy intensity of industries in China and the United States? Drawing inferences from per GDP statistics requires a fair amount of insight into a country's industrial base. Great care must be taken in choosing the correct statistic for any analysis. Statistics can be used to gain an understanding, or they can be used to make a pre-meditated point.

China is the world's largest energy consumer, the largest generator of electricity, and the largest producer, consumer, and importer of coal.

China has a two-part energy challenge: how to build new capacity that keeps pace with its economy and how to secure reliable energy supplies both domestically and abroad. China's primary energy consumption has grown 178% since 2000. The growth between 2011 and 2012 alone was the equivalent of primary energy consumption for countries such as Turkey, Ukraine, and Thailand. Moreover, for several decades to follow, China must continue to build new capacity beyond the scale of many countries. Limited domestic fossil fuel resources mean that it must secure significant portions from exporters. It follows, then, that domestically China is working to develop every possible economically viable energy source.

A dominant 67% of primary energy is sourced from coal, 37% above worldwide averages. So, the issue with coal in China is its relatively high contribution within the primary energy mix. Similar to the case in the United States and India, China's fossil fuel wealth lies in its coal deposits. Fossil fuels, led by coal, provide more than 90% of primary energy, with hydro contributing another 7%. Nuclear and nonhydro renewables are relatively small contributors in China's current energy mix. In 2012 China imported 57%, 26%, and 3% of oil, gas, and coal consumption, respectively. To a large extent the disproportionately high use of coal reflects China's natural desire to be energy-independent, and producing 97% of consumption domestically has been very helpful toward that goal.

Its petroleum imports ranked second in the world, behind the United States. Territorial disputes between China and Japan in the East China Sea and its Association of Southeast Asian Nations neighbors in the South China Sea can be traced in part to access to petroleum resources. All countries covet domestic energy independence, and consequently any ambiguity in territories bearing petroleum resources gives rise to regional tensions. Strategically, China has acquired interests in foreign oil and gas assets not only to help secure supplies but also to secure technology know-how to further develop its own resources. For example, in 2013 China purchased the Canadian oil company Nexen for $15.1 billion, its largest overseas acquisition at the time. Many of China's oil fields are considered mature, so continued extraction will require enhanced oil recovery methods acquired through these types of foreign acquisitions.

What We Can Anticipate (PECF = 1.0 × FFWF = 0.4)

Don't blink your eyes when looking at the PECF figure for China, because it is rising rapidly. China, for now, is on an economic fast track and has joined the ranks of energy-vulnerable countries. For decades, manufacturers the world over have shifted operations to China primarily because of the low cost of labor. More recently, however, low labor rates aren't the whole story. For example, China and Taiwan manufacture a large share of the world's solar PV panels. The manufacturing process is highly automated, so what, then, is the impact of lower-cost labor in this situation? Large manufacturing investments in solar PV operations by China and Taiwan created superior economies through scale; labor was not the leading factor. Sustained prosperity in China will depend now on a new economy competing beyond low-cost labor, an economy where stable and affordable energy will play a pivotal role.

The framework correctly anticipates China's current situation: high dependence on imported fossil fuels and an energy mix significantly influenced by economics. Climate change concerns will be subordinate to economic development goals. China is unique, developing and yet economically prosperous, and its strong financial standing allows it to make long-term investments and suffer short-term pains where other countries cannot. So, how does China plot its future?

Beijing Directs the Plan

Every five years China publishes a plan that broadly sets the strategic direction for the country. Annual energy policies are then issued that provide specific actions in alignment to the five-year plans. A verbatim text from the energy plan preface sets the stage: "Energy is the material basis for the progress of human civilization and an indispensable basic condition for the development of modern society. It remains a major strategic issue for China as the country moves towards its goals of modernization and common prosperity for its people."[1]

The 11th Five-Year Plan targeted a 20% reduction in China's energy intensity (energy per GDP). The 12th Five-Year Plan calls for an additional 17% reduction.[2] The first plan target was largely achieved by closing inefficient power units and low-hanging efficiency gains from top industrial consumers. The current plan continues this effort with incrementally more-difficult-to-achieve targets. Together these plans will help China achieve an impressive reduction in energy intensity (baseline 2005) by 2020.

Closely related to and following the first goal, China targeted a 16% reduction in CO_2 emission intensity (per GDP) between 2011 and 2015. This was accomplished with increased supply from nonfossil energy sources. Nonfossil energy sources that accounted for 8% of China's total energy in 2011 will increase to 15% by 2020. The plan includes large new capacities in nuclear and wind. Financially supporting these goals, China plans to invest RMB 5.3 trillion in nonfossil energy sources.

Returning briefly to coal, China plans to cap the contribution of coal in total energy at 65% by 2017. This sounds good until one realizes that capping on a percentage basis still means that coal will increase in order to meet the country's growth in consumption. Figure 23.1 shows the outlook from China for electricity generation. Basically, declines in the mix of coal are offset by increases in wind and nuclear power. China's electricity

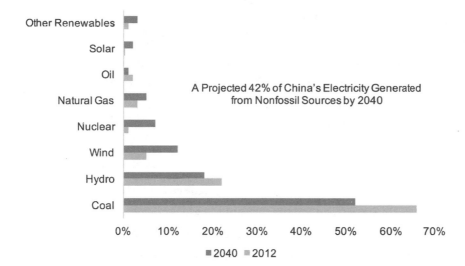

A Projected 42% of China's Electricity Generated from Nonfossil Sources by 2040

generation increased 45% from 2008 to 2012; projections are that it will increase an additional 49% between 2020 and 2040.

What is surprising to note is that although China has strategically established solar PV manufacturing, the technology is projected to play only a relatively minor role through 2040, contributing no more than 2% of its electricity. Remembering the scale of China, however, 2% is still a huge amount of solar PV. Continued advances in solar PV efficiency will need to occur for accelerated adoption. To underscore that point, wind energy is already considered to be on par with coal generation, while solar PV is not expected to achieve this tipping point until sometime beyond 2020. As with the rest of the world, in China when the desire for low-cost energy confronts the desire to reduce CO_2 emissions, economics wins the day.

More noteworthy, however, is China's investment in solar thermal capacity. Through 2014 China had deployed 290 GW of solar thermal capacity, representing 70% of the world's total. It ranks first in the world in hydropower, adding 22 GW in 2014 for a total of 280 GW, representing 27% of the world's capacity. Its wind capacity increased 22.2 GW in 2014 for a total capacity of 115 GW, also ranking first in the world. Although solar PV capacity isn't a top priority now, China still ranks second in

Figure 23.1. China's changing electricity energy mix. *Source:* U.S. Department of Energy, *China–Energy Brief* (Washington, D.C.: U.S. Department of Energy, Energy Information Administration, 2015).

the world, with 28 GW of capacity. China had only 14.7 GW of installed nuclear capacity in 2013 but much bigger plans. Thirty-one new reactors in various stages of planning and construction will boost nuclear capacity to 58 GW. Lest we think China is going completely clean, its coal consumption is still projected to increase through 2025.

Every Energy Source Remains an Option

China's energy policy is similar in several ways to that of the United States. Changes in its energy mix are largely determined by economics. China is fully committed to deploying wind because of the economic attractiveness of the technology and secondarily for its clean profile. Similarly China has set goals to improve its energy intensity for economic reasons—goals that, by the way, also deliver reduced CO_2 emissions. It remains fossil fuel–centric, with contributions from alternatives to help meet the rapid growth in the demand for energy. China has more than 14,000 mi of crude oil pipelines and 32,000 mi of natural gas lines feathered out to gather fuel. Nature drains her waters via rivers to the sea, while humankind has built intricate webs of pipelines delivering fuel to churning and burning power plants that deliver on-demand energy. As we see elsewhere, the petroleum web is still weaving threads in China. In summary, China is invested in extracting as much fossil resources as technology and economics allow. Internal pressure to improve poor air quality is increasing and may lower the projected use of coal in the future.

Unlike the United States, however, China is articulating nuclear energy as an important nonfossil alternative in its energy plan. On December 15, 2014, the Chinese Experimental Fast Reactor was brought to full power for three days.[3] The reactor is a sodium-cooled fast reactor constructed with assistance from Russia. The plant has a thermal capacity of 65 MW, able to produce 20 MWe of electricity. Xu Dazhe, head of the China Atomic Energy Authority and the State Administration of Science, Technology and Industry for National Defense, is quoted as saying, "The achievement has laid a solid foundation for fast reactor technology development, commercialization and nuclear fuel cycle technology development."[4] As in Russia, nuclear power is viewed as an attractive energy option for the future, and beyond 2020 fast-breeder reactors together with fuel cycle technologies will become the standard.

Notes

1. Chinese Government, *China's Energy Policy 2012* (Beijing: Information Office of the State Council, People's Republic of China, October 2012), http://www.iea.org /media/pams/china/ChinaEnergyWhitePaper2012.pdf.
2. KPMG China, "China's 12th Five-Year Plan: Sustainability," April 2011, https:// www.kpmg.com/CN/en/IssuesAndInsights/ArticlesPublications/Documents /China-12th-Five-Year-Plan-Sustainability-201104-v2.pdf.
3. "Chinese Fast Reactor Completes Full-Power Test Run," *World Nuclear News*, December 19, 2014, accessed April 15, 2015, http://www.world-nuclear-news .org/NN-Chinese-fast-reactor-completes-full-power-test-run-1912144.html.
4. Ibid.

24

India

No power is as expensive as no power.
—Dr. Homi Bhabha

India's energy environment calls to mind an image of an ocean wave curling above and threatening to overtake the surfer who hopes valiantly to ride it to shore. India is working hard to stay ahead of its rising energy appetite. Rising consumption, an untenable energy infrastructure, and a massive population accessing energy require a nearly impossible balancing act.

In 2012, a major blackout left 700 million people without power in 20 of India's 28 states. Unfortunately, there was no single cause that could easily be fixed. Deficiencies exist throughout India's energy sectors, from reliable fuel supplies to generation to transmission, and all of them need investments. A fix will require a systemic overhaul of India's energy infrastructure.

Current economic activity is hindered by energy insecurity, and future growth is placed at risk. India's GDP was the world's tenth-largest in 2013 at $1.877 billion. But on a per capita basis its 2013 GDP was 22% that of China's, so there are great aspirations for continued economic development. Since 2006 China has added 84 GW of power capacity per year to fuel its economic activity, while India has added only 14 GW per year, falling short of its own targets. An energy pinch point has arrived in India that will force corrections or realize Dr. Bhabha's warnings.

Energy security in India means supplying a lifeline of energy to its citizens independent of their ability to pay, and this lofty idea unfortunately spawns subsidies that often miss their mark. It commits the state to providing this energy at all times at competitive prices. Elsewhere in the world energy security refers to self-sufficiency or securing stable supply partners, but for India it is different; it means both access and affordability. Figures from the World Bank for 2010 indicate that one in every four of India's residents, roughly equivalent to the entire population of the United States, lacked access to electricity.[1] This sheds light on India's energy security priority.

As in China, coal plays a disproportionately high role in India's energy mix and will likely continue to be a major source for the near term. Domestic petroleum resources are limited, while India's hard coal reserves are the third-largest in the world. Consequently, it should not come as a surprise that India is the world's third-largest producer and consumer of coal. Here, though, lies a challenge. The coal sector is a bit of a monopoly, managed by two state-owned companies. There is little private participation. Perpetually poor management within these state-owned companies results in a sector unable to achieve new capacity targets, plants that routinely operate below optimum levels, and regular supply shortages. The energy that these plants deliver is state-subsidized, and those subsidies are demanding more and more of the government's budget. How does India find a way to attract investment to address these gaps? It seems that it must rationalize subsidies and privatize the sector.

Symbolic of the need for systemic changes is the astounding degree to which losses occur in India's power sector. India incurs aggregated transmission and commercial losses of 31% of its generated electricity, an amazing figure when compared with rates in other countries. Aggregated transmission and commercial losses in China, South Korea, and Brazil were 5%, 4%, and 17%, respectively. Losses in the transmission of electricity are unavoidable, so this figure will never be zero. But the aggravating part of aggregated losses for India is that most of the loss is occurring through theft, nonbilling or underbilling, administrative errors, and nonpayment.

Figures from the World Bank for 2010 indicated that one in every four of India's residents, roughly equivalent to the entire population of the United States, lacked access to electricity.

India's sheer size, population density, and language and cultural differences add special challenges and shape a complex environment.

What We Can Anticipate (PECF = 0.3 × FFWF = 0.2)

The framework describes India as dually challenged given that both factors are below 1. Some of us have it easier than others, but India has more than her share of challenges. India is home to a huge population (>1.2 billion), and by 2025 it will overtake China as the world's most populous country. Bestowed with few domestic energy resources and a determination to develop economically, India will need to work extremely hard, and it won't come easy. The model suggests that India will strive to leverage every possible resource it has to support economic progress. Climate change concerns may be appreciated, but acting on those concerns will challenge economic development and limit energy access. India will remind the world that it uses one-fourth the per capita energy of the rest of the world and that calls to limit its use of coal are unfair. The top priority in India is economic development, powered by an energy mix that maximizes that potential. Climate concerns will be addressed only indirectly through economic concerns that drive improved energy efficiency.

Where Climate Concerns Conflict with Energy Access

India's 2011 census revealed that an amazing 62.5% of rural households still use firewood as the primary fuel for cooking, 12.3% use crop residue, and 10.9% use dung. Still, India's largest energy source is coal, followed by petroleum. Despite having the world's third-largest reserve of hard coal, India had to import 13% of its demand in 2013 because of the inefficiencies in its state-run coal operations. Petroleum imports were 72% of its consumption. Even with investment, the domestic production of its fossil fuel resource cannot keep pace with demand—primary energy demand has been growing 5% per year since 2000 and will continue to grow even faster.

The power sector is the largest and fastest-growing sector, representing 38% of primary energy demand in 2012 and expected to increase to

47% by 2035 according to the International Energy Agency.[2] The transportation sector was only 8%, well below levels elsewhere in the world and therefore likely to grow. Fossil fuels provided nearly 82% of India's electricity, with coal well above worldwide averages. Nonhydro renewables accounted for only 3% of electricity energy. This is India's baseline. Where does she go from here?

Energy independence is an ideal sought by every country around the world. For India, this means increasing production from domestic fossil fuel resources, expanding its nuclear program, and investing in certain renewable energies to bring access to its citizens. Energy independence also means that India will need to institute pricing reforms and reduce subsidies. Near term the government intends to incentivize wind and solar development. Administrative and regulatory barriers will be streamlined to support investments in capacity and infrastructure. The transportation sector is expected to grow, and the government plans to mandate some level of biofuels and alternative vehicles.

India currently has six nuclear power plants, with a generation capability of 4.8 GW, contributing roughly 3% of India's electricity. For decades India's nuclear industry was isolated due to its unique pursuit of a thorium-based program and objection to signing the Nuclear Nonproliferation Treaty. Then in 2005, a deal was struck between the United States and India that effectively welcomed India back into the world's civil nuclear fraternity.[3] India now has access to technology and supplies for its nuclear projects. This agreement breathed new life into the hopes for India's nuclear industry. As of 2014, 4.3 GW of new reactors were under construction, with a long-term goal of providing 25% of electricity via nuclear power by 2050. Like China, India expects the fast reactor to become the standard beyond 2050.

Overall India ranked sixth in the world in nonhydro renewable capacity, with wind power at 22.3 GW. Through the end of 2014, its 44.9 GW of hydropower ranked sixth in the world. As mentioned previously, even though only 3% of India's land area has annual solar insolation above 2,000 kWh/m^2/y, it still represents a potent opportunity for domestic energy production.[4] Between renewables and nuclear power, India has the potential to achieve its goal of energy independence.

Energy-Challenged

India is in the most challenging position of all the countries studied. By 2025 India will be the most populous country in the world, desperately working to emerge economically. It already imports roughly 70% of its petroleum and 35% of its natural gas consumption, it has energy subsidy levels of nearly 20%, and it struggles to fund necessary infrastructure that could improve its prospects. India is riding a wave that has little margin for error.

India's primary energy consumption was 92% carbon-based in 2013, 5% higher than the world average. Its use of coal, though, is 25% higher than worldwide averages. The plan outlined begins the journey to non-carbon energy sources, but by 2050 India will still rely primarily on fossil fuels, a reminder that if worldwide carbon is to be reduced, more developed countries will have to do more than their share. Hopefully nuclear and other renewables will come on line faster than projected to help reduce import reliance and avoid health and environmental impacts from the overreliance on domestic coal. Until then, India will remain dually challenged.

Notes

1. World Bank, "Access to Electricity (% of Population)," accessed May 1, 2015, http://data.worldbank.org/indicator/EG.ELC.ACCS.ZS.
2. Sun-Joo Ahn and Dagmer Graczyk, *Understanding Energy Challenges in India* (Paris: Organisation for Economic Co-operation and Development/International Energy Agency, 2012), https://www.iea.org/publications/freepublications/publication/India__study__FINAL__WEB.pdf.
3. U.S. Department of State, "U.S.–India: Civil Nuclear Cooperation," accessed March 27, 2015, http://2001-2009.state.gov/p/sca/c17361.htm.
4. REN21 Renewable Energy Policy Network for the 21st Century, *Renewables 2015 Global Status Report.*

25

South Korea

South Korea seems less concerned with diminishing fossil fuel
supplies and climate change than with finding a strong position
in an emerging alternative energy marketplace.

South Korea has little access to domestic fossil fuel resources, and yet
this hasn't deterred it from establishing the world's 14th-largest nomi-
nal GDP economy. You would think that a country beholden to the rest
of the world for its energy supplies would be articulating the rapid establishment of domestic sources. It is not!

Renewable energy contributions to Korea's total primary energy supply rank among the lowest in developed countries at roughly 2%. Either
South Korea is taking a unique approach in prioritizing its energy plan,
or it is simply dodging the inevitable shift from fossil fuels. As in the rest
of the world, South Korea's electric power paradigm has historically been
built to meet demand with little to no supply adjustments. That system
has created a spoiled overfed child. Since 2008, electricity consumption
in South Korea has increased 4.6% annually, while production is struggling to keep pace. The result is an energy intensity factor above compa-
rable counterparts around the world. In specific terms, South Korea has
an energy intensity of 10,726 (BTUs per unit of GDP), more than twice
that of Japan and Germany.

South Korea, though, is taking smart steps to avoid the increased
costs in power generation by avoiding extremes in demand levels. Among
other functions, smart grids will intelligently shift demand to avoid

expensive peak loads, while in-place advanced pricing programs send consumers a strong signal to both conserve and shift energy demand. Together smart grids and advanced pricing help reduce the costs of delivering electricity.

It was compelling to discover that although South Korea imports 98%–99% of its fossil fuel requirements, it exported 1.2 million barrels per day of refined petroleum products such as jet fuel. South Korea's export of refined petroleum product makes it a quasi-exporter without having measurable natural resources. Building petroleum refining capability in a fossil fuel–poor country might seem counterintuitive, but for South Korea there were two strategic reasons to build refineries: an economic opportunity to add value and a means to secure supplies of petroleum. One of its neighbors to the south, Singapore, has done the same thing. Benefiting from their refining capability, South Koreans enjoy some of the world's lowest prices for gasoline. To further reduce supply risk, South Korean companies engage in many overseas energy projects and provide financial support for some of those projects.

Even more interesting, three export industries in South Korea—electronics, nuclear, and shipbuilding—are targeting new opportunities related to the world's changing energy mix. Leveraging its electronics acumen in building smart grid technologies is a natural fit, as is customizing ships for offshore wind farm construction. South Korea is already one of the world's top suppliers of nuclear reactor technology. Its Ministry of Trade, Industry and Energy in 2010 announced intentions to export 80 nuclear reactors worth $400 billion by 2030.[1] Already under way are four nuclear reactors in the United Arab Emirates worth $20 billion. Even if the Fukushima disaster has dampened some of the world's market potential, there are still substantial numbers of planned nuclear reactors, and South Korea's share is projected to be 20%.

What We Can Anticipate (PECF = 3.1 × FFWF = 0.0)

The framework classifies South Korea as vulnerable given its combination of a high PECF and a low FFWF. One would expect an urgency to shift energy dependence to domestic sources. The framework would also predict that the country's developed economy would provide the financial resources and skills to undertake an energy transition. Until this happens, South Korea would employ all means to secure reliable import supplies. Under scrutiny, South Korea strays from the framework's predictions. Planning to supply only 11% of primary energy from renewables by 2030 lacks the urgency the model predicts. Let's look at why this is the case.

Fossil Fuels and Nuclear Energy and Feeling OK

South Korea's use of fossil fuels both for primary energy and within the power sector is above worldwide averages. Surprisingly, this overdependence isn't overly influential in its energy planning. Primary energy consumption grew by 3.8% between 2008 and 2012 and is expected to continue to increase. The negligible contribution of renewables in power generation came in equal measures from wind, solar, and biomass and waste. A shift is occurring in the primary energy mix through 2017 that will reduce petroleum consumption by 7% and replace this with some combination of liquid natural gas, coal, and nuclear energy.

Nuclear power from 23 reactors (20.7 GW) across four sites contributed 29% of South Korea's electricity. New plant construction is under way, which will increase its nuclear capacity.

Nuclear power from 23 reactors (20.7 GW) across four sites contributed 29% of South Korea's electricity. New plant construction is under way, which will increase its nuclear capacity. Like Japan, South Korea is also assessing overall nuclear plant safety and addressing the impacts of false safety certifications from foreign-made components. So, although nuclear capacity is increasing, the rate of increase has been slowed while safety standards are evaluated. South Korea has been developing fast reactor technologies for several decades and hopes to get a demonstration reactor operating by 2025, followed by commercial start-ups by 2050.

Three main targets have been established in South Korea's energy plan. First, on the demand side, it plans to reduce 15% of planned electricity consumption by 2027. Second, additional nuclear capacity of 12.2 GW will contribute one-third of the country's electricity by 2022. Finally, renewables will contribute 6% of primary energy requirements by 2020. No wonder South Korea's plan to cut 30% from its GHG emissions by 2020 has no firm commitments. Indicative of its measured approach to renewable technologies, South Korea is planning and/or conducting several pilot studies. These include concentrating solar power, a new tidal turbine, and a sodium-cooled nuclear fast reactor.

As noted, South Korea's top priority is its plan to establish a comprehensive nationwide smart grid system by 2030. It encompasses technologies for large-capacity energy storage and charging stations for vehicles. South Korea also plans to export its smart grid technologies. Where the rest of the world lunges forward in building temporal capacity ahead of storage capacity and smart grids, South Korea is prioritizing and validating this important supportive technology.

For the medium term as it continues to rely on fossil fuels, the government is investing in combined cycle electricity generation and carbon capture and sequestration to address both efficiency and climate change concerns. Unfortunately, the carbon capture and sequestration has no commitment dates, a common thread around the world. South Korea has replaced a feed-in tariff system with renewable portfolio standards to guide new capacity decisions. The RPS milestone is 10% of electricity by 2022, a meager target for a country so dependent on fossil fuels.

A few recent achievements further exemplify the country's slow and methodical approach. In 2013 South Korea added 0.4 GW of solar PV, a modest start. Now, the government has mandated biodiesel blend levels to 2.5%, with plans to increase this to 3%. Breaking out a little, South Korea opened its Sihwa ocean energy plant (254 MW), the largest of its kind in the world.[2] This is a lot of power, harnessed from the ocean's tidal range.

Perplexingly Slow to Adopt Meaningful Alternatives

A nationwide smart grid with advanced energy storage systems along with vehicle recharging stations lays an important and sound foundation. Renewables representing a paltry 11% of primary energy in 2030 still leaves South Korea vulnerable to supply disruptions, although substantial nuclear power coming on line will provide some buffer. South Korea is moving slower than expected in deploying domestic energy sources. Prioritizing a nationwide smart grid may prove to be more sensible in the long term. By targeting technologies that represent opportunities to establish an export business, South Korea seems less concerned with diminishing fossil fuel supplies and climate change than with finding a strong position in an emerging alternative energy marketplace. South Korea appears confident in its fossil fuel supply chains and continued production from its partners. Meanwhile, the pace of developing renewable capacity is inconsistent with the depletion window established earlier in this book.

Notes

1. World Nuclear Association, "Nuclear Power in South Korea," January 2016, accessed April 15, 2016, http://www.world-nuclear.org/information-library/country -profiles/countries-o-s/south-korea.aspx.
2. Daewoo Engineering and Construction, "Sihwa Lake Tidal Power Station: World's Largest Clean Maritime Energy Development Project," in 2011 Sustainability Report, September 2012, http://www.daewooenc.com/eng/contribution /download2015/2012   daewooenc%20SR__en.pdf.

26

Country Roundup

We traveled around the globe looking at country energy plans with an eye trained to look for notable efforts to migrate from fossil fuels to the more exclusive use of alternative sources. The 13 or so countries we chose to study are home to 50% of the world's population, with diversities in both their economies and their access to conventional energy sources. These select countries tell a tale of motivations, ideas, technologies, policies, and urgency in shifting to alternatives.

A glass-half-full outlook would report that renewables delivered 21.9% of worldwide electricity in 2013. On closer examination, though, we see that renewables are holding position and not making inroads, and that in 1980, coincidentally, renewables also delivered 21.9% of the world's electricity. Undaunted in our optimism, let's consider that in 1980 hydropower was the dominant renewable but today other renewables are making noticeable contributions and for the first time their combined capacities exceed that of hydropower. We would also note that worldwide demand for electricity has grown 3.1% annually since 1980, and renewables have carried their share of that increased demand. And now reality: If fossil fuel use is to decline, progress like that of the last 35 years will be grossly insufficient.

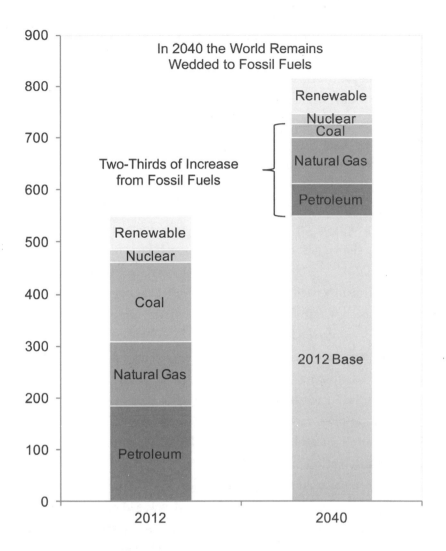

Figure 26.1. Primary energy increase (quadrillion BTUs) by source—48% increase between 2012 and 2040. *Source Data:* U.S. Energy Information Administration, *International Energy Outlook 2016: With Projections to 2040*, report no. DOE/EIA-0484(2016) (Washington, D.C.: U.S. Energy Information Administration, Office of Energy Analysis, U.S. Department of Energy, May 2016), http://www.eia.gov/forecasts/ieo/pdf/0484(2016).pdf, 165, Table A2.

A Global Energy Projection Portrays a Big Gap

When the world's energy plans are pulled together, what will our energy mix look like in 2040? The Energy Information Administration published an *International Energy Outlook* report with energy consumption projections by energy source through 2040 (Figure 26.1). The depiction is spectacularly unsatisfactory if fossil fuels are to be phased out in the next 50 to 100 years. We will continue to deplete fossil fuels at higher and higher rates over the next 25 years without much progress in building alternative sources. If the EIA's outlook forecast proves correct, there is little chance that the transition will be anything but disruptive. This is a trajectory we should all want to rethink!

According to the EIA, we will be burning far more fossil fuels in 2040 than we do today, while renewables will have made only modest gains. This is alarming, even apart from the depletion issue. The fossil fuel category is projected to provide most of the world's increased demand even when we are 25 years further into the depletion window. To publish this EIA energy consumption outlook without a distressing title is operating with a high degree of detachment, for the future will judge us poorly for our lack of progress. This *International Energy Outlook* makes one wonder whether the biggest obstacle is technology, economics, inertia, or a flaw in our DNA.

To build capacity in alternative fuels, the first priority is to cap fossil fuel at a global level, not regulate CO2 emissions.

The energy mix needs to be different. We are taking the path of least resistance, as shown in Figure 26.1, and exposing our unwillingness or inability to peer into the future. The bar to the right shows the growth by fuel type over the 2012 baseline. A higher sense of urgency would have us hold fossil fuels to 2012 levels and invest in alternative capacity to meet the world's 48% increase in consumption by 2040.

In order to pivot toward alternatives, the first and highest priority would be capping fossil fuel at a global level, not capping CO_2 emissions. The distinction is that the accurate measurement of CO_2 emissions is more elusive than the quantification of fossil fuel consumption. This places a particularly heavy burden on developed countries to lead the energy transition and eliminate fossil fuel use in favor of alternatives for the global math to work. It assumes that maximum effort is applied to building alternative sources, not to changing the mix of fossil fuels, and that altogether replacement of fossil fuels is the top priority. Finding ourselves well down the transition path will minimize conflicts in our global community while also reducing our exposure to pricing volatility.

The sections that follow will describe the soft technologies of policies that, along with alternative energy technologies, can help meet the challenge. The other benefit is that exporters can focus on adapting their economies as opposed to increasing their dependence on the revenues from a finite resource.

Logically following from the EIA chart, most national energy plans envision a minor mix of renewables. On the bright side, beyond just a few countries articulating a large role for alternatives, there are many communities around the world setting 100% renewable targets. The world's energy transition will follow a ground-up rather than top-down pattern.

If we return to our time line of needing to end the substantial use of fossil fuels this century and agree that it will take 30+ years of single-minded focus to build alternative capacity, we really can't afford to wait too much longer. Hopefully by 2050, renewables will be the primary source for electricity generation, and we take advantage of fossil fuels to navigate a smooth transition. If we plan well, we will be fortunate to have sufficiently affordable fossil fuels to serve as a backup; otherwise, making the jump across all the other energy sectors will be much more difficult.

Selected Highlights: First, Hydrogen

Japan's and Germany's articulations of a new-age fuel, hydrogen, take us most realistically to the time when fossil fuels are no longer used energetically, a time when other sources of energy will power our world. Many parts of the world have built energy policies fixated on a sustained blend of fossil fuels along with marginal alternative energies. The only long-term planning aspect of many energy plans is the delay in the deployment of alternatives. Much of the world hasn't come to terms with life absent fossil fuels and seems determined to fight nature's life cycle law.

One of the first challenges in harnessing new sources of energy is adapting our power sector to electricity that is generated at the whims of nature. Electricity is the mainstay of how the world delivers energy—so much so that the World Bank regularly monitors the percentage of a country's population with electricity access, considered a measure of social health. However, because electricity has to be used the instant it is produced, the world needs to accumulate energy. Hydrogen for both Japan and Germany is both of these things: a means to deliver and a means to accrue energy. If we're still not sure that we need to cache energy, then consider a world where there is none. A still and cold cloudy day means no heat or lighting; it means taking a train that stalls on the track because the current harvest of energy is too low. Strategic reserves are also accumulated energy, a nation's stockpiling of energy to mitigate the risk of long-term disruptions in supply and production. Modern society isn't engineered to forgo energy for even a second; we need a flow of energy at all times, and serving this absolute requirement is long-term storage. It's simply not enough to note that wind and solar production periods complement each other; we need energy storage to ultimately replace all the energy sectors.

Japan's energy plan highlights the critical role hydrogen will play in the future. It speaks about building a "Hydrogen Society" using hydrogen much like we use electricity today. Japan hopes to be the world's "solutions provider" for advanced energy storage and retrieval technologies. Interestingly, we are already seeing practical applications of this vision in hydrogen fuel cell vehicles and the ENE-FARM stationary fuel cell systems.

What a country envisions in its energy plan reveals its view of the future and the solutions necessary to thrive. Germany, if not Japan, is steering toward a future where large amounts of temporal energy are harnessed and storage and retrieval technologies are vital. Germany has dispatched fossil fuels from its long-term thinking in order to plan its path forward. Elsewhere several other countries reveal modest roles for hydrogen in comparison.

Nuclear Energy: Silently Building Capacity

With headlines pronouncing the progress of renewables, it is interesting to note that projected increases in nuclear capacity make up 33% of projected growth from all the renewables together. It may come as a surprise to realize the number of countries adding nuclear power capacity (Table 26.1). Of the countries studied, only Germany is purposely halting all nuclear operations by 2022. All the other countries studied are constructing and/or planning nuclear power projects. We've seen that two countries, China and Russia, have already moved fast reactors into final performance testing phases. Plans for the disposition of spent nuclear fuel are integrated into the closed fuel cycles that several countries are articulating. The fraternity of countries planning to practice

Table 26.1. Status of Nuclear Power for Countries in the Study Group.

COUNTRY	EXISTING NUCLEAR	NUCLEAR PLANS
Brazil	☑	☑
Canada	☑	☑
China	☑	☑
Germany	☑	☐
India	☑	☑
Japan	☑	☑
Russia	☑	☑
Saudi Arabia	☐	☑
South Korea	☑	☑
United States	☑	☑

nuclear power is increasing; nearly one in five countries will generate power from nuclear plants.

Smart Grids

How did our current grids fall so far behind in today's digital world? The world's million-plus miles of electric transmission and distribution lines were designed in the twentieth century, long before smart electronics came to be. The better question is, How does our current grid system function so well without smart equipment? Waste! South Korea has prioritized the development and deployment of smart grids. It has established an aggressive goal to build a nationwide smart grid by 2030. Many countries speak to the need for funding, while South Korea is field-testing and fine-tuning solutions. An important philosophy resident within its smart grid concept is the decentralization of energy sourcing and the role of online intelligence in mastering the supply and demand relationship. By comparison, much of the rest of the world seems spellbound by the centralized model and demand orientation.

Taking to the High Wire

Germany should be applauded for taking to the high wire as it marches into its energy transition. Right or wrong, it made a value decision to discontinue both nuclear power and the use of fossil fuels. Germany is the first fully developed economy that was largely fossil fuel–dependent to make the break toward renewables. Germans are already one of the most efficient consumers of energy and plan to reduce consumption even further. Germany's "Energiewende" is a field test of high order for the rest of the world. The lack of domestic fossil fuel resources has given Germany a glimpse of the world's energy future. Guided by environmental custodianship, a desire for energy security, and technological confidence, Germany has embarked on a path we will all be required to take someday.

The Realistic Role of Biofuels

Brazil is producing 450,000 barrels of biofuel per day. While the rest of the world's transportation sector remains highly dependent on petroleum and regulates emissions, Brazil is using more biofuel than gasoline. The rest of the world slowly funds electric recharging infrastructure, while most refueling stations in Brazil offer flex-fuel. Brazil didn't wait on the future; it acted swiftly in embracing biofuels for transportation. This strategy doesn't preclude Brazil from adopting alternative fuel vehicles in the future, but it allowed it to act in the present. Where most of the world waits for better economics or some technological breakthrough, it is refreshing to see a country move in the present with tools as they exist.

Seeing Past the Moment

When Saudi Arabia, home to one of the largest petroleum reserves, makes plans to diversify its energy mix, it reflects long-term thinking absent most everywhere else. Saudi Arabia has targeted big investments in nuclear, solar, and wind energy to come on line by 2040. So, like Germany, Saudi Arabia at least has begun planning its energy transition. To aspire and to achieve are two different things, though, and we shall watch

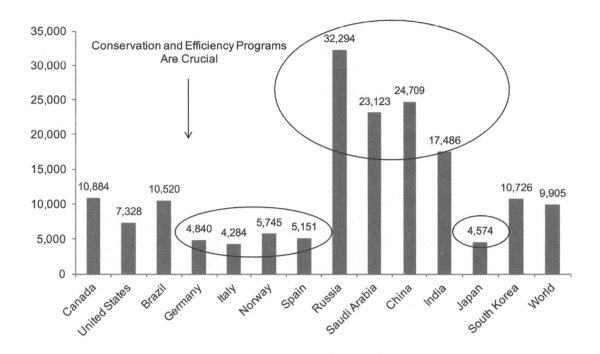

this closely. Credit must be given to Saudi Arabia for thinking beyond petroleum. As natural as renewables are in a fossil fuel–poor country, renewables are odd in such a petroleum-rich country. It is a test of time awareness that most countries fail.

Conservation and Efficiency

Conservation and efficiency are common themes in most country plans. Figure 26.2 displays the energy intensity for our country study group. Energy intensity is the amount of primary energy per unit of GDP stated in terms of British thermal units. What is startling to learn is the range of energy intensity for the countries studied. No doubt some of the variation is due to each country's industrial base and climate, but a large portion is waste and therefore an opportunity.

On one extreme is Russia, with energy intensity over 30,000 BTUs per GDP, while Germany and Italy are at the opposite extreme, both below 5,000. Supply- and demand-side technologies are converging on the rampant use of energy. It is truly sad to consider the waste incurred

Figure 26.2. Energy intensity (BTUs per GDP) for countries in the study group. *Source Data:* U.S. Energy Information Administration, "Energy Intensity: Total Primary Energy Consumption per Dollar of GDP (Btu per Year 2005 U.S. Dollars (Market Exchange Rates)) for 2011," accessed May 7, 2015, http://www.eia.gov/cfapps/ipdbproject/IEDIndex3.cfm?tid=92&pid=46&aid=2.

over so many decades. Respectful consumption will occur, as happens with most things precious, so starting now is a good strategy, and continuing with alternatives is even better. Germany and Italy are leading the world by showing us how efficient we can be and how much energy we really need!

Offshore Wind Farms

Most of the world's wind energy is gathered onshore, while a lot of the world's raw potential is offshore. Development in offshore environments is more challenging and expensive but necessary to harness much of the wind's practical potential. Germany's offshore wind farms are typical of most designs; the wind turbines are bolted onto a platform anchored into the seabed. Japan is developing a design that goes in a different direction. The wind turbines are mounted on floating platforms that are anchored at sea, avoiding the eco-disruption involved with drilling and underwater construction. Tackling the challenges of economical offshore farming will allow new capacities of wind to come on line. Japan's unique approach may present a solution. This same approach is possible with floating solar PV platforms.

We Are Lingering Too Long in the Fossil Fuel Era

Figure 26.3 summarizes the percentages of electricity generated from nonfossil sources. The most striking observation is the diversity utilized across the study group. Norway leads, with more than 98% of its electricity derived from noncarbon sources, with most of this coming from hydropower. Brazil and Canada, like Norway, are also beneficiaries of relatively abundant hydro resources. We see that Western Europe leads the group in nonhydro contribution, while Saudi Arabia remains completely dependent on fossil fuels. South Korea currently chooses between fossil fuels and nuclear, and Spain has notable contributions from all three types of nonfossil energy sources—perhaps the best way forward. Overall, the world still relies on fossil fuels for two-thirds of electricity generation, but decade by decade this figure will shrink.

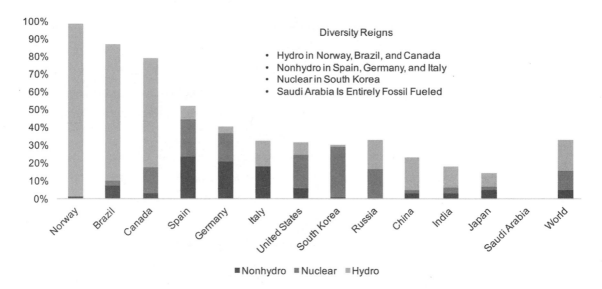

Figure 26.3. Portions of electricity generated from nonfossil sources for countries in the study group. *Source Data:* U.S. Energy Information Administration. "International Energy Statistics," accessed May 7, 2015, http://www.eia.gov/cfapps/ipdbproject/iedindex3.cfm?tid=2&pid=alltypes&aid=12&cid=regions&syid=2008&eyid=2012&unit=BKWH.

Again, if we accept that we are entering the depletion window, then worldwide nonfossil sources seemingly stuck at 33% in the power segment is distressing. You would think that climate change and the end of easy and cheap fossil fuel would see the world rally to new solutions. Although we have achieved advancements unimaginable only a generation ago, we seem to suffer the same deficiencies as our ancestors in facing the inevitable limits of a finite resource. Symptoms include an unwillingness to confront the obvious limits of the resource, an inability to see across generations, and failure to adapt.

Former high school biology students might recall a lab in which they inoculated specimens containing bacteria onto an agar plate. The agar is a limited ecosystem. The colonies grow unchecked, consuming the nutrients, and the bacteria finally perish when the nutrients are consumed. Certainly we can see that the agar plate has its limits, but can we apply that same awareness to our accessible energy supplies?

It would be nice to think that we can avoid the bacteria's fate. Most country policies embed an assumption that fossil fuels are abundant and that the risks in supply are self-generated. In general, fossil fuel—wealthy countries only see production limitations created by regulation and limited access. Country plans also cite economics as the main obstacle for change, as opposed to addressing supply and environmental risks. The supply of fossil fuels is not infinite as we treat it but, rather, fixed and determined for us by forces millions of years old. We're acting like the trolls in J. R. R. Tolkien's *Hobbit*, arguing too long into the night without making a decision, only to turn to stone when the morning comes. So, why can't we incentivize an end to the use of fossil fuels as an energy source? We can!

Another trait we demonstrate is an ability to rationalize just about anything. Several of the countries we investigated are proudly pursuing the transition to renewables domestically, acknowledging their green value. These same countries are investing in and exploiting fossil fuels as export leaders. Marrying those two positions together is typically human.

The cigarette industry and consumer can offer some warnings for our use of fossil fuels. The health effects of smoking are downstream, and it took a massive accumulation of data to compile statistics that finally couldn't be ignored. The progression is slow, but the diagnosis is quick and fatal. As with smoking, we can continue to burn fossil fuels for decades, but eventually we will receive a frightening diagnosis.

On a positive note, according to REN21's *Renewables 2015 Global Status Report*, renewable investment in the power sector outpaced investment in fossil fuels for the fifth consecutive year. This is good news! Perhaps, then, the next 30 years will see substantial increases in the market share of renewables from the 21.9% we've seen for the last 35 years. We can only hope that many of the projections we've seen are wrong and that they understate the path the world will in fact take in building alternatives.

PART IV

History, Inertia, Interventions, and Adaptation

Energy Policy Conundrums

27

Our Response to Limited Resources

A certain way of understanding human life and activity has gone
awry, to the serious detriment of the world around us.
—Pope Francis

So fossil fuels are accessibly finite, energy is absolutely essential to our
way of life, and global consumption is rising. Now we come face-to-face
with the question of which is our greatest challenge: surmounting the
technical obstacles of replacing fossil fuels with alternative sources or
altering humankind's propensity to want to walk head down into the
future. Is it possible to mobilize ourselves and come to terms with fossil fuels as economically finite and avoid societal disruption through the
development of alternatives? Can we relax, confident that necessity will be
the mother of invention, or should we be petrified at the looming depletion of economically viable fossil fuels and the prospect of repeating past
and often tragic mistakes?

Our short history provides many not-so-pleasant examples of how
societies deal with stress caused by a limited resource and their failure to
adapt. Across history famine and drought have killed millions, scattered
others, and in extreme cases caused the total collapse of civilizations.
Energy now is nearly as important as food and water. Unlike unfore-
seen famines unleashed by harmful weather patterns, the limits of fossil fuels are obvious, so we can be spared tragedy if we earnestly build
alternatives. Absent seeing the conspicuous and acting with urgency, it's
a safe bet that we will not everywhere transition our energy sourcing in
a peaceful manner. We've seen country energy plans that perpetuate the

It is a safe bet that we will not transition our energy sourcing worldwide in a peaceful manner.

status quo behind excuses—either the technology itself or the economics of the technology is lacking. Our nature needs to be considered as we confront this enormous transformation. In the spirit of self-awareness, we would do well to take a look at cases where humankind confronted shortages and what we may learn from these occasions. If we know our anthropological bias, we can better guard against tendencies that often betray survival. Here are a few clues.

The History, Mystery, and Theory of Easter Island

Easter Island's history reveals a society that was unable to see across generations in a brewing catastrophe, the complete deforestation of their island. Every successive generation harvested a share of the forest, failing to adapt in spite of the inevitable end result of their practices. Easter Island provides a grim reminder of the consequence of not reckoning with the limits of a vital resource.[1]

In the South Pacific Ocean, far away from other habitable lands, hundreds of huge carved statues dot Easter Island's coastline, providing evidence that an advanced society once existed. More intriguing than the mystery shrouding these statues, however, is the story of the rise and collapse of the society on this remote island.

When the islands were visited in the eighteenth century, there were approximately 3,000 inhabitants living in squalid conditions. The islanders themselves had degraded along with their habitat. Visitors noticing the immense statues spread around the island couldn't imagine the current residents being capable of crafting and moving these monument-sized stone structures. Subsequent research offered a tale of self-inflicted destruction.

It is estimated that the island was first settled around A.D. 500 by Polynesians. The earliest settlers may have numbered 20–30 people. They inhabited an island with no freshwater streams, no mammals, few fish nearby the island, and soil that presented few choices for crops. The diet that sustained the islanders was primarily chicken and sweet potato. The island, however, was heavily forested. The inhabitants used the abundant forest to build homes, heat, and cook. They also cut down trees to move their impressive statues, rolling them from the island's stone quarry on beds of tree trunks. Generation by generation, trees were taken faster than nature could replenish them, but the deforestation happened incrementally over 1,000 years. Each generation took a piece of the forest without the vantage point of time. Surely they would have seen the inevitable deforestation if only they could see across generations. The islanders' way of life sequentially became limited; they went from living in wooden homes to primitive stone and cave dwellings. Ceremonial statues, once considered important enough to warrant construction on an immense scale, now lay unmoved in the quarry, lacking sufficient trees to transport them around the island. As time went by the islanders forgot the past; without trees, they even lost their ability to build canoes and were stranded by their own hands. As the inhabitants' behavior digressed (i.e., cannibalism), the island's population declined precipitously, and in 1877, the Peruvians removed all but 110 of the residents. Today sheep graze the island amid the eerie stone statues.

How does this happen? Easter Island rose and fell in the course of 1,300 years, and it had nothing to do with drought, natural disaster, or marauding invaders. The reckless use of a limited resource and the failure to adapt resulted in societal ruination. The few survivors had to be mercifully removed. Are we doing the same thing, consuming fossil fuels at an increasing rate because it seems there is still "plenty" left and leaving future generations to adapt? Is there a lesson we should take given that Easter Island's tree resources took 1,300 years to deplete and we are on pace to consume economically viable fossil fuels over the course of just 200 years?

Our history, patterns, and nature pose questions about the long-term planning that must be applied to energy policy. Are we able to act now for the benefit of future generations, or are we lingering too long at the well and underestimating the time and effort needed to adapt?

Like the water of Owens Lake, so, too, oil and gas flow through an extensive system of pipelines with valves that can instantly divert supplies and wreak havoc.

Water Wars at Owens Lake

In our relatively more recent past, a small farming community in the western United States proved powerless against the city of Los Angeles,

California, which faced an acute need for water resources. The story of Owens Lake illustrates how a scarce resource can inspire desperate measures and that "to the victor go the spoils."[2] It is quite fathomable that similar events could unfold with oil and gas.

At the beginning of the twentieth century, Owens Valley in southeastern California was home to a ranching and farming community. This community was located near Owens Lake, which formed nearly 800,000 years earlier and covered 200 mi^2. Coincidentally, at the same time, the city of Los Angeles was growing in size beyond the Los Angeles River's ability to supply its population. Los Angeles needed water.

Effectively, L.A. "planners" and "developers" anonymously purchased water rights along the Owens River—the same river that fed and sustained Owens Lake. This anonymous group then built a 250-mi aqueduct from the river to the city of Los Angeles. The taps, 50 mi upstream of Owens Lake, were opened in 1923, diverting the water to Los Angeles.

The ecosystem of Owens Lake, which had flourished for 8,000 centuries, became a dust bowl within decades. The dry bed of Owens Lake became the largest source of particulate pollution in the United States, and of course the unsuspecting Owens Valley farming and ranching community was decimated. Los Angeles secured its future through the destruction of another community.

In the western United States, and in many places around the world, water is treated as a precious and limited resource. Where water is scarce, conservation measures and sustainability are the only ways forward. It is time to think preemptively about economically finite fossil fuels as a scarce resource, since we still lack practical alternative capacity. Although supplies seem anything but scarce today, low-cost suppliers are scrambling to convert their reserve assets to revenues while demand is high and alternatives are minimal. If we act wisely, as these low-cost supplies are depleted they will give way to more attractive alternatives. No matter the future source of energy, as with water, conservation and sustainability will be valuable skills.

What happened to Owens Lake is a tale that should warn us all of the global implications of desperation when powerful suppliers choose to redirect the flow of fossil fuels. Like the water of Owens Lake, so, too, oil and gas flow through an extensive system of pipelines with valves that can instantly divert supplies and wreak havoc. But unlike the Owens Lake community, where there simply wasn't an alternative to its water supply, countries around the world can develop domestic alternatives to mitigate supply risk. Retaining dependence on fossil fuels for the vast majority of the world's energy when alternatives exist is reckless policy.

Organization of the Petroleum Exporting Countries and Politics

The Oil Embargo of 1973–1974 caused a sudden shock to the U.S. economy and to the national sense of independence.[3] There are at least two lessons to be taken from the experience. First, the United States overnight went into emergency response by rationing, forgoing nonessential travel, and other measures. Second, the embargo ended before a sustained behavior change could be embedded, and the country quickly reverted back to old habits.

The Arab-Israeli war was under way, and the Organization of the Petroleum Exporting Countries (OPEC), dominated by Arab members, moved to ban exports of oil to all countries supporting Israel. The United States had pledged its full support to Israel. Throughout the course of the embargo, the United States resorted to the use of its reserves along with strict rations while longer-term solutions were sought. This sudden oil shortage tested the societal ability to adapt to a sudden and dramatic loss of crude oil, a basic commodity. Long multiblock lines for fuel were the new norm. Theft by siphon shed a new light on our vulnerabilities, and the locked gas cap was born. The United States experienced high inflation, a national 55 mph speed limit was mandated to help lower gasoline consumption, and new fuel economy standards for cars were legislated. As the embargo was crippling the national economy, the United States pursued relatively long-term solutions. It sought to reduce dependence on OPEC by increasing domestic petroleum production and by applying greater scrutiny to the management of the nation's petroleum reserves. The fuel economy of motor vehicles has been improved dramatically since

1973, although commitments to fuel efficiency have fluctuated in the intervening years.

The United States has invested in domestic oil and gas production since then, but we are as vulnerable today as we were then because we are just as reliant, if not more so, on a limited energy source. It's like a rubber band that you stretch for just a moment: it readily returns to its original shape. Gasoline engines still power our transportation vehicles. In 40+ years we've not made any substantial structural changes that acknowledge fossil fuels as economically limited in the twenty-first century. Inflation-adjusted gasoline prices in the United States have risen 63% since 1990, and yet speed limits are now up again. Bugatti Veyrons are among our automobile choices, delivering more than 1,000 metric horsepower and an estimated 8 mpg. We've relaxed and reverted quickly to our old ways. Whether the supplies are domestic or imported, crude oil reliance has not changed, and any future disturbance seems likely to thrust us backward to the shock of the 1970s.

Contrast this with the sustained experience of people living during the Great Depression. Unlike the Oil Embargo, which was settled in a year, the effects of the global depression were felt over a decade. My mother was 10 years old when the depression hit, and its impact embedded values and norms that stuck with her the rest of her life. Little was wasted, and pride was taken in repurposing items. Things were saved on the chance they might be useful later. Socks were darned, hand-me-downs were expected, and a single new dress was rare and special. It seemed to her that "everyone was affected so there was no time to despair. We adjusted, we made sacrifices, and we appreciated what we had." Now 97, my mother still holds values shaped by the Depression, values that stand out as odd today. She is perplexed by new generations' consumption and waste. The lessons of her generation have not transferred themselves to following generations (whoops, that's me). They have not fundamentally affected those of us who did not experience the hardships directly. In anthropological summary, we resist adaptation until it is thrust upon us. Clearly, we need some kind of powerful force to alter course, despite the impending consequences of a sustained dependence on fossil fuels.

Fossil Fuels and Politics in the Twenty-First Century

Just a few years ago, in January 2006, negotiations on a natural gas price increase between Russia and the Ukraine failed.[4] Prior to the election of pro-Western leadership in Ukraine, Russia had subsidized natural gas to the Ukraine at $50 per 1,000 m³. Now in response to the Ukraine's new political direction, Russia was proposing pricing of $220. For the Ukrainians this sharp price increase from a supplier of one-third of their natural gas consumption was as threatening to their economy as the 1973 Oil Embargo was to the United States. Negotiations failed, and within its rights, Russia turned off supplies to the Ukraine.

Whether intended or not, a consequence of Russia's action was reduced supplies to the Western European countries whose gas flowed through the Ukraine. Was this due to a drop in pipeline pressure and flow, or was the Ukraine still taking supplies? Fingers pointed in all directions as the Ukraine and Western Europe confronted supply uncertainty in the middle of winter. It was the natural gas version of the OPEC embargo. Energy-exporting nations such as Russia and OPEC members can and have intertwined politics into their supply positions. They can make these decisions overnight and thrust tremendous stress on dependent countries.

Western Europe was a bystander to the political rift between Russia and the Ukraine but immediately felt the ripple effect. Europe considered the future and wondered about the sanity of a continued dependence on Russia. Would Russia use the pipeline for political influence? Importing countries are beginning to realize that there are just a very few fossil fuel–wealthy nations and continued dependence could undermine their ability to prosper. Like the Oil Embargo of 1973, the Russian natural gas crisis of 2006 left an imprint in Western Europe's psyche. Partially in response to this national threat, Western Europe has built the world's highest renewable energy capacity on a per capita basis.

Territorial Tensions Related to Oil and Gas

China's island building in the South China Sea has escalated tensions in one of the world's busiest sea-lanes, which also happens to contain meaningful oil and gas reserves. The South China Sea holds 3% and 1% of the world's recoverable natural gas and oil, respectively, roughly equal to all of China's reserves. Stress is building between China and its Association of Southeast Asian Nations neighbors who are geographically far closer to those assets and might be wondering whether they could go the way of Owens Lake. As we have seen, China needs to build new energy capacity every year on a scale that is larger than the total long-term requirements of many entire countries. The island building could be viewed singularly as China seeking greater regional influence, but to dismiss the oil and gas factor entirely would be off the mark. Similarly, the Falkland Islands located in the South Atlantic have significant oil and gas reserves that no doubt inflamed tensions and increased the stakes between the United Kingdom and Argentina, on opposite sides of the Atlantic Ocean.

For countries dependent on fossil fuel imports these examples should provide impetus to develop domestic energy sources to mitigate risk from these types of situations. Many importing countries in the short term are reducing risk by building multisource fossil fuel supply networks while looking to build domestic energy sources in the longer term.

Why and How Elements and Raw Materials Fit into the Story

Considering our options when accessible fossil fuels are depleted naturally makes one wonder whether the world is keeping track of endangered raw materials in the same way that we monitor endangered species.[5] Has any element/raw material perished? Are there any near extinction? The answer to both is yes; there are endangered raw materials, and the 20 most critical are shown in Table 27.1.

Our response to limited supplies includes one or all of the following: hoard supplies, locate new sources, find substitutes, increase recycling, and/or restrict supplies. Hoarding and finding new supplies are the most popular. Responses requiring more fundamental behavior adaptations are largely absent, even though we can readily understand that balancing our consumer desires with sustainable practices is really the only approach that will work for the long haul.

Table 27.1. European Union's Top 20 Critical Raw Materials.

RAW MATERIAL
Antimony
Beryllium
Borates
Chromium
Cobalt
Cooking coal
Fluorspar
Gallium
Germanium
Heavy rare earth
Indium
Light rare earth
Magnesite
Magnesium
Natural graphite
Niobium
Phosphate rock
Platinum group metals
Silicon metal
Tungsten

Source: European Commission, "Critical Raw Materials," accessed April 16, 2015, http://ec.europa.eu/growth/sectors/raw-materials/specific-interest/critical/index__en.htm.

As this discussion shows, everything is intertwined; many of the endangered raw materials are crucial in renewable technologies. They are used in wind turbines and in pure electric and hybrid vehicles. Often the concentration of these elements in our waste streams is higher than the concentration in the ore from which they were initially recovered! Part of the long-term solution is establishing highly efficient recycling programs for these precious materials. A sustainable future is possible for products that include limited raw materials when we match our consumption to what we recycle. Future utilization is not possible if we approach the consumption of critical raw materials with a "new discovery" mentality.

Without systematic and comprehensive recycling of critical raw materials, renewable energy sources will be as limited as fossil fuels.

Predatory Pricing: A Glimpse of the Risk with Dependence on Just a Few Suppliers

We have a recent example of how countries can instantly apply predatory pricing and disrupt dependent industries overnight. Rare earth elements (REEs) are on the E.U. top 20 critical raw materials list, and periodically they make the news. The last time they garnered media attention, China controlled 97% of the world's rare earth supplies. When demand soared, prices soared as well, jumping from $3,100 per metric ton to $110,000, setting off alarms around the world. The discovery of new sources alleviated supply pressures, prices dropped, and today consumption continues unabated, as if we've solved the problem for eternity. Absent a more complete response, countries that hold critical supplies are certain to again remind us what finite means, and the prices will only go higher.

As it relates to energy, the solution to the finite supplies of fossil fuels is in plain sight. Renewable energies provide a means for countries to become energy self-sufficient and insulate themselves from potentially disruptive fossil fuel supply relationships. The deployment lead time for new energy sources is decades, so when we fail to plan far enough ahead we risk a turbulent transition in the future. Our behavior patterns warn us that we need to work in accordance with time scales that span multiple generations if we are to avoid revisiting previous fates. We are used to thinking of "long-term" on a much shorter scale. A successful future is not only a technological challenge but also a behavior challenge. We waste much and consume more than necessary, and our business models consciously make things obsolete to churn sales. This may sound like an economic manifesto, but it is absolutely true, and like everything else our pattern of living and working should be subject to adaptation.

Without meaningful alternative capacity, the price and supply of fossil fuels will inevitably become disruptive for all of the reasons we have discussed—supplies concentrated in an ever smaller group of countries, production costs increasing, and consumption on the rise. The world will become unstable, people will suffer, and conflicts will occur. Scarcity brings to the surface our worst character traits, and energy is something we all need.

A Sustainable Habitation with Alternative Energy Sources

Whether our energy is harnessed from fossil fuels or renewables, no amount of technology can allow us to avoid the need for changes in our behavior. Unless we institute systematic and comprehensive recycling of critical raw materials, renewable technologies will be as limited as fossil fuels.

Rare earth elements, mentioned above, are essential constituents in permanent magnets used within wind turbines and some electric motors. Although permanent magnets constitute the largest end use of REEs, other applications including catalysts, metal alloys, and polishing powders also place significant pressure on sustainable supplies.[6] Potential shortages of dysprosium, a heavy REE, used in the construction of neodymium permanent magnets, is of particular concern according to the U.S. DOE. Dysprosium allows the magnet to maintain its properties at high temperatures, a requisite for these products.

REEs are found collectively in ores and need to be separated for specific downstream applications. Hence the supply of any single REE reflects the concentration of that element in the mined ore and the separation efficiency. Every ton of rare earth minerals that we extract comes with a ton of radioactive waste that includes uranium and thorium. It would be comforting, if optimistic, to envision that the world might integrate its thinking into simultaneously processing that uranium and thorium waste for use in nuclear power.

The worldwide supply of REEs is overly reliant on just a few sources, with China at the top of the list. Current production is able to meet demand, but forecasted increases in demand principally from "clean energy" will likely lead to shortages in the next decades. A wind turbine requires 0.17 kg/kW of neodymium for permanent magnets. This means that a 2-MW wind turbine requires 750 lbs of neodymium. The sustained use of REEs will require a robust recycling program to meet demand beyond the next 100 years.

Lithium (Li) is an element found in many of today's electric and hybrid vehicle batteries. Supplies of lithium are found in only a small number of countries. Bolivia has the largest amount of the resource, followed by Chile and China. Bolivia hasn't translated much of its resource into extractable reserves as yet, while Chile, Australia, China, and Argentina are the leading producers. Projections for the number of electric vehicles and the kg/vehicle of lithium suggest that known supplies will be exhausted in less than 50 years. Absent efficient recycling programs, more efficient use of lithium, and/or alternatives to lithium batteries, shortages will occur. Fortunately, as with REEs, it is imaginable that recycling programs can be established to meet demand well into the future. It is important to remember that unlike fossil fuels or atomic fuels, REEs and lithium are not consumed within renewable applications. However, one day nuclear fusion may also require lithium (130 metric tons per TWy), where it is consumed like a fuel in the deuterium-tritium reaction. To put that into perspective, it takes more than 1 billion metric tons of coal to produce a TWy of thermal energy. We would be balancing thousands of years of nuclear potential against uses that allow recyclable applications.

Another critical element that will play an important role across energy sectors is platinum (Pt), used as a catalyst in hydrogen fuel cells. A catalyst by definition isn't consumed but, rather, brokers a chemical reaction. Like lithium and REEs, the supplies are limited, and known resources can satiate demand for less than 100 years without recycling.

Finally, solar PV promises to deliver vast amounts of energy that will require massive amounts of material. Amazingly there are more than 20 different types of solar PV technologies. Within these technologies there are many types of commercial semiconductor materials, and others are in development. Examples include silicon-based material, cadmium telluride (CdTe), and copper indium gallium selenide (CIGS). To some extent this diversity of materials helps mitigate risk with near-term substitutes, while long-term supplies will absolutely require recycling. Thin-film systems that use tellurium, indium, or germanium in particular will be limited to gigawatt-scale capacities, while the world demand will be a thousand-fold higher.[7] Silver, used in electrodes, is less critical but still limited to the multiterawatt-scale manufacture of PV panels.[8] Efforts to use less silver and/or find alternatives are attempting to overcome the constraint.

Interestingly, with fossil fuels we have been concerned with economical depletion, while in the future we will be concerned with the efficiency of our global recycling programs. Just because a technology is renewable doesn't mean that we can take for granted that it will be available for an "eternity." So as we are trying to dodge economical depletion of fossil fuels we should be mindful that we may have only one shot to get renewables right. Our behavior will need to change; we will need systematic and effective recycling in order to continue to secure energy.

As we move from fossil fuels to alternatives, we will still be rightly concerned with our impact on the environment. We needn't worry about the long-term well being of the planet—the earth will be fine. Humans, however, must confront the repercussions of our habitation and whether or not the world can maintain an equilibrium with conditions conducive to our survival. Switching sources of energy without changes in behavior that acknowledge this truth will only solve one set of problems and allow others to fester. The smallest possible footprint should always be a goal.

Notes

1. Clive Ponting, "The Lessons of Easter Island," Eco-Action, accessed April 16, 2015, http://www.eco-action.org/dt/eisland.html.
2. Kirk Siegler, "Owens Valley Salty as Los Angeles Water Battle Flows into Court," NPR, March 11, 2013, accessed April 16, 2015, http://www.npr.org/2013/03/11/173463688/owens-valley-salty-as-los-angeles-water-battle-flows-into-court.
3. U.S. Department of State, Office of the Historian, "Milestones: 1969–1976. Oil Embargo, 1973–1974," accessed April 16, 2015, https://history.state.gov/milestones/1969-1976/oil-embargo.

4. Andrew E. Kramer, "Russia Cuts Off Gas Deliveries to Ukraine," *New York Times*, January 1, 2009, accessed February 27, 2015, http://www.nytimes.com/2009/01/02/world/europe/02gazprom.html.

5. European Commission, "European Critical Raw Materials Review."

6. MIT, Mission 2016: Strategic Mineral Management, "Rare Earth Elements Supply and Demand," accessed June 4, 2015, http://web.mit.edu/12.000/www/m2016/finalwebsite/problems/ree.html.

7. Shakuntala Makhijani and Alexander Ochs, "Renewable Energy's Natural Resource Impacts," in *State of the World 2013: Is Sustainability Still Possible?* (Washington, D.C.: Worldwatch Institute/Island Press, 2013), 88.

8. Jacobson and Delucchi, "Providing All Global Energy with Wind, Water, and Solar Power, Part I."

28

Global Energy Price and Cost Points

Every time history repeats itself, the price goes up.
—Anonymous

Energy pricing is a local choice that reflects policies of access, taxation, conservation, and global competitiveness. Differently, direct costs are unavoidable, and capturing indirect but traceable costs is hard but necessary for sound decision making.

Most consumers around the world purchase electricity and gasoline, so examining the worldwide pricing for these staples will be interesting. Most of us are familiar with the price of gasoline but are much less tuned to the price of electricity. Surprisingly, the worldwide variability in pricing for both of these staples goes way beyond the cost of supply. Local energy taxes and subsidies convey the choices governments make to address priorities. Taxes can be levied to fund welfare programs, subsidies can share with citizens the benefits of a rich domestic energy source or support global competitiveness, and higher price points can be used to spur conservation and efficiency measures.

Gasoline Prices Range from Giveaway to Overly Taxed

Given the price range in Figure 28.1, we would expect to see a world of difference in consumer behavior between Venezuela and Norway, and in fact there is. Both of these countries are rich in petroleum reserves, and

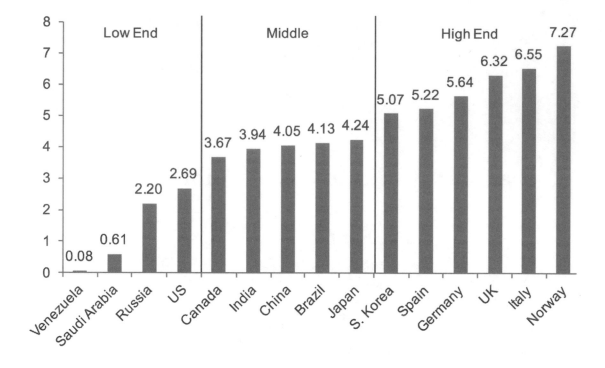

Figure 28.1. Gasoline prices (US$/gal), March 2015.
Source Data: International Energy Agency, "Key World Energy Statistics," accessed, October 23, 2016, https://www.iea.org/publications/freepublications/publications/Kenworld.statistics_2015.pdf.

yet one country nearly gives away gasoline, while the other has the most expensive prices in our study group. Venezuela has a price per gallon of gasoline of US$0.08, while Norway has a price per gallon of US$7.27. In Venezuela, giveaway pricing has created a cross-border smuggling trade, while in Norway drivers purchased more electric vehicles than gasoline-powered vehicles in 2014. One could argue that Norway's disproportionately high tax on petroleum subsidizes other parts of its economy. Elsewhere, many of the wealthy oil and gas countries provide subsidies for domestic use; some are close to the brink of free. At the other extreme, Europeans pay considerably more for their petrol, twice the price paid by the U.S. consumer. In the United Kingdom the tax on gasoline is more than the price of gas in the United States, 61% of the consumer price or US$3.86/gal, while in the United States the tax is US$0.19/gal. A price increase at the pumps of US$2.00/gal will cause more consumer pain in the United States than in Europe, where the increases would be 74% and 32%, respectively. The U.S. consumer is not particularly incentivized to purchase more fuel-efficient vehicles, which punctuates the downside of low prices. Such a wide range in global gasoline pricing affects the types of vehicles purchased and the usage of those vehicles; the lower the pricing, the less it drives behavior conducive to conservation and efficiency.

Global Prices for Electricity Have a Fivefold Range

Not surprisingly, the price per kWh of electricity also varies around the world, and the variation isn't entirely due to domestic power costs. Figure 28.2 shows the household prices for electricity in darker shades and the industry prices in lighter shades for most of the countries we've studied. The price of electricity varies a lot, ranging for industry from 3.5¢/kWh in Norway to 16.2¢/kWh in Japan, nearly a fivefold difference. Household prices ranged similarly, with South Korea at the low end and Denmark at the high end. Norway, where more than 97% of electricity is generated from hydropower, has higher residential prices, while its industry prices are lower, presumably to foster global competitiveness. Across all the countries we can see that industry is able to purchase electricity at a discount below the cost for households, with the exception of Mexico, where the rates are roughly the same. And it seems that households in Germany and Denmark pay an especially disproportionately higher price as compared with industry.

A fully amortized and functional fossil fuel plant will be hard to displace without other incentives.

Surprisingly, South Korea has low electricity pricing in both the household and industrial sectors. You would expect South Korea to mirror the pricing of resource-poor countries such as Japan and Germany. So what accounts for its low pricing? South Korea may have one of the most advanced pricing mechanisms in the world, designed to shift consumer demand and lower the cost of electricity generation.[1] Interestingly, in South Korea pricing is directly related to the variable costs of electricity generation, as opposed to a bidding price that is often unrelated to costs. In South Korea, the consumer bill includes a demand charge and an energy charge within six nonlinear pricing levels. For instance, using twice as much electricity causes a 2.5× increase in the energy charge and a 4× increase in demand charges. The price signal has been so strong that in spite of overall growth in demand, capital investments in capacity and transmission infrastructure have been avoided because of consumers' demand shifts. Recall that for the United States, 10% of electricity

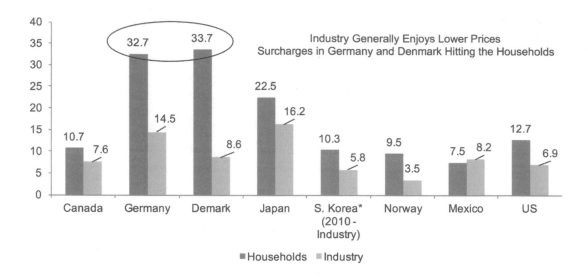

Figure 28.2. Electricity pricing including taxes (US¢/kWh), 2011. *Source Data:* U.S. Energy Information Administration, "International Energy Statistics," accessed February 26, 2015, http://www.eia.gov/cfapps/ipdbproject/IEDIndex3.cfm?tid=6&pid=29&aid=12.

generation assets are used only 5% of the time. South Korea employs advanced pricing to avoid such grossly inefficient investments. We have already seen that South Korea is targeting smart grids as an export industry; it has realized the positive effect of a strong price signal, and in combination with smart electronics it expects even greater efficiencies.

The relatively lower electricity pricing for industry helps attract and retain business. Japan's industrial prices are triple those of nearby South Korea, creating a challenge for its energy-intensive industries. Industry reasonably seeks globally competitive energy pricing, while residential energy pricing competes with other local family expenses. Consequently, governments more readily send strong price signals for efficiency and conservation to households that lack the means to shop elsewhere.

Unsubsidized Levelized Cost of Electricity Generation Is Quite Revealing

Examining the relative cost of power generation for the different types of energy reveals an economic gap between fossil fuels and renewables that is blurring, and the trends favor renewables in the longer term. The term *levelized cost of energy* (LCOE) is used as a means to compare power plants of different generation technologies and cost structures. LCOE is calculated on a net present value basis, taking the sum of the discounted investments and costs over the economic lifetime divided by the sum of the electricity generated. Because subsidies are always local, the following values have excluded subsidies in order to draw global conclusions. All the levelized costs are taken from the utility perspective for comparison purposes. Residential energy costs are discussed separately and compared with retail electricity prices. There are many points of data needed to determine LCOEs, including investment costs, capacity factors of the plants, operating costs, and fuel costs for each type of technology. Unfortunately, each of these data points comes with a range, as will the LCOEs. Nevertheless, the analysis allows us to see how alternative energy costs compete with those of conventional generation technologies. It also reveals the differences among the conventional sources. The projected levelized costs with conventional sources are increasing as lower carbon emissions and improved air quality are prioritized. The LCOE is a useful tool for investment decisions but less helpful for plant retirement determinations. A fully amortized and functional fossil fuel plant will be hard to displace without other incentives.

The information in Figure 28.3 is taken from a 2014 study conducted by Lazard. Since the perspective is the utility scale, the energy costs are based on MWh. The types of conventional generation include gas combined cycle (most efficient method), coal, and nuclear. The high end for gas and coal incorporates costs to capture CO_2 emissions, while the lower end reflects most of the world's current installations. New fossil fuel capacity faces increased pressure to reduce CO_2 emissions, and consequently its costs are shifting to the right. The renewable technologies are solar PV and thermal (CSP), onshore wind, geothermal, and biomass. We've discussed the temporals (i.e., solar and wind), so it is also useful to see the LCOEs for hydrogen fuel cells and batteries, both of which are necessary to mitigate imbalances in supply and demand.

What we learn from the data is startling. First, the unsubsidized levelized cost of onshore wind is already less than that of current conventional sources, while solar PV is on par. Conventional gas combined cycle remains cost-effective even when smokestack CO_2 is comprehended. This

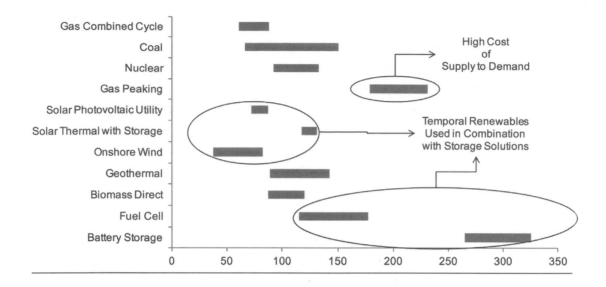

Figure 28.3. U.S. unsubsidized levelized cost of energy (US$/ MWh), utility view. *Source Data:* "Lazard's Levelized Cost of Energy Analysis—Version 8.0," Lazard, September 2014, accessed March 9, 2015, https://www.lazard. com/media/1777 /levelized_cost_of _energy_-_version_80 .pdf, 2.

explains the world's heightened investments in gas production, development, and power capacity. Going forward, biomass and geothermal power are cost-competitive with coal (with CO_2 capture), and nuclear power is more expensive than gas. The nuclear costs reflect the relatively high regulatory compliance expenses associated with safety and the fact that fossil fuel costs have yet to monetarily account for their tolls. There is a regulatory imbalance that artificially distorts our energy economics; true fossil fuel costs lie to the right. Gas peaking is worth noting because it highlights the high cost incurred to provide peak loads. It underscores the importance of shifting our demand behavior to avoid wasteful capital investment and a threefold increase in generation costs.

When temporal solar PV and onshore wind are producing, their costs are competitive with those of conventional plants, but their costs increase much like those of peaker plants when the production needs to be stored and later retrieved. Presuming that temporals and storage systems are used in combination to provide round-the-clock electricity, the overall cost for solar and wind moves to the right. This is because some portion of the energy would be used directly and the balance would be converted to hydrogen or stored in a battery for subsequent use. Two things need to occur as the world migrates to high contributions from temporals. First, the infrastructure to store and retrieve needs to be built, and second, the cost of storage needs to be refined.

What Figure 28.3 also implies is that we are approaching a tipping point for the adoption of renewables, and a successful transition is well within sight if we have the political and social will. The degree to which we can shift consumption patterns or otherwise store and retrieve energy cost-effectively will be crucial. Although the present costs of fossil fuels don't accurately account for carbon abatement, if we were to assign a value to the carbon energy sources, the tipping point would obviously move forward. A $50/ton CO_2 emission charge is cited in one evaluation report on Europe's emissions trading system. The report believes that a figure in this range will send the necessary signal to effect changes.[2] Still $50/ton represents only one-third of the true costs of fossil fuel emissions by some estimates. A report in Germany projects that the LCOE for carbon fuels will increase through 2030, while renewable costs will continue to decline.[3] The cost lines have already begun to cross, and that trend will continue in the next decades. Considering that the useful life of a new power plant is 30 to 40 years, this sets the pace for converting the power sector to alternatives and is the reason that planning and incentives need to be enacted now before further investments in the status quo are made. A long planning horizon would see a vast majority of new capacity powered by alternatives.

Also to be considered is the fact that significant renewable capacity will require land and ocean spaces to harness the energy. The costs in the graph mask these hard choices. For instance, the utility costs for solar do not comprehend the land requirements needed to harvest the necessary quantities of energy. Obviously, energy decisions will be based largely if not solely on economics, but they will also reflect an acknowledgment that we gather our staples at the whim of our blue island in the sky. Since every energy choice carries a charge, maximizing conservation and efficiency in lowering our energy appetites is vital to minimizing the tolls.

The costs of conventional energy have taken 150 years to arrive at their current levels. While the costs of renewables are still declining, conventional production costs can hope to remain the same but will certainly increase as enhanced technology is applied. Impressively, the utility-scale

LCOEs for onshore wind have decreased 58% in the last five years, while solar PV costs have decreased 78%.

Distributed Energy Makes Energy Personal

Figure 28.4 includes the range in the retail price of electricity compared with renewable costs both subsidized and unsubsidized. The subsidized categories are marked with an "S". Like onshore wind in the utility sector, micro turbines are already cost-competitive with retail pricing in most markets. The cost of solar PV is gradually falling, and market by market around the world solar PV is becoming less expensive than current retail prices for electricity. German households, for instance, pay less than half what U.S. residents pay per watt of solar PV capacity. If U.S. consumers had access to the prices in Germany, solar PV would provide savings against current utility prices for electricity. It makes one wonder whether the system is rigged to retain the existing centralized architecture and its investors. Having suffered and recovered from a cynical moment, though, if one assumes that the pricing in Germany heralds what can be expected, residential solar PV investment will accelerate in the relative near term.

Trends Favor Renewables

While the global pricing for energy staples varies significantly by choice, the costs of alternatives are favorably competing with the costs of conventional sources. Many of the new technologies can be locally sourced, and that enhances energy security. Increasing pressure to curb CO_2 emissions and the declining cost of alternatives are already causing cost lines to cross in some markets, with many more to follow in the decades ahead.

Peering into the future, as opposing forces converge on the use of fossil fuels, the world is likely to become socially and economically unstable if we do not have alternative energy sources to turn to. Despite all the rhetoric that opening pipelines or drilling in sensitive areas is critical to our national security, there is no better way to strengthen global security than for countries around the world to build domestic capacity with

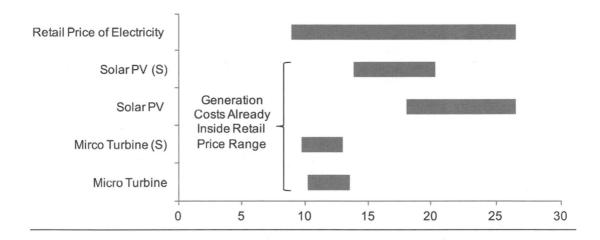

alternative technologies and become energy self-sufficient. In the long run, global security is undermined by continued dependence on fossil fuel sources that are held by just a few nations.

Notes

1. Kamil Bojanczyk, "The Price Is Right: Advanced Electricity Pricing in South Korea," Greentech Media, June 20, 2012, accessed June 7, 2015, http://www.greentechmedia.com/articles/read/the-price-is-right-advanced-electricity-pricing-in-south-korea.

2. Jos Sijm, Hector Pollitt, Wanter Wetzels, Unnada Chewpreecha, and P. Koutstaal, Splitting the EU ETS: Strengthening the Scheme by Differentiating Its Sectoral Carbon Prices, report no. ECN-E-13-008 (Petten, the Netherlands: ECN, 2013).

3. Christoph Kost, Johannes N. Mayer, Jessica Thomsen, Niklas Hartmann, Charlotte Senkpiel, Simon Philipps, Sebastian Nold, Simon Lude, Noha Saad, and Thomas Schlegl, Levelized Cost of Electricity Renewable Energy Technologies (Freiburg: Fraunhofer Institute for Solar Energy Systems, November 2013), accessed June 8, 2015, https://www.ise.fraunhofer.de/en/publications/veroeffentlichungen-pdf-dateien-en/studien-und-konzeptpapiere/study-levelized-cost-of-electricity-renewable-energies.pdf.

Figure 28.4. Residential view of electricity cost/price (US¢/kWh). PV = photovoltaic; S = subsidized.
Source Data: "Lazard's Levelized Cost of Energy Analysis—Version 8.0," Lazard, September 2014, accessed March 9, 2015, https://www.lazard.com/media/1777/levelized_cost_of_energy_-_version_80.pdf, 3.

Energy Policy Tool Kit

It has become appallingly obvious that our technology has
exceeded our humanity.
—Albert Einstein

It's clear after examining country energy plans that priorities are as unique as fingerprints. Population, geography, resources, industry, politics, and values all interplay in forming a blend of priorities that shape energy plans. Addressing global climate concerns, such as those advocated by the Intergovernmental Panel on Climate Change, is inherently difficult given such diversity, and to some countries these concerns directly conflict with higher priorities. Across the countries we've examined, top priorities ranged from economic development to affordable access, national security, petroleum development and production, health and environment, and alternative energy development.

Most often, the time horizon for energy planning is only a few decades. It seems that the longer the planning period, the more likely country plans would share common priorities since we all share the same world. Instead, the pressures of the day are forcing a focus on the present and undermining efforts to plan with a grander view. Effective tactical plans are understandably short term but ideally born out of thinking undertaken with a longer view. As discussed earlier, one of many risks with a short view can be tactical actions that work against a natural shift to alternative energies.

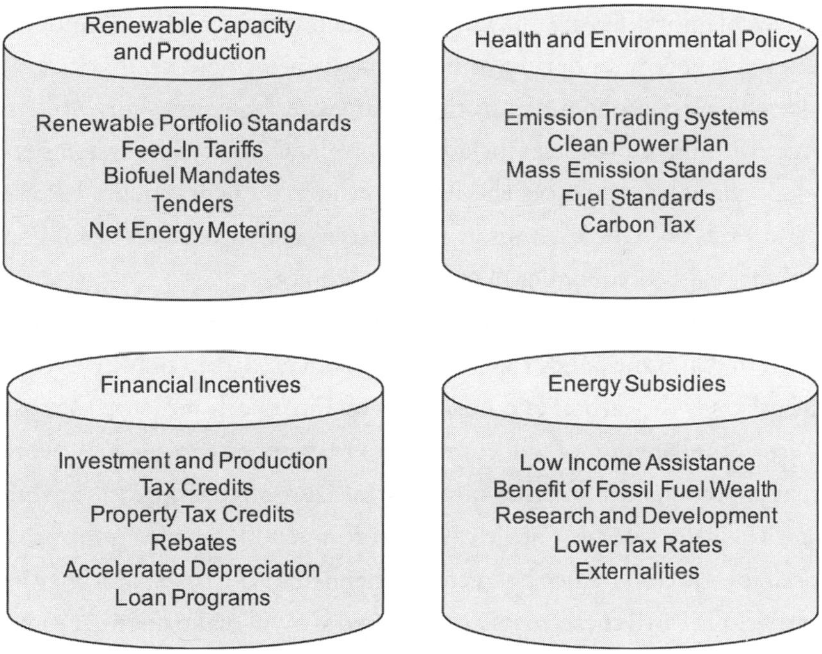

Renewable Capacity and Production

Renewable Portfolio Standards
Feed-In Tariffs
Biofuel Mandates
Tenders
Net Energy Metering

Health and Environmental Policy

Emission Trading Systems
Clean Power Plan
Mass Emission Standards
Fuel Standards
Carbon Tax

Financial Incentives

Investment and Production
Tax Credits
Property Tax Credits
Rebates
Accelerated Depreciation
Loan Programs

Energy Subsidies

Low Income Assistance
Benefit of Fossil Fuel Wealth
Research and Development
Lower Tax Rates
Externalities

The Policy Choices Are Not the Problem—It's the Lack of a Long-Term View in Policy Making

Figure 29.1. Policy soup.

Up until now we've talked about the potential of renewables from a technology perspective. But how do we actually engineer change? How does any community go about the difficult task of altering its sources of energy? How do we sit across the table from the energy industry and stimulate, incentivize, and effect change? What tools do we have? And most fundamentally, can we agree on the problem that needs solving?

There are two classes of policy: those that encourage renewable capacity and production and those that address the health and environmental consequences from our sourcing of energy. And then there are subsidies. Figure 29.1 displays renewable policies on the left and health and environmental along with subsidy policies on the right. Renewable policies are of two main categories: capacity and production programs and financial incentives. The capacity and production group includes renewable portfolio standards and feed-in tariffs, biofuel mandates, renewable energy tendering, and net energy metering. The financial category includes economic incentives that encourage capital investment. The environmental policies include carbon allowance trading systems and regulation. The carbon tax should be considered a hybrid policy because it addresses both the environmental concerns with greenhouse emissions and the relative economics of our energy choices.

It is very common to find a combination of these policies concurrently in play. One policy tool that stands out as critical to fully leverage the others is the carbon tax. As long as we incorrectly measure the costs of fossil fuel energy we will continue to make poor choices. Our world turns on economics, and the scales that measure our energy costs are broken. There are traceable health and environmental costs from the use of fossil fuels that have not been comprehended in the costs, and they are significant. Until these costs are captured we will make poor decisions. The other policy tools are diluted in their effectiveness unless we first correct the basic flaw in our evaluation of energy costs. Let's push forward, though, and describe these other policies before we return to the most controversial of all, the carbon tax.

Policies should be judged on the basis of the behaviors they induce and their alignment with long-term goals. Regulation to some extent indicates an acceptance of an industry and applies stipulations on its continued operation. Yes, we want to regulate the development and use of fossil fuels, but inevitably we need to replace them. The following example clarifies the distinction between regulation of an industry and policy that fosters the regeneration of an industry.

Within the power segment, clean air regulations indeed reduce emissions and encourage shifts from a dirtier fossil fuel—coal—to the cleaner fossil fuel—natural gas. Alternatively, properly valued carbon charges can stimulate shifts to alternative energy sources by correcting the economics of the choices. There can be no doubt that economics ring loudest in our world today. If we want to guide behavior in the twenty-first century toward a wholesale change in energy sourcing, economics will be the most potent policy tool, not regulation.

So what specific policies are available for countries to employ? Some were helpful early on and are giving way to new policy inventions as the energy transition progresses. Following are the major policies leadership has to effect change.

A Stake in the Ground: Renewable Portfolio Standards

Renewable portfolio standards are both a policy and a goal. A state, province, or country sets a goal for the contribution of energy from renewable sources by a target date. Utilities pay a penalty on every unit of electricity in their shortfall. The program can prioritize a specific technology such as solar PV, or it may disallow certain technologies to suit preferences. Most often this policy is applied in the power sector. Around the world there were 79 states, provinces, and countries with RPS policies in place at the end of 2013.[1] Recently, the state of California tried unsuccessfully to legislate a contribution level for alternative vehicles, a transportation sector version of RPS.

The Initial Spark: The Feed-In Tariff

The feed-in tariff (FIT) is often a tactical complement to an RPS policy. The design of the system includes setting the tariff, capacity targets, and qualifying technologies for the program. The tariff is really an elevated price for electricity guaranteed to investors for a term of typically 10 to 20 years. The program is funded through electricity surcharges. If the price or tariff is too high, the program costs increase unnecessarily; too low a price may not incentivize enough investment to meet the capacity goals. Germany was particularly successful in jump-starting its capacity in renewables with this program. Over the years as the levelized cost of electricity from renewables has declined so has the tariff and presumably the surcharge. At the end of 2013, there were 79 states, provinces, and countries with FIT policies, the same number as those with RPS policies.

Tendering Schemes for Renewable Energy Sources:
A Refinement of FIT Policies

The FIT policy was first engineered when renewable capacity was negligible and more costly than conventional sources. Now more than a decade later renewable capacity is increasing, and the cost differences are much less pronounced. So, too, the policy tool kit needs to evolve. Ideally you want the development of energy sources to reflect the locality's natural potentials and value systems. Tendering programs can be highly specific regarding where the renewable energy is needed and what technologies are desired.

Tendering is a tool that can be considered a natural evolution of the FIT policy. The advantage is that tenders can align to a long-term blueprint for energy production. The blueprint specifies the degree of decentralization, preferred types of renewables based on the local value system, and where they are best located. The tendering process can be administered over a long horizon in alignment with the blueprint. If 20 MW of solar PV capacity are needed in a certain location, a tender can be written to precisely characterize that requirement. Bids are accepted, and the qualifying bid with the lowest price wins the tender. Payments may take the form of an FIT at this point. This policy derivation rapidly increased from use in nine states, provinces, and countries in 2009 to 55 in early 2014.

Simple Has Its Limits: Net Energy Metering

If you believe that simple is better, you'll appreciate net metering. The cost of the participant's renewable energy investment is recovered through a lower net utility bill. The utility agrees to pay wholesale price for any surplus generated by renewable users and charge retail price for electricity those users take from the grid. Surprisingly the purchase of surplus electricity at wholesale prices is controversial, but it is actually quite generous. At low levels of renewable capacity this program works well because the excess is immediately consumed elsewhere. However, it is easy to see limitations in this program as consumers install higher and higher capacities of temporals. Naturally there will be more and more occasions when total production exceeds total demand and the surplus is either stored or lost. The value of the consumer's surplus is lower for the utility because it is unpredictable and potentially necessitates storage. Paying the consumer at the wholesale price is a defendable position.

As scalable energy management systems (storage) emerge, the consumer's appreciation for the grid connection and net metering will evolve. Consumer installations will link storage capacity with temporal capacity and strive to achieve grid autonomy. The ideal is a combination of production and storage to meet consumers' 24-hour requirements. Retaining the grid connection becomes a comforting backup for extended periods of poor production or seasonal variation. Just as the purpose of a grid connection matures, the real value of net metering will change, and consumers will find solace in their storage systems.

CO2 regulatory limits are aimed at lowering carbonization, but what is urgently required is the overall decarbonization of the atmosphere, something inherently incompatible with the continued use of fossil fuels.

Biofuel Mandates Linked to the Internal Combustion Engine

Simply, biofuel mandates establish blends for renewable ethanol and biodiesel, thereby reducing the transportation sector's dependence on petroleum. In early 2014 there were 63 states, provinces, and countries with biofuel mandates.

Attracting Money: Investment Incentives

Several types of financial incentives are used to spur the capacity and production from renewable energy sources. They include investment and production tax credits, property tax credits, and rebate programs. Larger projects have access to loan programs and accelerated depreciation schedules to help stimulate alternative energy development.

The Missing Piece

We know that ever-increasing contributions from temporals will require complementary technologies and infrastructure. Early on, FITs and other policies were helpful in initiating renewable capacity. But further capacity requires smart grids to smooth supply and demand peaks and advanced storage systems to avoid losing energy. Where are the incentives for infrastructure and storage technologies? We will hit a temporal capacity limit absent parallel investments in these areas. Breaking through the current storage barrier will be vital to the replacement of fossil fuels.

Health and Environmental Policies

Referring back to Figure 29.1, let's examine a few indirect policies positively affecting the relative attractiveness of renewables—tools designed to safeguard human health and the environment including climate change. The tools fall into one of two categories: regulation and CO_2 emission–based charges. Regulation seems to be the preferred tool to address air quality and many other environmental concerns associated with the development, production, and consumption of fossil fuels. Regulatory bodies typically assess technology and best practices when establishing targets for the regulation. This is the strength and weakness of applying regulation to the challenge of driving a wholesale change in energy sourcing. The limits, often publicly characterized as "stringent," usually reflect what can be done as opposed to what needs to be done. In the case of CO_2, the established regulatory limits are aimed at less carbonization, but the decarbonization of our atmosphere is what is urgently required, something inherently incompatible with the continued use of fossil fuels absent the deployment of robust carbon capture and sequestration technology. The carbon charges are then the best method to advance climate policy because they incorporate a cost for CO_2 emissions. This allows noncarbon alternative energies to compete equitably for future investments. Where regulation accepts the negative impacts but also then imposes costs on current operations, the carbon charges help correct our economic lens for better decision making.

Emissions Trading: Carbon Allowance Systems (Cap and Trade)

A carbon allowance program places a cap on CO_2 emissions and progressively reduces that cap toward a goal. Carbon allowances are typically applied to the power sector. As an example, utilities are required to have allowances for each metric ton of carbon they emit. For reference, 1 ton of coal emits about 2 tons of CO_2. They acquire these allowances through auctions where a price is established. Setting the right price is understandably an important element of a successful program. A price set too low won't incentivize a change in behavior, while a price set too high will create pain points before responsible behavior can be implemented and results can be achieved. Individual utilities can buy or sell allowances as necessary to meet their obligations. At the end of a compliance period, a utility is required to provide allowances for each metric ton of CO_2 emitted.

Allowances are a form of currency; a utility can either bank excess allowances or sell them in the market. Noncompliance varies by program, but in the case of California's program, a utility must surrender four allowances for each metric ton in excess of the cap. Over time the number of allowances is reduced, until finally the program has achieved its CO_2 reduction goal.

Europe's Emission Trading System experience highlights an important lesson: an optimum program can be elusive. Just as Europe was auctioning a round of allowances, a recession occurred. This resulted in an oversupply and devaluation of the credits. Since there was no floor price for the allowances, they took on the characteristic of junk bonds. The cost of noncompliance created little economic pain and little change in behavior.

Another lesson is a reminder of the pitfall of unintended consequences. Within the power segment there were calls to define low- and high-carbon trading schemes for industry and residential segments, respectively. Those in industry argue that because they compete in global markets, a high local fee can render them uncompetitive. In that situation they may be forced to relocate, referred to as CO_2 leakage. The unintended consequence is conceding lower prices to industry, which means that households are asked to carry a disproportionate share of carbon emission costs. So, a good program on paper can turn out to be far more complex to implement and ultimately ineffective. Well-intentioned programs can easily turn into tangled webs of administrative confusion.

Carbon allowance systems are designed to arrive at the end goal through two types of influences. Carbon emissions are a function of how much and what types of energy we use. They can encourage lower demand through conservation and efficiency and secondarily influence the adoption of lower-carbon-emitting sources. A high carbon price would force utilities to adopt noncarbon sources, while lower pricing would shift the onus of the program to the consumer, who would be expected to conserve. Thus far programs have established low carbon prices, and the residential consumer has paid the bulk of the programs' cost. This seems counterintuitive if the real solution is a reduction of fossil fuel resources in the utility mix.

Several emissions trading programs are in place today, and Table 29.1 shares the key elements of each. As mentioned earlier, the targets are all short term (i.e., 2020), and the low price floor for the allowance (i.e., $10) does not send a strong enough signal to cause a change in the energy mix. However, California's emissions trading program is generating revenue that can be used to fund other state energy projects. The initial sale of 29 million allowances raised revenues of $525 million, which supported air quality programs within the state.

The establishment of a process for the speculative buying and selling of carbon allowances that represent just a portion of a power plant's CO_2 emissions seems convoluted at best. The policy seems to be attacking the symptom as opposed to the problem, the use of carbon-based fuels.

Carbon Regulations

In 2009, the U.S. National Research Council issued a report titled *Hidden Costs of Energy: Unpriced Consequences of Energy Production and Use.*[2] The U.S. Congress requested this report, and it estimated that nonclimate (i.e., ground-level ozone pollution) and climate (greenhouse gas) damages were at $120 billion in 2005. Economists, in this context, use the term *externalities* to refer to costs and benefits not reflected in the current price of fossil fuels. The basis of exposing externalities is that energy decision making will be flawed if these actual costs are high but are not considered. Beyond carbon allowance systems, two other policies

Table 29.1. Comparison of Carbon Cap and Trade Programs.

	CALIFORNIA'S GREENHOUSE GAS (GHG) CAP AND TRADE PROGRAM	EUROPEAN UNION'S EMISSION TRADING SYSTEM	QUEBEC'S CARBON MARKET
Population	38 million	500 million	8 million
Jurisdiction	California	27 E.U. members plus Norway, Iceland, and Lichtenstein	Quebec
GHGs covered	CO_2, CH_4, N_2O, SF_6, perfluorocarbons (PFCs), NF_3, and other fluorinated GHGs	CO_2, N_2O, and PFCs	CO_2, CH_4, N_2O, SF_6, PFCs, NF_3, and other fluorinated GHGs
Sectors covered	Electricity, industry, ground transportation, and heating fuels	Fossil fuel power plants	Electricity, industry, ground transportation, and heating fuels
Emission thresholds	Emitters of at least 25,000 CO_2 annually	Any combustion installation over 20 MW	Emitters of at least 25,000 CO_2 annually
Target	17% below 2013 levels by 2020	21% below 2005 levels by 2020	20% below 1990 levels by 2020
Carbon allowances (millions)	162.8	2,039	23.7
Price floor	$10 per metric ton, rising 5% per year starting in 2014	No price floor	$10 per metric ton, rising 5% per year starting in 2013

Source: C2ES Center for Climate and Energy Solutions, "California Cap-and-Trade Program Summary," January 2014, http://www.c2es.org/docUploads/calif-cap-trade-01-14.pdf.

are designed to address if not capture those externalities—emissions regulations and taxes—but their results can vary dramatically based on the stringency of the targets and strength of the economic signal. Let's start with regulation.

Transportation Sector Example: Vehicle Mass Emissions and Fuel Standards

Mass emission regulations of vehicle tailpipe exhaust went through several phases as they became law and industry complied. In the first phase, industry objected, warning of significant costs and the loss of competitiveness. In the second phase, the regulations were accepted, and industry applied existing technology to achieve the goals. In the final phase, benefits greater than expected and costs less than anticipated were realized. Good to know.

Vehicle emissions standards regulate known toxins including carcinogenic formaldehyde (CH_2O), carbon monoxide, nitrous oxides, nonmethane organic gases, and particulate matter. Carbon monoxide (CO) is produced during the combustion of fossil fuels when there is insufficient oxygen. The ancient Greeks observed the toxic effects of carbon monoxide when burning coal and applied this observation to a method for performing executions, and today we are still well aware of the dangers of CO poisoning from vehicle emissions. Nitrous oxides and nonmethane organic gases react in the presence of sunlight to cause ground ozone: smog. The harmful consequences of these toxins have been established, and the U.S. EPA was authorized to protect human health. As mentioned earlier, most of petroleum is carbon and hydrogen, with variable amounts of sulfur and other impurities. Sulfur is the instigator of acid rain and must be removed during the refining process; hence the EPA incorporated limits within the clean fuel standard. Similarly for lead, the EPA regulated the pollutant at the refining stage. Car manufacturers were given five years to introduce lead-free models, and the transition away from lead-containing gasoline was made.

The mass emission regulations were successful for several reasons. First, the regulations were stringent enough to deliver good results, and yet they were within the technology wheelhouse of auto manufacturers.

Second, the point of regulation, the auto industry, applied pressure where the solutions lay. The same was true for fuel regulations; the point of regulation (directed to the refiners) delivered maximum impact. Third, the standards were regulating known toxins—formaldehyde and carbon monoxide as examples—that were not subject to limitless debate.

As noted above, the regulations applied to the transportation industry have been extremely effective in improving air quality. Regulations increase costs to the manufacturer, but since the auto industry is global and everyone must comply to participate in the global market, there wasn't a loss of competitiveness. This is an important distinction because power sector regulations that are locally enforced can increase costs and undermine local industry's global competitiveness, and consequently advancing health and environmental quality is, then, a question of who on the world's stage blinks first.

Altering energy sources will incur immediate costs, with benefits realized only later. This is the test of leadership.

The end result of the mass emission regulation has been cleaner carbon-based vehicles, not zero-carbon vehicles. Will the CO_2 emission regulations have similar results in 30 years? Will the power industry remain essentially carbon-based with lower carbon emissions while atmospheric CO_2 levels increase? It is a valid question with a predictable answer.

CO_2 Regulation Is Not Enough: Restating the Case for a Carbon Tax

Scientists link the ppm levels of atmospheric CO_2 with changes in climate including average global temperatures and sea levels, and the higher the number, the greater the change. Atmospheric CO_2 levels were 317 ppm in 1960, and in 2015, a mere 55 years later, the world crossed the perilous 400 ppm threshold, a quantity not seen in the last 400,000 years. The slope of the line shows no sign of altering. Without a dramatic change in the world's energy sector we are on our way to 500 ppm and beyond. Fortunately, the depletion of economically viable fossil fuels places a lid on the harm we can self-inflict.

As previously discussed, the *International Energy Outlook* projects continued increases in worldwide use of carbon fuels up to 2040. Coal, natural gas, and petroleum consumption is poised to increase 17%, 70%, and 34%, respectively, above 2012 levels. Absent broadscale CCS, the world's

2040 energy mix will usher atmospheric CO_2 to much higher levels. The U.N. Intergovernmental Panel on Climate Change recently concluded that the world needs to decarbonize by the end of the twenty-first century. This is the right goal, but the *International Energy Outlook* portrays the gap between goal and planning. If this outlook is unacceptable, then current policies must be reconsidered.

Since 1980, our study group of just 13 countries emitted two-thirds of the worldwide energy-related CO_2. The United States alone is responsible for nearly 22% of worldwide emissions. If we placed the highest burden to reverse atmospheric levels on the largest contributors, the U.S. commitments to cap GHG emissions at 26%–28% below 2005 levels are grossly insufficient. Only Germany, with a goal to reduce GHG emissions by 80% to 1990 levels, seems aligned to the problem. Developed countries (such as the United States) most responsible for the 90-ppm increase in atmospheric CO_2 during the last 55 years need to lead the world's retirement of fossil fuels.

Climate regulation is dilutive and distracting when the world's underlying requirement is the complete discontinuance of fossil fuels. Regulation is the wrong tool for the task, unless it regulates a schedule for the replacement of fossil fuels this century, and that's why the Clean Power Plan is well intentioned but wrong. The liabilities of fossil fuels related to health and the environment can no longer exist outside the price. Economic policies with loud and clear price signals are needed to change the world's energy course. The correct goals need to be stated in terms of replacing carbon sources of energy and, finally, a cessation date for fossil fuels. This avoids the economic depletion of fossil fuel and honestly strikes a path toward reducing atmospheric CO_2.

To State the Obvious: Initiate a Meaningful Carbon Tax

The carbon tax is in a class by itself in that it could accomplish in one program what many other policies nibble at: the prompt withdrawal from the use of fossil fuels. The use of the word *tax* is no doubt polarizing, but it is honest. Revenues from carbon cap and trade programs are taxes with a different name. If a utility has no choice but to purchase allowances, let's be candid—it's a tax. No matter whether we call them use fees, taxes, or carbon taxes, the use of fossil fuels needs to bear a charge, and the tax revenues need to be directed toward breaking the hold of these prehistoric carbon fuels.

Progressive taxes on carbon emissions would begin to account for the externalities to which the economists refer. You apply the tax to sources of CO_2 emissions. For instance, one could tax gasoline to account for CO_2 emissions, and people would pay as they go; or you could apply the tax to new vehicles, where the choice of energy is made. A carbon tax of $50/ton for a fuel-efficient gasoline-powered vehicle would add $2,000 to the sticker price (20 lbs CO_2/gal, 35 mi/gal, and 150,000 mi lifetime). If the policy goal is to change the transportation sector's energy mix, applying the tax to new vehicles would make more sense, and it automatically grandfathers prior decisions, which makes it inherently progressive. The tax revenue would be used to support the necessary infrastructure for increasing the capacity of alternative energies. The carbon tax also corrects our economic eyesight and begins to level the playing field for alternative energies. It can also be a means to stimulate immediate investment in carbon abatement. Fully amortized fossil fuel power plants are sticky and difficult to displace without a carbon tax, and consequently this sector should not be grandfathered. Simpler in administration, a tax places trust in a functioning marketplace and allows renewables to enter the market on their good merits. It encourages utilities to invest long-term in noncarbon alternatives rather than a short-term strategy of shifting fossil fuels. Best of all, carbon emissions will naturally decline as the market shifts to these noncarbon alternatives, making the carbon tax transitional. Between renewable and health and environmental policies, the carbon tax is best aligned with a top priority of replacing fossil fuels, but political leadership may lack the courage to tackle it. The Intergovernmental Panel on Climate Change would be well advised to make this a top priority for global adoption.

A progressive carbon tax in developed countries that collects $50 per metric ton of CO_2 emissions would produce real change. This translates into a 5.1¢ and 2.3¢ tax per kWh of electricity from coal and gas, respectively. This tax can be avoided in one of two ways. First, CCS technology could be implemented in order to avoid a portion of the carbon tax. Most likely, though, renewables would become the more attractive option. Most importantly, the scale of the tax will render the status quo unattractive, causing a fundamental change in the energy mix or at minimum spurring investment in carbon abatement.

Global Energy Subsidies

Opponents as well as proponents of renewable energies often cite various subsidies in making a premeditated point. If we take a closer look, we can see that energy subsidies take many distinct forms that address different priorities, including low income energy assistance and economic development. In the United States, for instance, the Energy Policy Act of 2005 funded nearly $14.9 billion from 2005 to 2016 for low-income home energy assistance.[3] The United States has also suppressed fuel prices by restricting petroleum exports. And we've seen petroleum-rich countries such as Venezuela and Saudi Arabia subsidize gasoline prices at home as a tangible benefit of their carbon wealth. Another form of subsidy is the support that the United States and other countries provide for research and development programs ranging from more efficient jet engines, to nuclear technologies, to renewable technologies.

According to data from the International Energy Agency, global fossil energy subsidies were $548 billion in 2013.[4] This figure is fourfold the subsidies for both renewable energies and efficiency programs. Oil accounts for more than half the fossil fuel subsidies.

Well-intentioned programs often have undesirable side effects and miss their initial goal. Oil-exporting countries that choose to subsidize domestic markets unleash a number of negative responses, from cross-border smuggling to wasteful consumption and hastened decline of exports. Importers similarly experience waste but also encounter budget pressures in sustaining subsidies. Estimates surprisingly reveal that the wealthy take advantage of 40% of these subsidies, and only 7% make their way to the poor.[5]

As with the fossil fuel subsidies, renewable subsidies can also miss the mark. Ill-advised renewable projects can overpay for capacity or invest

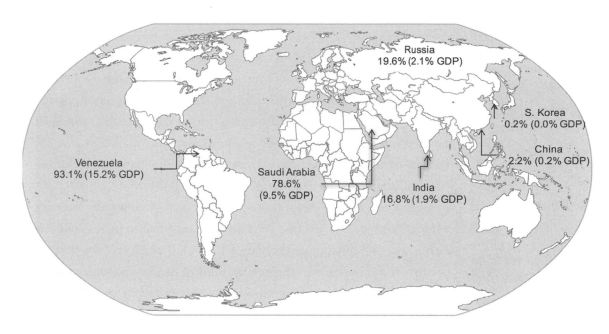

in bad technology. For this reason, many countries are looking to ratio-nalize, not eliminate, subsidies. They understand that writing a check to the poor could be far more effective than many of the programs in place. Indonesia has recently embarked on "right pricing" its petroleum with a 40% increase for petrol.

Figure 29.3 shows a map of the 2014 global fossil fuel subsidies for fire of the countries in our study group (with the addition of Venezuela) for which the International Energy Agency provided data. Subsidy levels are stated in terms of the proportion of cost bypassed. The global range in the data set went from an incredible 93% in Venezuela to 0.2% in South Korea. Many countries continue to provide generous subsidies above the 50% level despite the negative behaviors these subsidies spawn. These include the north African countries of Algeria, Libya, and Egypt and the Middle Eastern countries of Bahrain, Kuwait, Oman, Qatar, United Arab Emirates, Iran, and Iraq. Stated as a percentage of GDP, the subsidies ranged from 0.0% in South Korea to a startling 19.3% in Iran. Figure 29.3 shares the percentage of GDP for several countries in our study group.

The subsidies thus far discussed are of the pretax form, the easiest to estimate. Pretax subsidies occur when the consumer price is below

Figure 29.2 Fossil fuel consumption subsidies as a percentage of the full cost of supply, 2014. *Source:* International Energy Agency. "Energy Subsidies" (Energy Subsidies on-line database), accessed June 2, 2015, http://www .worldenergyoutlook .org/resources /energysubsidies/.

the costs of supply. The International Monetary Fund recently published a working paper on energy subsidies discussing both pretax and post-tax subsidies.[6] Its estimates of pretax subsidies are strikingly similar to the International Energy Agency figures. The report estimates posttax subsidies to be nearly 10-fold the size of pretax figures, at $5 trillion. It includes instances where energy consumption taxes are below efficient levels and the price fails to charge for environmental damages including local pollution and climate change.

Global fossil energy subsidies were $548 billion in 2013. The wealthy take advantage of 40% of these subsidies, and only 7% make their way to the poor.

While projections for pretax subsidies show declines as countries rationalize programs, posttax subsidies are expected to increase in line with increasing consumption of fossil fuel. A fair portion of the posttax estimate is attributable to the externalities of health and environment previously discussed. Posttax subsidies provide a means to cross-check the $50 per ton of CO_2 carbon tax proposal previously mentioned. Apportioning coal's portion (45%) of the International Monetary Fund—reported posttax externalities results in a value of $130 per ton of CO_2 for coal. The proposed carbon tax would capture $50 of the $130, a great first step in properly pricing fossil fuel.

Explicitly Setting Milestones to Decarbonize Our Energy Sourcing

In summary, if the top priority is replacing fossil fuels for all the reasons laid out throughout this book, then we should set a schedule to end the use of fossil fuels energetically at least. We should establish energy mix base camps that deliver an increasing percentage of alternatives, with the summit being zero use of fossil fuels by the end of the century. "Zero" in this context means that the vast majority (>90%) of the world's energy will be harvested from nonfossil sources, with fossil fuels continuing to decline.

A progressive carbon tax is the most forthright way to capture fossil fuel externalities and give renewables a fair economic playing field while providing the revenue for the requisite infrastructure to build high levels of renewables. A master plan would guide investment over the time period, specifying technology mix and sourcing locations. The plan would incorporate the decentralized merits of renewables while also building complementary energy storage requirements. Renewable portfolio standards become the goal-setting base camps; tenders including nuclear technology would be the stepwise execution of the master plan.

Today, policy makers are operating inside a maze and trying to make it through without hitting dead ends. If they had the benefit of an aerial view prior to entering the maze, they would stand a better chance.

The 2015 Paris Climate Accord

The Paris meeting of the U.N. Framework Convention on Climate Change (UNFCCC) galvanized the countries of the world (or at least their representatives) around a goal of limiting global rise in temperatures to 2°C above preindustrial levels. Or stated differently, the UNFCCC now interprets the evidence of climate change as a terminal sentence unless urgent collective action is taken. After decades of research, the relationship between the concentration of atmospheric greenhouse gases and temperature is sufficiently—if not perfectly—understood to define a threshold the world cannot exceed without forcing the earth to a new equilibrium, one that is certain to be hostile to human biology. A global gathering untethered from the challenges of the day shouldn't be dismissed as "whimsical," as some "realists" have described it. One could easily view the usefulness of the meeting cynically and discount its relevance, but that would be stumbling our way into the future.

The Paris agreement is a breakthrough, but there are two Herculean tasks that lie ahead for the world. The first is for countries to deliver on their commitments at home. The second is that we must take meaningful action that decarbonizes our energy sources and that translates into moving away from fossil fuels. Granted fossil fuels and CO_2 aren't the entirety of the problem; however, they are the primary cause of the problem and a necessary focus if we are to sidestep the consequence of climate change. Is there sufficient binding energy from Paris to deliver a course correction? Can the science of climate change translate into a hard-held certainty that our species and many of our cohabitants will perish unless we act? Is this belief powerful enough for countries to set and deliver commitments? Let's hope so—but be ready with other policies if the forces of inertia hold us hostage to fossil fuels and imperil our future.

The hard work for each country representative following the Paris meeting is traveling home and balancing climate and local priorities that are often in direct conflict. If you are a developing nation, providing affordable access trumps climate concerns, and it would be extremely hard to turn a blind eye to domestic supplies of coal. If you are one of the fortunate countries with large portions of the world's fossil fuel reserves, you will be determined to extract and export the fuels before they lose attractiveness. Nobody wants to hold an asset that is eroding in value, so exporters will only increase their pace of production. Universally, altering energy sources will incur immediate costs with benefits realized only in the longer term, and this is the test of leadership. When opportunities to improve the present quality of life of citizens or continue a thriving petroleum export business conflict with complying with climate obligations, local priorities will be difficult to subordinate. The strength of the Paris commitments must be forceful enough to overcome local priorities. Beyond pledges, the UNFCCC will need policies to elevate the climate priority across all classes of countries. The meeting was a turning point; the language is far from perfect, and the commitments are soft, but brokering a global agreement on climate change is a major milestone in acknowledging our global imperative.

The Paris effect must go beyond the symbolism of 195 countries finding common ground on the threat of climate change, however. The world has moved in a few short decades from studying, to synthesizing, to agreement that the earth's climate is changing, but alignment on the risks is still forming. We need global harmony, but there can be no doubt that a much smaller group of countries will need to lead the way, or the goal will not be achieved. Based on fossil fuel consumption levels for 2012–2013, the left portion of Figure 29.3 shows the proportion of global CO_2 emissions from fossil fuel consumption for countries that represent 80% of total emissions. How these 20 countries, home to 62% of the world's population, respond in the near term will determine whether or not we can remain within the 2°C level. And within this group the opposing local energy priorities vary a lot. No single policy tool will work for this diverse group of countries. The right portion of Figure 29.3 takes these same data but translates them into a per capita measure, and this helps

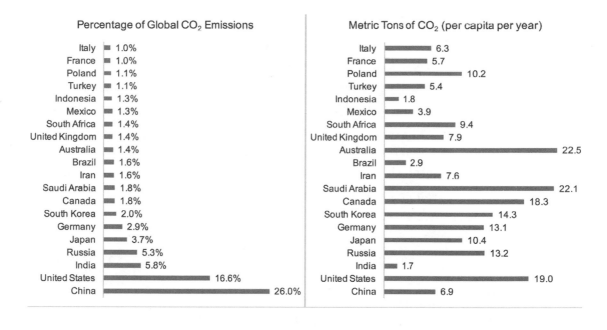

Figure 29.3. CO_2 emissions from the consumption of fossil fuels.

surface a select group of countries. There is an exclusive club of nine countries each emitting greater than 10 metric tons of CO_2 per capita per year! This group has benefited the most from the use of fossil fuels and should help lead the retreat. Every country needs to do its part, but without abstention this group must take prompt action. Leadership means tangible and quantifiable decarbonization of a country's primary energy mix.

The second pitfall in carrying out the Paris commitments is the mathematics of what we are choosing to measure. Greenhouse gas emissions may be the problem, but they are not the best choice for monitoring our course correction. There are many quantifiable inflows and outflows of GHGs in our atmosphere, but many factors such as seabed methane and agriculture, for example, are less well defined, and this will render the mathematics faulty. This will lead to a disagreement between the true atmospheric GHG levels and what the sum of country-level action plans would otherwise project. It's like taking one's temperature with a hand to the forehead rather than using a thermometer. If one accepts the urgency required to hold to 2°C, then our choice of measure cannot be flawed.

Measuring the consumption of fossil fuels is much more direct. It is more precisely the culprit in climate change and unavoidably needs to be curtailed to halt the progression of CO_2 in the atmosphere. The use of fossil fuels must recede and give way to the use of noncarbon sources of energy. It is helpful to improve efficiency where we use fossil fuels and important to conserve, but cumulatively this will not be enough. We can grasp at the notion of carbon capture and sequestration but should only consider its broad implementation as relevant. We can similarly grasp at the promise of new technologies but should only respond with the commercialized solutions. We need to rapidly decarbonize, and this is the best action and measure.

Optimistically, Paris represents a patient coming to terms with a diagnostic warning and taking healthful steps. The risk of climate change due to GHG emissions is like the loss of the earth's ozone and chlorofluorocarbons. The science detected, studied, and conclusively linked the depletion of ozone with humankind's use of chlorofluorocarbons. The world responded with substitutes and moved on. Regulation was the key tool in the case of preserving the earth's ozone, while the precise tool for climate response is yet to be determined. It is noteworthy that we found substitutes for chlorofluorocarbons, just as we will need to find substitutes for fossil fuels.

Paris was a starting point, and now the period of implementation begins. The accord proposes a Green Climate Fund of US$100 billion a year to underwrite the global consequences of climate change—an expensive Band-Aid while policy addresses the root cause. Unfortunately, there is a high likelihood that the aspiration alone of the UNFCCC won't translate into limiting and subsequently reducing atmospheric GHG levels. Stronger policies will need to be ready to ensure that we act, and the carbon tax is the most conspicuous.

There is a limit to our ability to physically adapt as the world around us changes, and the projected pace of climate change is too fast to avoid mass extinctions. There are many sources of anthropological emissions of greenhouse gases, but none contributes more to climate change than the energetic use of fossil fuels. If we are to evade our demise, our choice and use of energy need to be altered abruptly.

Notes

1. REN21 Renewable Energy Policy Network for the 21st Century, *Renewables 2015 Global Status Report*.
2. National Research Council, *Hidden Costs of Energy: Unpriced Consequences of Energy Production and Use* (Washington, D.C.: National Academies Press, 2010).
3. U.S. Department of Energy, "Energy Policy Act of 2005," PsN:PUBL058, August 8, 2005.
4. International Energy Agency, "Energy Subsidies," January 1, 2013, accessed April 16, 2015, http://www.worldenergyoutlook.org/resources/energysubsidies/.
5. Shelagh Whitley, *Time to Change the Game: Fossil Fuel Subsidies and Climate* (London: Overseas Development Institute, November 2013), accessed November 1, 2014, https://www.odi.org/sites/odi.org.uk/files/odi-assets/publications-opinion-files/8668.pdf.
6. International Monetary Fund, "IMF and Reforming Energy Subsidies," May 1, 2015, accessed June 9, 2015, http://www.imf.org/external/np/fad/subsidies/.

30

Four Repressive Forces on the Use of Fossil Fuels

There is nothing more deceptive than an obvious fact.
—Arthur Conan Doyle

After examining the energy plans for a representative group of countries, we find that both the interest and urgency to adopt alternative sources range from high to why. Just as annual solar insolation depends on where you live, the activation energy that will spark an energy transition depends on the aggregation of forces, and they vary a lot.

Imagine yourself at the Kentucky Derby; the horn sounds, the gates open, but only a few of the horses are off and running to win. Others canter forward, feeling a slight pressure to run, and frustratingly to the spectators, a final group of horses refuses to step out of the gate. Examining country energy plans is a bit like watching that race from the stands. A few countries are in a headlong sprint to the zero fossil fuel finish line, a time when fossil fuels are no longer used as a source of energy. Others are proceeding with just a dash of urgency, not trying to win but still running. They are setting alternative capacity goals but are satisfied with a mix including fossil fuels. Perplexingly the remaining appear to have no interest in the race or its objective, even though sooner or later all the horses will need to make the trek to reach the paddock and there are clear rewards for getting there early.

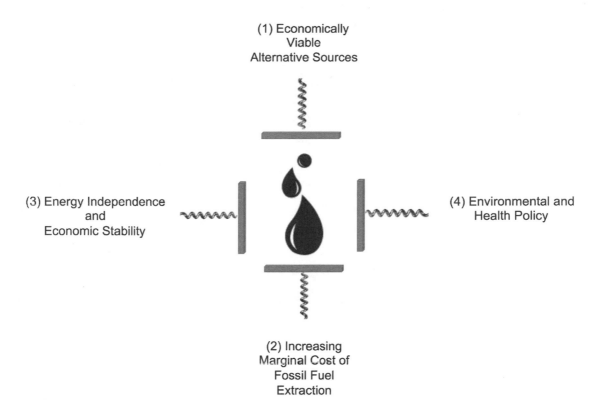

Figure 30.1. Twenty-first-century forces driving the decline of fossil fuels.

The depletion window for accessible fossil fuels is near, and alternate energy sources have the potential and technologies to power our future. Certainly the technology will improve, as will the economics, but perfection can get in the way of transitioning. What most countries seem to lack is a planning horizon that peers sufficiently far ahead to see the implications of the forces at play. Many countries believe that fossil fuels can last a lot longer, while others think that the problem to solve is climate change as the world continues to extract and burn fossil fuels. Only a select few countries are moving to place their entire dependence on alternate sources and advance their long-term national security through energy independence. Fossil fuels have been so convenient and plentiful; it's just too bad they're undeniably limited, unevenly distributed, and wastefully consumed in a manner harmful to our health and our environment.

"Four Forces" Have Fossil Fuels in a Vise Grip

Only a select few countries are moving to place their entire dependence on alternate sources and advance their long-term national security through energy independence.

Four forces are bearing down on the energetic use of fossil fuels. Everywhere these forces will exert steady repressive pressure on the continued use of fossil fuels: first, the emergence of cost-competitive renewable technologies; second, the increasing marginal costs of fossil fuel production; third, country energy plans that establish self-sufficiency goals for importers and economic transitioning goals for exporters; and fourth, environmental and health policies incompatible with fossil fuels (Figure 30.1). The intensity of any one of these forces and their collective strength depend on where you live. Much like the telltale signs of the seven signals from chapter 1, these four forces are converging, and recognizing this will allow us to drive behavior consistent with the complete cessation of fossil fuel use by the end of the century.

As we have seen in the technology section, the marketplace has already delivered economically attractive alternatives. The fast pace of innovation is in its early phase and set to deliver even more solutions. Nothing is static in the energy sector; renewable technologies are rapidly becoming more efficient and less expensive, while fossil fuel costs will increase as low-cost fields give way to higher-cost projects. Disruptive technologies are breaking the monopolistic hold of fossil fuels. The declining costs of alternative energy will for the first time since the dawn of the petroleum era place a limit on the acceptable costs of extracting fossil fuels. Those pursuing higher-cost projects will have to come to terms with the real possibility that they may never extract the reserve and that lost jobs may not return. Absent meaningful alternative capacity, though, the consumer pricing for energy staples will become erratic. Inflation-adjusted gasoline prices in the United States have had six instances since 1974 where the year-to-year price change was greater than 20%. There were no such instances prior to 1974 dating all the way back to 1918.[1] The resource is in the ground, but more advanced and expensive extraction technologies are needed to get it above ground. The production of fossil fuels will just get more expensive and contentious throughout the remainder of the century.

The desire for energy self-sufficiency relates to the disproportionate distribution of fossil fuels around the globe, long a source of regional tensions and insecurity for importers. Affordable access to energy is considered a fundamental right and passageway to development. Energy is unlike other industries such as electronics or automobiles in that the ideal most countries strive for is self-sufficiency, and that is counter to normal marketplace principles. The fossil fuel export trade has been the privilege of just a few select countries. In contrast, renewable energies for most of the world will permit domestic energy production that bestows national and economic security. The combination of viable alternative technology and the inherent desire everywhere to be self-sufficient will tip the scales in favor of domestic solutions. This poses a bigger challenge for the major exporters of the world: how to transition their economies to a post–fossil fuel era. Retooling an economy to attract new types of jobs and build the necessary skills is difficult, and it will take time. We've seen that Norway, for example, has established an $860 billion post–fossil fuel fund to support the inevitable adaptation of its economic base.

The fourth force, safeguarding environmental and human health, is also placing a limit on our consumption of fossil fuels. In addition to their prominent role in creating greenhouse emissions, fossil fuels are also a large source of air pollutants. The concern is the cumulative damage to human health and environmental health from the extraction, development, and use of fossil fuels. The health impacts are somewhat mitigated by the regulation of pollutants, but one can easily conclude that without economic forces things won't change fast enough on the climate front. Carbon trading systems seem to be convoluted schemes that are overly complicated and ineffective where implemented. A carbon tax, the most direct economic tool, may remain just a good idea due to high resistance and lack of political courage. But there are hopeful signs that our faulty economic lens is getting an adjustment. More than 400 firms around the world are setting internal prices on carbon emissions. These prices range from $6.00/tonne for Microsoft to $60/tonne for ExxonMobil.[2] These firms apply these prices internally to value projects and help make long-term investment decisions. The pricing reflects a belief that carbon abatement will ultimately be a necessary condition to operate with fossil fuels.

The forces stemming from environmental and human health concerns will also increase as evidence unfortunately builds and conditions deteriorate. Opening access to environmentally sensitive areas will be contested, and with each debate all parties will be scarred and battered, but finally the oil and gas industry will be less determined. But even as developed countries are capping CO_2 levels and in some cases decreasing emissions, the developing world will continue to increase the global use of fossil fuels, and pollution will only intensify.

When we began our inquiry, the initial concern was the physical depletion of fossil fuels. What will happen is vastly different. The four forces described here will render fossil fuels depleted in an economic sense rather than a physical sense. A strong impetus for change will be the continued commercialization of cost-competitive renewable technologies and each nation's desire for self-sufficiency. When the pressure hits a certain point the transition will move quickly, a natural micro-economic phenomenon resulting from the convergence of these forces. There remains a high risk, though, that policy will fail to adapt or even recognize the signals. Waiting too long will amplify the social, environmental, and economic disruptions that are certain to occur if we overly rely on fossil fuels to feed a growing population and accelerating per capita energy demand. Failing to internalize the negative impacts on health and climate in the cost of fossil fuel energy is the largest obstacle to changing energy sourcing and improving health and environmental conditions. This man-made economic failure is inhibiting our positive response and undermining our future. The sooner we correct this distortion, the faster we will decarbonize our energy mix and reap the benefits.

If we lacked viable alternative technology, this would be a distressing realization, and it might be understandable that we would choose to ignore the dangers with the hope that another generation would figure something out. But we have learned that this isn't the case. We do have options, and we can do something about it now. We should be celebrating the arrival of these innovations rather than diminishing their importance. As with removing a Band-Aid, the slower the process, the more painful. A fast transition is appealing on all levels! Unfortunately, the use of fossil fuels is projected to increase over the coming decades, even as the forces for decline are exerting terminal pressures. Developed countries that lack fossil fuel reserves will be the first to replace fossil fuels, starting with the power and transportation sectors. Once this happens the pace of fossil fuel decline will accelerate as countries around the globe convert to noncarbon sources, and we now have a good idea of which countries will lead and which will be caught lagging behind.

Notes

1. Tim McMahon, "Inflation Adjusted Gasoline Prices," InflationData.com, accessed April 16, 2016, http://inflationdata.com/articles/inflation-adjusted-prices /inflation-adjusted-gasoline-prices/.
2. "Carbon Copy," *Economist*, December 14, 2013, accessed September 22, 2015, http://www.economist.com/news/business/21591601-some-firms-are-preparing -carbon-price-would-make-big-difference-carbon-copy.

31

The Only Path Forward Departs from the Use of Fossil Fuels

"Hope" is the thing with feathers—/that perches in the soul—
—Emily Dickinson

There are many reasons to be both hopeful and demoralized. Let's go with hopeful. We have examples of small communities, large cities, and entire countries navigating a wholesale change in their energy sourcing, displacing the use of fossil fuels. For the rest of the world these early movers will demonstrate that the impossible is possible. We've wrapped our arms around a lot of material, and now it's time to summarize the key observations and offer a few opinions.

Rediscovering Ancient Technologies

Renewable technologies are far older than petroleum technologies, dating back thousands of years. The development of windmills and tide mills in the ninth century is not surprising, but the discovery of the 2,000-year-old Parthian Battery and the knowledge that 200 years ago a French scientist documented the photovoltaic effect remind us that renewable technologies have long been part of our scientific knowledge base and that only the recent mass extraction of ready-to-use and transportable fossil fuels caused them to recede in relevance. Now these old discoveries are central to the development of new solutions. To a certain extent fossil fuels took us on a brief detour, and we are getting back on course.

So slowly—but relentlessly—is the age of fossil fuels ending that it is difficult to detect. Over the course of this century, alternative sources will be superior in cost and viewed as better for our environment. The greatest challenge for energy planning is overcoming political and economic shortsightedness and corporate special interests that obscure the obvious. When we acknowledge the inevitable, it will mark the end of one era and the dawn of another. A grand plan for the next period of energy sourcing can only be grasped with a longer view. The dead ends and wasted opportunities of plans designed with decades-long thinking mired in the continued use of fossil fuels can be disastrous. The consequence is that price increases associated with fossil fuels will happen too fast for an orderly adjustment to occur. Most national energy plans seem to hold onto the security of fossil fuels, refusing to draw up a plan absent their contribution. The reality is that fossil fuels as a backbone of energy will be a small tick in our long history. The seven signals from chapter 1 portray a world in an energy flux, and projecting the recent pace of change into the future will understate the real shift that lies in store.

In the future, non-carbon-related jobs in the energy sector will be more stable than jobs directly tied to fossil fuels.

Job Creation

The International Renewable Energy Agency calculates that in 2015 there were 7.7 million renewable energy jobs worldwide, an 18% increase from the prior year.[1] Solar PV, with 2.5 million jobs, leads all renewables, followed by biofuels, hydro, and wind at 1.8 million, 1.5 million, and 1 million, respectively. Five of the top 10 countries with employment in the renewable energy field are located in Asia. The agency also projects that renewable capacity will double by 2030, and higher employment will follow.

In the United States, between 2008 and 2012, direct employment in the energy field was shifting in large numbers. Coal, for instance, lost nearly 50,000 jobs, while gas and renewables more than made up for that loss, reporting gains of 94,000 and 79,000, respectively.[2] Although jobs in gas have increased, the volatility in this sector leaves an uneasy feeling in those lucky to have jobs. Behind the numbers, communities reliant on coal are facing a downward spiral in employment opportunities absent adaptation to new opportunities in the energy field or beyond. Whether or not we can realize "clean coal" will determine the short-term fate of the jobs in these communities. It is unlikely despite best efforts.

The good news is that there are plenty of opportunities for energy employment in our near future. The power grids need to be extended and upgraded with smart electronics. Major home appliances need new features that support smart grid integration. Advanced storage technologies, including batteries and hydrogen, will mean manufacturing and installation jobs across several large customer segments. The renewable technologies themselves will create jobs in raw material sourcing and recycling, manufacturing, and installation and service. Conservation and efficiency measures will also create jobs in the development and manufacture of new construction materials and tools that reduce energy intensity. Looking into the future, overall non-carbon-related jobs in the energy sector will be more stable than those directly tied to fossil fuels.

Nuclear power can be a perfect complement to temporal energy sources, a desirable baseload with inherent storage. The risk profiles of fossil fuel and nuclear energy have flipped in the last few decades.

Domestic Energy Helps Global Security

A small share of the world owns the fossil fuel wealth, while the potential to gather renewable sources gives most countries an opportunity to gain a measure of energy independence, dramatically improving prospects for global stability. Energy is nearly as fundamental to a nation as food and water, and self-reliance is a healthy and natural instinct. Access to energy supplies can be a cause for conflicts with limited fuels, while it can become a stabilizing force with domestically available renewable sources. Claiming that opening access to restricted lands for domestic drilling is in our national security interests is partially true, but we have superior choices.

Nuclear Power Needs a New Introduction

Nuclear power has largely receded from public discussion, and yet the fraternity of countries practicing civil nuclear power is expanding. The United States made tremendous strides in leading the development of the fast reactor technology that addresses liabilities with today's thermal-neutron designs. For reasons likely related to the lack of political will, the United States severely cut funding decades ago, but China, Russia, and others have continued the work of validating and commercializing this technology. Specifically, Russia has already built and commissioned the BN-800 fast reactor.

Nuclear power has lacked a voice to educate the public on the value of these new capabilities. Instead we continue to operate more than 400 dated thermal-neutron reactors around the world that continue to unnerve the public, while many countries are building and planning capability that portends an expanded role for nuclear energy by the end of the twenty-first century. Alternatively, a closed fuel cycle incorporated with a fast reactor is a means to consume the highly dangerous long-lasting radioactive waste from today's plants, decarbonize our atmosphere, and simultaneously provide a fuel source for hundreds of years. There is a way forward. The future of nuclear power shouldn't be condemned by the mistakes of the past, which are comparatively few when matched against the hundreds of toxic incidents related to fossil fuels. We should learn from them and allow these innovations to create new choices to power the world. It is essential that fast reactor technology become the standard fission solution for nuclear power to play a safe and relevant role. The status quo of nuclear energy is unacceptable, and so is that of fossil fuel. But nuclear has options.

Strategically, nuclear power can be a perfect complement to the temporals, a desirable baseload with inherent storage. We may hesitate to consider it, but we should know that the risk profiles of fossil fuel and nuclear have flipped in the last few decades. The accumulation of negative health and environmental evidence with the use of fossil fuels has increased, while closed fuel cycles and fast reactor technology have lowered nuclear risks.

Building Our Energy Infrastructure Is Like Building the Internet

The path to large-scale adoption of temporals is blocked by underinvestment in necessary support technologies and infrastructure, and this is less a question of money and more a statement of priorities. Without the empowerment of the Internet, our smart devices wouldn't be so smart, and we wouldn't be so interested. We need to break the storage barrier. The challenge isn't the cost-effective gathering from alternative sources but, rather, the accumulation of energy to satiate round-the-clock demand. Increasing contributions from temporals can only come with improvements in electrical grids and economical storage and retrieval solutions.

Safeguarding the Environment

Efforts to extract more fossil fuels are environmentally destructive. The economic pressures are not in place, and regulatory pressures alone won't provide the necessary protection; the world needs alternative energies and increased economic pressure to counter the forces of inertia that increase the world's fossil fuel dependency. Take the case of the potential to reduce CO_2 emissions through the implementation of carbon capture and sequestration technology or coal's integrated gasification combined cycle with precapture. Neither economic nor regulatory pressures have caused industry to implement these capabilities, even though we know how (CO_2 capture). If one wants to argue that CCS technology isn't up to the mark, that position should be qualified to be fair. Stating that CCS technology isn't ready really means that neither the strength of the economics nor the muscle of regulatory pressures is sufficient to make the case for investment. Recall the CCS installation in Canada's Saskatchewan province that captures 90% of CO_2 and 93% of sulfur oxide. The technology exists, but in this case it was economics that drove implementation. Saskatchewan's carbon capture installation underscores that most often economics, not technology, prevent the adequate protection of the environment. Absent strong economic and appropriate regulatory pressures, the environment that supported the flourishing of the human species is exposed to further harm.

Climate policies that marginally lower emissions but allow us to remain fundamentally fossil fuel–dependent work in opposition to policies that seek to replace fossil fuels altogether. Reversing the atmospheric concentration of GHGs is critically important, and the fastest way to achieve this is by replacing the use of fossil fuels with noncarbon sources. Try as we may, the universal capture and sequestration of CO_2 from fossil fuel sources is elusive and expensive. More helpful are environmental regulations that protect sensitive environments from petroleum extraction access, most notably frontier locations. Challenging petroleum at the source will prove much more effective than climate policy. It helps us all by focusing our collective energies on longer-term solutions.

Properly Valuing the Benefits of Decentralization

Today our power paradigm is based on centralized generation delivered through millions of miles of transmission and distribution lines to billions of points. On the other extreme we have one billion motor vehicles, each with its own carbon power plant, representing the essence of decentralization. Transportation will remain decentralized, and our vehicle's power plant will change from gas to alternatives. But what becomes of the power sector? A major freedom offered by renewable energy is the capability to scale the technology, from large centralized operations to small distributed installations. Long-distance transmission of electricity loses energy, so local production is better wherever possible. Inherently, most renewables can be decentralized, and that will force a new type of power grid, a system of micro-grids that manage local demand and supply like individual neurons in our brain that collectively provide function. In a decentralized model, individual communities, one by one, are in a position to become energy self-sufficient. Renewables allow resource-poor countries to access affordable energy and are likely the paths to first-time access for insular communities. We are already seeing the use of the term *net-zero* for homes and businesses designed to stand alone energetically. To remain wedded to today's centralized model will limit the advantages of the renewable technologies. Localities around the world should draw up and execute a blueprint for energy self-sufficiency. Every country studied covets energy security, and no other energy source can deliver that security for the long term more than domestic renewables. Today's centralized power architecture was determined by Tesla's alternating current equipment and fossil fuels. Now distributed energy sourcing will naturally result in decentralization. Adding force behind this trend is the evolution of hydrogen-fueled micro—combined heat and power systems.

The generation of electricity will no longer be the exclusive domain of centralized power plants.

A Vote for a Worldwide Carbon Tax

Wrongfully, fossil fuel pricing does not comprehend significant and traceable costs related to health and the environment. Consequently, countries everywhere would be wise to implement a progressive carbon tax to foster better decision making. A worldwide carbon tax is the ideal policy tool, and realistically a group of influential countries will need to exercise leadership and step out ahead for this to happen. The review of all the policy inventions raises the concern that we may be outsmarting ourselves. We are creating administrative burdens and cross-purpose policies that seem far from simple and straightforward. Furthermore, the *International Energy Outlook* report suggests that these policies aren't working. We have seen examples of wealthy countries subsidizing domestic energies resulting in wastefully high energy intensities. A tax can help conservation and efficiency efforts, but it can also properly direct long-term capital investments. It would take a special brand of leadership to propose a carbon tax, but smart grids, hydrogen infrastructure, and alternative energy fueling stations all require funding and investment. Lack of timely investment is blocking higher capacities of temporals. Let the market respond to a true and transparent energy cost when investing in the energy sector, and good things will happen.

Noblesse Oblige

Convening task forces on climate change is addressing the symptoms of our current energy choice, not leading change. Most parts of the world don't have the technological resources, only the need for what just a few countries can develop. The developed countries, then, have an obligation to lead the energy transition on behalf of the underdeveloped. They have the resources, they have the benefit of full access to energy, and they have the moral obligation, having been responsible for a large majority of the GHGs already in the atmosphere.

Adopting the Right Energy Mix

Even in a post—fossil fuel world, there is no way to hide the effects of unchecked energy consumption. The disproportionate and excessive use of energy we see today will need to give way to a more prudent use of energy in the future. Our review of renewable technologies reveals conclusively that the practical potential of all renewable energies other than solar can deliver only a portion of the world's projected energy requirements. We will need a substantial capacity from a solar and nuclear mix to satisfy tomorrow's energy demands without continuing to bend nature in unsustainable ways. The right price of energy can reduce our energy intensity without undermining our quality of life. Every action has an equal and opposite reaction, and that is true of energy. Every source exerts an equal and opposite toll for deployment.

Early hydroelectric projects didn't comprehend, or at least grossly underestimated, their toll. Being classified as a green technology shouldn't obviate the need to assess a potential toll. Hard choices and trade-offs will need to happen. Power densities are vastly lower with renewable sources, and some technologies such as concentrated solar power require an exclusive use of large tracts of land. Comparing a natural gas power plant with a solar PV plant, the former has power densities of 1,000 W/m^2, while solar PV's is just 9 W/m^2. Thus, at a utility scale, the land use required for solar PV is two magnitudes that of natural gas. The good news is that solar PV can capture energy in multiuse areas such as parking lots, rooftops, and hopefully even roadways. Utilities should extend their definition of a power plant to include distributed sources. We should prioritize multipurpose land use as much as possible. Building the future's energy mix will force the world to make value judgments as to the risk of nuclear, surface use for renewables, and our energy intensity.

The Future's Renewable Fuel

We will always need to accumulate energy for supply risk mitigation. Electricity is the ultimate form of transportable energy, but it is instantly fleeting; we still need a fuel. A new-age fuel will be a fundamental element of our new energy structure. The main criticism of temporals is overcome with an integrated plan to store, transport, and retrieve energy as required. Hydrogen becomes a perfect complement to renewables, and unlike fossil fuels, the hydrogen cycle is reversible and sustainable. There is no doubt that batteries will play a big role in the power and transportation sectors, but there is also no avoiding the need for a renewable fuel. High-temperature industrial processes operating at thousands of degrees will require a fuel, and the hydrogen cycle discovered in the nineteenth century provides the scientific basis for the future's renewable fuel.

Biofuels Will Go the Way of the Combustion Engine

Biofuel's relevance is intertwined with the use of the internal combustion engine, which will decline in use. Biofuels are the least efficient method of harvesting solar energy because they have the lowest power densities of the renewables. They have been a low-investment means to reduce dependence on petroleum but an expensive use of land and food resources. The United States and Brazil produce most of today's biofuels. We should de-emphasize the use of biofuels as other renewables come on line, but not until then.

A Final Message of Urgency and Hope

We can be thankful that millions of years ago the earth incubated such a convenient and massive amount of energy, but we cannot defer building alternative capacities and thus deliberately rob future generations of the adequate time to transition. The lead time to build an energy network utilizing the benefits of renewables is half a century. Let's not wait 50 years to start. The seven signals from chapter 1 tell us that it is time to act now.

Disproportionate investment directed toward fossil fuels is wasting time, money, and inventive effort and missing the opportunity for job growth in the Energy Renaissance. We shouldn't take the bait and prioritize further petroleum for extraction but, rather, protect those remaining areas considered too environmentally sensitive or valuable to withstand extraction. We must place our top-priority investments behind solutions for an eternity as opposed to decades.

A truth has emerged from this journey, and it is related to inertia. The obstacle to a major energy transformation is a mind-set masked in economics, not technological challenges. To linger between fossil fuel dependency and a promising but unrealized alternative energy mix is dangerous. While some petroleum companies recognize the future and are planning to adapt to remain relevant, others are dangerously binding us to an end-stage energy source. We are fortunate that innovators are casting their creative minds to new ways in which to harvest energy.

The "four forces" discussed in chapter 30 will collectively drive petroleum demand into the ground. Like geothermal hot spots, there are new-age renewable hot spots. They occur where access to fossil fuels is expensive and consumption is high, where longer-term planning horizons are used, and generally wherever the collective pressures of the four forces are the greatest. The complete transition away from fossil fuels isn't a theory for these hot spots; it is a necessary move to secure a stable future. How these communities overcome inertia, redirect investments, employ technology, and adapt behavior will be exciting to follow.

Notes

1. International Renewable Energy Agency, "Renewable Energy Employs 7.7 Million People Worldwide, Says New IRENA Report," press release, May 19, 2015, accessed June 4, 2015, http://www.irena.org/News/Description .aspx?NType=A&mnu=cat&PriMenuID=16&CatID=84&News__ID=407.
2. Katherine Tweed, "Coal Loses Nearly 50,000 Jobs, Wind and Solar Add 79,000," Greentech Media, April 3, 2015, accessed April 16, 2015, http://www.greentechmedia .com/articles/read/coal-loses-nearly-50000-jobs-wind-and-solar-add-79000.

Appendix A

PRIMARY ENERGY CONSUMPTION FACTOR
AND FOSSIL FUEL WEALTH FACTOR

Tables A.1–A.4 group countries according to their Primary Energy Consumption Factors and Fossil Fuel Wealth Factors. The groups are organized by countries above and below the worldwide average for energy consumption and those above and below the worldwide average for fossil fuel reserves. Countries that have PECFs above 1 and FFWFs below 1 are categorized as "vulnerable" (Table A.1). Absent domestic alternative sources, this group's high energy consumption and import dependence pose significant national security risks. Countries grouped under "dually challenged" have less and use less than the world averages (Table A.2). Most countries fall into this category, and where the lack of fossil fuels has arguably held them back, alternatives can unleash their development. A select group, classified as "fortunate," has been blessed with more than their share of energy supplies, and their consumption reflects this advantage (Table A.3). Finally, there are just a handful of countries referred to as "underconsumers" (Table A.4); they have more than their share of supplies, but this hasn't translated into higher energy consumption. The tables are intended to help in understanding a country's energy environment and how it might approach the future.

Table A.1. Fossil Fuel Wealth Factor and Primary Energy Consumption Factors for "Vulnerable" Countries.

COUNTRY	PRIMARY ENERGY CONSUMPTION FACTORS (>1)	FOSSIL FUEL WEALTH FACTOR (<1)
Bermuda	1.7	0.0
Antigua and Barbuda	1.2	0.0
Argentina	1.2	0.2
Aruba	1.8	0.0
Bahamas	1.7	0.0
Cayman Islands	1.8	0.0
Chile	1.1	0.1
Puerto Rico	1.4	0.0
Virgin Islands, U.S.	24.4	0.0
Austria	2.4	0.2
Belgium	3.2	0.0
Croatia	1.1	0.1
Cyprus	1.4	0.0
Czech Republic	2.1	0.5
Denmark	1.9	0.2
Faroe Islands	3.1	0.0
Finland	3.1	0.0
France	2.2	0.0
Hungary	1.4	0.8
Iceland	9.0	0.0
Ireland	1.8	0.0
Italy	1.7	0.0
Luxembourg	4.8	0.0

Malta	3.0	0.0
Netherlands	3.4	0.6
Poland	1.5	0.7
Portugal	1.4	0.0
Romania	1.0	0.1
Slovakia	1.9	0.2
Slovenia	2.0	0.5
Spain	1.8	0.1
Sweden	3.1	0.0
Switzerland	2.1	0.0
United Kingdom	1.8	0.1
Armenia	1.0	0.2
Belarus	1.7	0.1
Latvia	1.1	0.0
Lithuania	1.3	0.0
Bahrain	5.6	0.6
Israel	1.7	0.2
Seychelles	2.5	0.0
American Samoa	1.5	0.0
China	1.0	0.4
Guam	2.3	0.0
Japan	2.3	0.0
Malaysia	1.4	0.7
New Caledonia	2.2	0.0
New Zealand	2.7	0.7
Singapore	7.3	0.0
South Korea	3.1	0.0
Taiwan	2.8	0.0

Note: Countries in bold practice civilian nuclear power.

Table A.2. Fossil Fuel Wealth Factors and Primary Energy Consumption Factors for "Dually Challenged" Countries.

COUNTRY	PRIMARY ENERGY CONSUMPTION FACTOR (<1)	FOSSIL FUEL WEALTH FACTOR (<1)
Mexico	0.9	0.2
Barbados	0.9	0.0
Belize	0.5	0.0
Bolivia	0.3	0.6
Brazil	0.8	0.2
Colombia	0.4	0.7
Costa Rica	0.5	0.0
Cuba	0.5	0.1
Dominica	0.4	0.0
Dominican Republic	0.4	0.0
Ecuador	0.5	0.5
El Salvador	0.3	0.0
Grenada	0.5	0.0
Guatemala	0.2	0.0
Guyana	0.4	0.0
Haiti	0.0	0.0
Honduras	0.2	0.0
Jamaica	0.6	0.0
Nicaragua	0.2	0.0
Panama	0.9	0.0
Paraguay	0.9	0.0
Peru	0.4	0.1
Saint Lucia	0.5	0.0
Saint Vincent Grenadines	0.4	0.0
Suriname	0.9	0.3

Turks and Caicos Islands	0.8	0.0
Uruguay	0.8	0.0
Macedonia	0.8	0.7
Turkey	0.9	0.5
Estonia	0.8	0.0
Georgia	0.5	0.2
Kyrgyzstan	0.6	0.7
Moldova	0.5	0.0
Tajikistan	0.3	0.2
Uzbekistan	1.0	0.8
Jordan	0.6	0.0
Lebanon	0.9	0.0
Syria	0.6	0.2
Yemen	0.2	0.3
Angola	0.2	0.6
Benin	0.1	0.0
Botswana	0.5	0.1
Burkina Faso	0.0	0.0
Burundi	0.0	0.0
Cameroon	0.1	0.1
Cape Verde	0.2	0.0
Central African Republic	0.0	0.0
Chad	0.0	0.1
Comoros	0.0	0.0
Cote d'Ivoire (Ivory Coast)	0.1	0.0
Djibouti	0.3	0.0
Egypt	0.6	0.3
Eritrea	0.0	0.0

Ethiopia	0.0	0.0
Gambia	0.0	0.0
Ghana	0.1	0.0
Guinea	0.0	0.0
Guinea-Bissau	0.0	0.0
Kenya	0.1	0.0
Lesotho	0.1	0.0
Liberia	0.0	0.0
Madagascar	0.0	0.0
Malawi	0.0	0.0
Mali	0.0	0.0
Mauritania	0.1	0.1
Mauritius	0.7	0.0
Morocco	0.3	0.0
Mozambique	0.1	0.1
Namibia	0.4	0.2
Niger	0.0	0.0
Nigeria	0.1	0.5
Rwanda	0.0	0.0
Sao Tome and Principe	0.1	0.0
Senegal	0.1	0.0
Sierra Leone	0.0	0.0
Somalia	0.0	0.0
Sudan and South Sudan	0.1	0.2
Tanzania	0.0	0.0
Togo	0.0	0.0
Tunisia	0.4	0.1
Uganda	0.0	0.0
Zambia	0.1	0.0

Zimbabwe	0.2	0.2
Afghanistan	0.1	0.0
Bangladesh	0.1	0.0
Bhutan	1.0	0.0
Burma (Myanmar)	0.1	0.0
Cambodia	0.1	0.0
Fiji	0.0	0.0
French Polynesia	0.9	0.0
India	0.3	0.2
Indonesia	0.4	0.6
Kiribati	0.1	0.0
Laos	0.3	0.3
Macau	0.8	0.0
Maldives	0.6	0.0
Mongolia	0.5	0.0
Nepal	0.0	0.0
North Korea	0.5	0.1
Pakistan	0.2	0.1
Philippines	0.2	0.0
Samoa	0.2	0.0
Solomon Islands	0.1	0.0
Sri Lanka	0.2	0.0
Thailand	0.9	0.1
Timor-Leste (East Timor)	0.0	0.0
Tonga	0.3	0.0
Vanuatu	0.1	0.0
Vietnam	0.3	0.0

Note: Countries in bold practice civilian nuclear power.

Table A.3. Fossil Fuel Wealth Factors and Primary Energy Consumption Factors for "Fortunate" Countries.

Country	Primary Energy Consumption Factor (>1)	Fossil Fuel Wealth Factor (>1)
Canada	5.3	7.1
Greenland	2.0	14.6
United States	4.2	3.6
Trinidad and Tobago	9.5	2.9
Venezuela	1.5	9.4
Bosnia and Herzegovina	1.3	3.4
Bulgaria	1.5	1.5
Germany	2.3	2.3
Greece	1.6	1.2
Norway	4.9	4.2
Serbia	1.4	8.5
Kazakhstan	2.1	12.0
Russia	2.9	7.8
Turkmenistan	2.4	10.6
Ukraine	1.6	3.5
Iran	1.7	4.9
Kuwait	6.1	40.1
Oman	3.3	3.5
Qatar	9.4	99.1
Saudi Arabia	4.3	12.7
United Arab Emirates	5.4	17.3
Equatorial Guinea	1.2	2.1
Libya	1.3	10.6
South Africa	1.5	2.6
Australia	3.7	16.0
Brunei	5.0	9.9
Papua New Guinea	1.0	2.4

Note: Countries in bold practice civilian nuclear power.

Table A.4. Fossil Fuel Wealth Factors and Primary Energy Consumption Factors for "Underconsumer" Countries.

Country	Primary Energy Consumption Factor (<1)	Fossil Fuel Wealth Factor (>1)
Albania	0.5	1.4
Montenegro	0.9	1.0
Azerbaijan	0.8	1.5
Iraq	0.6	4.7
Algeria	0.7	1.2
Gabon	0.4	1.5

Appendix B

Table B.1 synthesizes energy platform positions for the Democratic and Republican parties since 1992. Expectedly, there are distinctions between the Democratic and Republican positions, but there are also several similarities. Themes include energy independence, endorsement of R&D for renewables, support for energy efficiency, and the promotion of environmentally sensitive exploration. Both parties have particularly stressed self-sufficiency since the terrorist acts of 9/11.

Generally, the Republican Party supports expanded access for fossil fuel exploration and development; cleaner technology for oil, gas, and coal; and reduced but environmentally responsible regulations. The Republicans over this period have generally supported nuclear power as an option for the future. Overall the Republican Party is more fossil fuel–centric. The Democratic Party supports strong environmental and health policy. Democrats further prioritize the development of renewable energy, as opposed to granting access for petroleum development. The Democratic Party's top priority seems to be environmental, where renewables have an inherent positive impact. President Obama, in 2008, set a renewable goal of 25% by 2025. He promised billions to support renewables research, creating up to five million jobs. The 2016 Party Platforms carry forward the same distinctions. The Republican Party wants a less intrusive federal government which means shifting the regulation of oil and gas development to the states and a philosophy that private not federal capital support energy investments. The Republican Party further wants to invest in nuclear research (i.e., thorium reaction) and technologies for the storage of electricity deemed "a breakthrough of extraordinary import." The Democratic Party more so than ever before is holding climate change as the challenge of our time and the single rationale for rapidly shifting away from the use of fossil fuels. The Democratic Platform has called for a price on greenhouse emissions and the continuation of responsible regulation to protect our health and the environment. Building on President Obama's goals, the Democratic Platform commits to reducing greenhouse gas emissions 80% below 2005 levels by 2050.

The distinctions above are highly charged and unfortunately polarizing. Absent alternatives and with our large energy diet, we are consuming more and more fossil fuels. Renewables provide the same percentage of our total energy demands as they did 30 years ago.

Table B.1. U.S. Presidential Party Platforms on Energy.

Nominee	Energy Platform Position
1992 Governor Clinton (D)	Energy Efficiency and Sustainable Development • Increase efficiency, reduce foreign dependence, and produce less toxic waste • Encourage alternative fuel vehicles • Increase reliance on natural gas • Promote clean coal technology • Invest R&D for renewables
1992 President Bush (R)	Power for Progress • Reduce regulation to foster domestic development of energy resources • Allow access to the coastal plains of the Arctic National Wildlife Refuge • Environmental policy should not overly restrict responsible exploration • Endorse the Superconducting Super Collider project • Hasten next-generation nuclear power plants • Endorse renewable energy sources
1996 President Clinton (D)	No energy policy in the platform

1996 Senator Dole (R)	Power for Progress

- Reaffirm position on deregulating to spur domestic development, reducing foreign sources while creating jobs
- Elimination of the Department of Energy to emphasize the need for greater privatization
- Advocate access to the Arctic National Wildlife Refuge
- Support and encourage the development of our domestic natural gas industry
- Encourage research for cleaner coal combustion technologies
- Foster alternative and renewable energy sources to assist in reducing dependence on unreliable foreign oil supplies
- Continue commitment to addressing global climate change in a prudent and effective manner that does not punish the U.S. economy

2000 Vice President Gore (D)	No specific energy platform position; platform included a section, "Protecting Our Environment," that spoke to climate change

2000 Governor Bush (R)	Energy

- Department of Energy has utterly failed in its mission to safeguard America's energy security
- Federal Energy Regulatory Commission has been no better, and the Environmental Protection Agency has been shutting off America's energy pipeline with a regulatory blitz
- Affordable energy, the result of Republican policies in the 1980s, helped create the New Economy
- Increase domestic supplies of coal, oil, and natural gas
- Improve federal oil and gas lease permit processing
- Provide tax incentives for production
- Promote environmentally responsible exploration
- Advance clean coal technology
- Maintain the ethanol tax credit
- Provide a tax incentive for residential use of solar power

2004 Senator
Kerry (D)

Achieving Energy Independence
- End America's dependence on Mideast oil
- Investments to harness the natural world around us—the sun, wind, water, geothermal, and biomass sources and a rich array of crops
- Support creating more energy-efficient vehicles, from today's hybrid cars to tomorrow's hydrogen cars
- Fund research to overcome the obstacles to hydrogen fuel and continue our other efforts to achieve energy independence
- Support balanced development of domestic oil supplies in areas already open for exploration
- Invest billions to develop and implement new, cleaner coal technology and to produce electric and hydrogen power
- Our commitment to conservation

2004 President
Bush (R)

Ensuring an Affordable, Reliable, More Independent Energy Supply
- Commit $2 billion over 10 years for clean coal research and development
- Commit $1.7 billion over five years to begin building hydrogen cars and the infrastructure to support them
- Strongly support removing unnecessary barriers to domestic natural gas production and expanding environmentally sound production in new areas, such as Alaska and the Rocky Mountains
- Support measures to modernize the nation's electricity grid
- Support renewable energy through extension of the production tax credit for wind and biomass
- Believe that nuclear power can help reduce our dependence on foreign energy and play an invaluable role in addressing global climate change

2008 Senator
Obama (D)

New American Energy
- Investment of billions of dollars over the next 10 years for green energy sector and create up to five million jobs
- Invest in research and development and deployment of renewable energy technologies—such as solar, wind, and geothermal, as well as technologies to store energy

- Dramatically increase the fuel efficiency of automobiles
- Commit to getting at least 25% of our electricity from renewable sources by 2025
- Direct the Federal Trade Commission and Department of Justice to vigorously investigate and prosecute market manipulation in oil futures

2008 Senator McCain (R)	Energy Independence and Security

- Increase our production of American-made energy and reduce our excessive reliance on foreign oil
- Simply must draw more American oil from American soil
- Support accelerated exploration, drilling, and development
- Encourage refinery construction and modernization and, with sensitivity to environmental concerns, an expedited permitting process
- Pursue dramatic increases in the use of all forms of safe, affordable, reliable—and clean—nuclear power
- Advocate a long-term energy tax credit equally applicable to all renewable power sources
- Look to innovative technology to transform America's coal supplies into clean fuels capable of powering motor vehicles and aircraft
- Extract more natural gas and do a better job of distributing it nationwide to cook our food and heat our homes
- Increase conservation through greater efficiency

2012 President Obama (D)	No updated energy platform

2012 Governor Romney (R)	Domestic Energy Independence: An "All of the Above" Energy Policy

- Let the free market and the public's preferences determine industry outcomes
- Ensure an affordable, stable, and reliable energy supply
- Encourage research and development of advanced technologies in this sector, including coal-to-liquid, coal gasification, and related technologies for enhanced oil recovery

- America's oil and natural gas reserves indicate an incredible bounty for the use of many generations to come
- Reasoned approach to all energy development
- Committed to approving the Keystone XL Pipeline and to streamlining permitting for the development of other oil and natural gas pipelines
- Timely processing of new nuclear reactor applications currently pending at the Nuclear Regulatory Commission
- Encourage the cost-effective development of renewable energy, but the taxpayers should not serve as venture capitalists

2016 Secretary Hillary Clinton (D)

Combat Climate Change, Build a Clean Energy Economy, and Secure Environmental Justice

- Committed to national mobilization and to leading a global effort to mobilize nations to address the climate change threat on a scale not seen since World War II
- Reduce greenhouse gas emissions more than 80% below 2005 levels by 2050
- Acquire 50% of electricity from clean energy sources within a decade
- Transform American transportation by reducing oil consumption through cleaner fuels and vehicle electrification, increasing the fuel efficiency of cars, boilers, ships, and trucks
- Eliminate special tax breaks and subsidies for fossil fuel companies as well as defending and extending tax incentives for energy efficiency and clean energy
- Carbon dioxide, methane, and other greenhouse gases should be priced to reflect their negative externalities
- Implement and extend smart pollution and efficiency standards, including the Clean Power Plan, while closing the Halliburton loophole, which stripped the Environmental Protection Agency of its ability to regulate hydraulic fracturing

2016 Donald
Trump (R)

A New Era in Energy

- Open public lands and the Outer Continental Shelf to exploration and responsible production
- Congress should give authority to state regulators to manage energy resources on federally controlled public lands
- Do away with the Clean Power Plan
- Support the development of all forms of energy that are marketable in a free economy without subsidies
- Encourage the cost-effective development of renewable energy sources—by private capital
- Support lifting restrictions to allow responsible development of nuclear energy, including research into alternative processes such as thorium nuclear energy
- Build on policies to find new ways to store electricity
- Aggressively expand trade opportunities and open new markets for American energy

Source: John Wooley and Gerhard Peters, "Political Party Platforms of Parties Receiving Electoral Votes 1840–2016," American Presidency Project, accessed July 27, 2016, http://www.presidency.ucsb.edu/platforms.php.

Selected References

A few sources have been used throughout the book and deserve acknowledgment for their immense usefulness. In particular, the U.S. Energy Information Administration provides an amazing amount of information. It archives data from 1980 by country, region, and the world on energy consumption and production by energy source. The Energy Information Administration website allows one to gain an international perspective. Specific to renewable energy activity, the REN21 Renewable Energy Policy Network for the 21st Century's *Renewables 2015 Global Status Report* was very useful. The International Energy Agency is also a very helpful source of information on country energy policies. The challenge in scanning the energy plans for 10 or more countries is acquiring translated versions of source documents. For instance, the English translation of the China policy was found through the Italian Embassy. I want to acknowledge that country briefs from both the Energy Information Administration and the International Energy Agency provided very helpful baseline information to complement other specific sources. Finally, fact resources such as the World Bank have been helpful throughout. Industry associations, whether petrochemical, renewable, or nuclear, have been invaluable sources of information.

General Information and Statistics

Boundy, Bob, Susan W. Diegel, Lynn Wright, and Stacey C. Davis. *Biomass Energy Data Book: Edition 4*. Oak Ridge, Tenn.: Oak Ridge National Laboratory, September 2011. http://cta.ornl.gov/bedb/pdf/BEDB4__Full__Doc.pdf.

British Petroleum. *BP Energy Outlook 2030*. January 2013. http://www.bp.com/content/dam/bp/pdf/energy-economics/energy-outlook-2015/bp-energy-outlook-booklet__2013.pdf.

Encyclopedia Britannica. "Moore's Law." September 22, 2013. Accessed April 17, 2015. https://www.britannica.com/topic/Moores-law.

ExxonMobil. *The Outlook for Energy: A View to 2040*. 2016. http://cdn.exxonmobil.com/~/media/global/files/outlook-for-energy/2016/2016-outlook-for-energy.pdf.

Institute for Energy Research. "IEA's World Energy Outlook 2014." November 21, 2014. Accessed March 26, 2015. http://instituteforenergyresearch.org/analysis/ieas-world-energy-outlook-2014/.

International Energy Agency. *Resources to Reserves 2013: Oil, Gas and Coal Technologies for the Energy Markets of the Future.* Paris: Organisation for Economic Co-operation Development/International Energy Agency, 2013. https://www.iea.org/publications/freepublications/publication/Resources2013.pdf.

NationMaster. "Transport. Road Density. Km of Road per 100 Sq. Km of Land Area. Countries Compared." Accessed April 17, 2015. http://www.nationmaster.com/country-info/stats/Transport/Road-density/Km-of-road-per-100-sq.-km-of-land-area.

N.C. Clean Energy Technology Center, Database of State Incentives for Renewables and Efficiency. "Renewable Portfolio Standards." Department of Energy 2013 Archive. Accessed February 25, 2015. http://www.dsireusa.org.

NextEra. *NextEra Energy Annual Report 2013.* Report no. CC191-1403. Juno Beach, Fla.: NextEra Energy, 2014.

Public Service Commission of Maryland. Renewable Energy Portfolio Standard Report 2014. Baltimore: Public Service Commission of Maryland, 2014.

REN21 Renewable Energy Policy Network for the 21st Century. *Renewables 2015 Global Status Report.* Paris: REN21 Secretariat, 2015. http://www.ren21.net/wp-content/uploads/2015/07/REN12-GSR2015__Onlinebook__low1.pdf.

Royal Dutch Shell. Annual Report (Year Ending December 31 2015). London: Royal Dutch Shell, 2016.

U.S. Department of Agriculture. *2012 Census of Agriculture.* Report no. AC-12-A-51. Washington, D.C.: U.S. Department of Agriculture, 2014.

U.S. Energy Information Administration. *Annual Energy Outlook 2013: With Projections to 2040.* Report no. DOE/EIA-0383(2013). Washington, D.C.: U.S. Energy Information Administration, Office of Integrated and International Energy Analysis, U.S. Department of Energy, April 2013. http://www.eia.gov/forecasts/aeo/pdf/0383(2013).pdf.

U.S. Energy Information Administration. *Annual Energy Outlook 2014: With Projections to 2040.* Report no. DOE/EIA-0383(2014). Washington, D.C.: U.S. Department of Energy, Energy Information Administration, April 2014. http://www.eia.gov/forecasts/aeo/pdf/0383(2014).pdf.

U.S. Energy Information Administration. *International Energy Outlook 2013: With Projections to 2040.* Report no. DOE/EIA-0484(2013). Washington, D.C.: U.S. Energy Information Administration, Office of Energy Analysis, U.S. Department of Energy, July 2013. Accessed March 25, 2015. http://www.eia.gov/forecasts/ieo/pdf/0484(2013).pdf.

U.S. Energy Information Administration. "International Energy Statistics." Accessed February 26, 2015. http://www.eia.gov/cfapps/ipdbproject/IEDIndex3.cfm?tid=6&pid=29&aid=12.

World Bank. "Access to Electricity (% of Population)." Accessed May 1, 2015. http://data.worldbank.org/indicator/EG.ELC.ACCS.ZS.

World Bank. "GDP (Current US$)." Accessed April 5, 2015. http://data.worldbank.org/indicator/NY.GDP.MKTP.CD?page=5.

World Bank. "Population Ranking." Accessed April 5, 2015. http://data.worldbank.org/data-catalog/Population-ranking-table.

Chapter 1. The Seven Signals Heralding an Energy Transition

ExxonMobil. *The Outlook for Energy: A View to 2040.* 2016. http://cdn.exxonmobil
.com/~/media/global/files/outlook-for-energy/2016/2016-outlook-for-energy.
pdf.

Gold, Thomas. *The Deep Hot Biosphere: The Myth of Fossil Fuels.* New York: Copernicus,
2001.

International Energy Agency. *Resources to Reserves 2013: Oil, Gas and Coal Technologies for the Energy Markets of the Future.* Paris: Organisation for Economic
Co-operation Development/International Energy Agency, 2013. https://www.iea
.org/publications/freepublications/publication/Resources2013.pdf.

"Shale Oil: In a Bind. Will Falling Oil Prices Curb America's Shale Boom?" *Economist*, December 6, 2014. http://www.economist.com/news/finance-and-economics
/21635505-will-falling-oil-prices-curb-americas-shale-boom-bind.

U.S. Geological Survey. "An Estimate of Undiscovered Oil and Gas Resources of the
World, 2012." Fact Sheet 2012-3042. U.S. Department of the Interior, U.S. Geological Survey, 2012. http://pubs.usgs.gov/fs/2012/3042/fs2012-3042.pdf.

Part I. The Sources of Energy We Draw Upon: Alternative Energies to the Front

Chapter 2. Fossil Fuels: Stored Solar Energy

American Public Gas Association. "A Brief History of Natural Gas." 2015.
Accessed April 9, 2016. http://www.apga.org/apgamainsite/aboutus/facts
/history-of-natural-gas.

Carnegie Mellon. "Environmental Decision Making, Science, and Technology."
Accessed April 9, 2016. http://environ.andrew.cmu.edu/m3/s3/09fossil.shtml.

Encyclopedia.com. "Coal." *UXL Encyclopedia of Science*, 2002. Accessed April 9, 2016.
http://www.encyclopedia.com/doc/1G2-3438100167.html.

Reitz, Rolf. "Reciprocating Internal Combustion Engine." Lecture, Combustion
Energy Frontier Research Center, University of Wisconsin, Madison, June 27, 2012.

Riva, Joseph P. "Petroleum." *Encyclopaedia Britannica.* Accessed April 20, 2015. http://
www.britannica.com/science/petroleum.

Chapter 3. Nuclear Energy and the Mass Defect

Ball, Philip. "Laser Fusion Experiment Extracts Net Energy from Fuel." *Nature*,
February 12, 2014. Accessed April 9, 2016. http://www.nature.com/news
/laser-fusion-experiment-extracts-net-energy-from-fuel-1.14710.

Georgia State University. "Nuclear Fusion." Accessed April 9, 2016. http://
hyperphysics.phy-astr.gsu.edu/hbase/nucene/fusion.html.

Goldschmidt, Bertrand. "Uranium's Scientific History 1789–1939." Paper presented
at the 14th International Symposium, Uranium Institute, London, September 1,
1989. http://ist-socrates.berkeley.edu/~rochlin/ushist.html.

World Nuclear Association. "The Cosmic Origins of Uranium." November 2006.
Accessed April 9, 2016. http://www.world-nuclear.org/information-library
/nuclear-fuel-cycle/uranium-resources/the-cosmic-origins-of-uranium.aspx.

World Nuclear Association. "Nuclear Fusion Power." February 22, 2016. Accessed April 9, 2016. http://www.world-nuclear.org/information-library/current-and -future-generation/nuclear-fusion-power.aspx.

World Nuclear Association. "Physics of Uranium and Nuclear Energy." September 2014. Accessed April 9, 2016. http://www.world-nuclear.org/information-library /nuclear-fuel-cycle/introduction/physics-of-nuclear-energy.aspx.

World Nuclear Association. "Supply of Uranium." September 2015. Accessed April 9, 2016. http://www.world-nuclear.org/information-library/nuclear-fuel-cycle /uranium-resources/supply-of-uranium.aspx.

Chapter 4. The Power behind Renewables

WIND ENERGY

Arent, Douglas, Patrick Sullivan, Donna Heimiller, Anthony Lopez, Kelly Eurek, Jake Badger, Hans Ejsing Jørgensen, Mark Kelly, Leon Clarke, and Patrick Luc-kow. *Improved Offshore Wind Resource Assessment in Global Climate Stabilization Scenarios*. Report no. NREL/TP-6A20-55049. Golden, Colo.: National Renew-able Energy Laboratory, 2012. http://www.nrel.gov/docs/fy13osti/55049.pdf.

Christopherson, Robert W. *Elemental Geosystems*. Upper Saddle River, N.J.: Prentice Hall, 2010.

Michand, Louis. *The Atmospheric Vortex Machine*. Sarnia: AVEtec Energy Corpora-tion, 2014.

Saeidian, Amin, Mojtaba Gholi, and Ehsan Zamani. "Windmills (ASBADS): Remark-able Example of Iranian Sustainable Architecture." *Architecture—Civil Engineering— Environment* (Silesian University of Technology) 5, no. 3 (2012): 19–30.

SOLAR ENERGY

"France Decrees New Rooftops Must Be Covered in Plants or Solar Panels." *Guardian*, March 19, 2015. Accessed April 10, 2016. https://www.theguardian.com/world/2015 /mar/20/france-decrees-new-rooftops-must-be-covered-in-plants-or-solar-panels.

Lindsey, Rebecca. "Climate and Earth's Energy Budget." NASA Earth Observatory, January 14, 2009. Accessed April 11, 2016. http://earthobservatory.nasa.gov /Features/EnergyBalance/.

Stackhouse, Paul W., Jr. "Surface Meteorology and Solar Energy." Atmospheric Science Data Center, NASA. Accessed April 17, 2015. https://eosweb.larc.nasa.gov/sse/.

Trieb, Franz, Christoph Schillings, Marlene O'Sullivan, Thomas Pregger, and Carsten Hoyer-Klick. "Global Potential of Concentrating Solar Power." SolarPaces Con-ference, Berlin, September 2009. http://www.solarthermalworld.org/sites/gstec /files/global%20potential%20csp.pdf.

U.S. Department of Energy. *2014: The Year of Concentrating Solar Power*. Report no. DOE/EE-1101. U.S. Department of Energy, May 2014. http://energy.gov/sites /prod/files/2014/05/f15/2014__csp__report.pdf.

U.S. Department of Energy, Energy Efficiency and Renewable Energy. "The History of Solar." https://www1.eere.energy.gov/solar/pdfs/solar__timeline.pdf.

HYDROPOWER

Energy.gov, Office of Energy Efficiency and Renewable Energy. "History of Hydropower." Accessed April 9, 2016. http://energy.gov/eere/water/history-hydropower.

Energy.gov, Office of Energy Efficiency and Renewable Energy. "Types of Hydropower Plants." Accessed April 10, 2016. http://energy.gov/eere/water/types-hydropower-plants.

International Energy Agency. "Renewable Energy Essentials: Hydropower." Paris: Organisation for Economic Co-operation and Development/International Energy Agency, 2010. http://www.iea.org/publications/freepublications/publication /hydropower__essentials.pdf.

"List of Largest Dams in the World." Wikipedia. Accessed April 10, 2016. https:// en.wikipedia.org/wiki/List__of__largest__dams__in__the__world.

Norman, Jeremy. "The Earliest Evidence of a Water-Driven Wheel (circa 250 BCE)." HistoryofInformation.com, March 19, 2015. Accessed April 10, 2016. http://www .historyofinformation.com/expanded.php?id=3509.

U.S. Census Bureau. "QuickFacts: United States. Housing." http://www.census.gov /quickfacts/table/PST045215/00.

Water-technology.net. "The World's Oldest Dams Still in Use." October 21, 2013. Accessed May 1, 2015. http://www.water-technology.net/features /feature-the-worlds-oldest-dams-still-in-use/.

GEOTHERMAL ENERGY

Allen, Richard M. "Convection in the Earth's Mantle." Lecture, EPS 122, University of California, Seismological Laboratory, Division of Geological and Planetary Sciences, Berkeley. http://rallen.berkeley.edu/eps122/lectures/L22.pdf.

Blodgett, Leslie. "Basics." Geothermal Energy Association, 2014. Accessed April 11, 2015. http://geo-energy.org/Basics.aspx.

Energy.gov, Office of Energy Efficiency and Renewable Energy. "Geothermal FAQs." Accessed April 11, 2015. http://energy.gov/eere/geothermal/geothermal-faqs.

Geothermal Education Office. "Geothermal Energy—Worldwide." Accessed April 11, 2015. http://geothermaleducation.org/geomap__1.html.

Hulen, J. B., and P. M. Wright. *Geothermal Energy: Clean Sustainable Energy for the Benefit of Humanity and the Environment*. Salt Lake City: University of Utah, 2001.

Kiehl, J. T., and Kevin E. Trenberth. "Earth's Annual Global Mean Energy Budget." *Bulletin of the American Meteorological Society* 78, no. 2 (1997): 197–208.

Williams, Quentin. "Why Is the Earth's Core So Hot? And How Do Scientists Measure Its Temperature?" *Scientific American*, October 6, 1997. Accessed April 11, 2016. http://www.scientificamerican.com/article/why-is-the-earths-core-so/.

SOURCES OF OCEAN ENERGY

Hautala, Susan, Kathryn Kelly, and LuAnne Thompson. "Tide Dynamics." Lecture, University of Washington, 2005.

Munk, Walter, and Carl Wunsch. "Abyssal Recipes II: Energetics of Tidal and Wind Mixing." *Deep-Sea Research Part I: Oceanographic Research Papers* 45, no. 12 (December 1998): 1977–2010. doi:10.1016/S0967-0637(98)00070-3.

National Oceanic and Atmospheric Administration National Ocean Service. "Tides and Water Levels." Accessed May 1, 2015. http://oceanservice.noaa.gov/education /tutorial__tides/welcome.html.

National Oceanic and Atmospheric Administration National Ocean Service. "Where Do I Get NOAA Tides and Currents Data?" July 16, 2015. Accessed December 20, 2015. http://oceanservice.noaa.gov/facts/find-tides-currents.html.

Ocean Energy Council. "Tidal Energy." Accessed May 1, 2015. http://www .oceanenergycouncil.com/ocean-energy/tidal-energy/.

Chapter 5. The Potential and Toll of Alternative Energy Sources

Archer, Cristina L., and Mark Z. Jacobson. "Geographical and Seasonal Variability of the Global 'Practical' Wind Resources." *Applied Geography* 45 (2013): 119–130. doi:10.1016/j.apgeog.2013.07.006.

Delucchi, Mark A., and Mark Z. Jacobson. "Providing All Global Energy with Wind, Water, and Solar Power, Part II: Reliability, System and Transmission Costs, and Policies." *Energy Policy* 39, no. 3 (2011): 1170–1190. doi:10.1016/j .enpol.2010.11.045.

Kursinski, Robert. "The Carnot Cycle." Lecture, ATMO 551a, University of Arizona, 2008.

Lu, Xi, Michael B. McElroy, and Juha Kiviluoma. "Global Potential for Wind-Generated Electricity." *Proceedings of the National Academy of Sciences of the United States of America* 106, no. 27 (2009): 10933–10938. doi:10.1073/pnas.0904101106.

Tsao, Jeff, Nate Lewis, and George Crabtree. "Solar FAQs." April 20, 2006. http:// www.sandia.gov/~jytsao/Solar%20FAQs.pdf.

Part II. Age-Old Technologies Find Application: Seeds of Innovation, Dormant for Centuries, Are Now Germinating

Chapter 7. Solar

Amos, Jonathan. "Solar Impulse: Global Flight Completes First Leg." *BBC News*, March 9, 2015. Accessed April 13, 2015. http://www.bbc.com/news /science-environment-31772140.

BrightSource Energy. "Ivanpah." Accessed April 13, 2015. http://www .brightsourceenergy.com/ivanpah-solar-project#.Vw56MWP7WFI.

Casey, Tina. "Earth to BrightSource: Give Up, the Media Will Never Get Ivanpah Right." *CleanTechnica*, November 18, 2014. Accessed April 13, 2015. http://cleantechnica .com/2014/11/18/earth-brightsource-give-media-will-never-get-ivanpah-right/.

Conca, James. "Thermal Solar Energy—Some Technologies Really Are Dumb." *Forbes*, November 11, 2014. Accessed April 13, 2015. http://www.forbes.com/sites/jamesconca/2014/11/11 /thermal-solar-energy-some-technologies-really-are-dumb/#4e7aeacd1574.

CSP World. "CSP World Map." http://www.cspworld.org/cspworldmap.

Energy.gov, Office of Energy Efficiency and Renewable Energy. "Dish/Engine System Concentrating Solar Power Basics." August 20, 2013. Accessed April 13, 2015. http://energy .gov/eere/energybasics/articles/dishengine-system-concentrating-solar-power -basics.

Feldman, David, Galen Barbose, Robert Margolis, Mark Bolinger, Donald Chung, Ran Fu, Joachim Seel, Carolyn Davidson, Naïm Darghouth, and Ryan Wiser. "Photovoltaic System Pricing Trends. Historical, Recent, and Near-Term Projections, 2015 Edition." SunShot, U.S. Department of Energy, August 25, 2015. http://www.nrel.gov/docs/fy15osti/64898.pdf.

Green Rhino Energy. "Solar Power Home." Accessed April 13, 2015. http://www.greenrhinoenergy.com/solar/#.

Harrington, Kent. "Molten Salt Gives Concentrated Solar a Unique Advantage." American Institute of Chemical Engineers, October 25, 2013. Accessed April 13, 2015. http://www.aiche.org/chenected/2013/10/molten-salt-gives-concentrated-solar-unique-advantage.

International Energy Agency. *Technology Roadmap: Solar Photovoltaic Energy—2014 Edition*. Paris: Organisation for Economic Co-operation and Development /International Energy Agency, September 2014. Accessed February 27, 2015. https://www.iea.org/publications/freepublications/publication /TechnologyRoadmapSolarPhotovoltaicEnergy__2014edition.pdf.

Japan Space Systems. "SSPS: Space Solar Power System." Accessed April 13, 2015. http://www.jspacesystems.or.jp/en__project__ssps/.

National Renewable Energy Laboratory, Concentrating Solar Power Projects. "Ivanpah Solar Electric Generating System." November 20, 2014. Accessed April 13, 2016. http://www.nrel.gov/csp/solarpaces/project__detail.cfm/projectID=62.

Overton, Thomas W. "Ivanpah Solar Electric Generating System Earns POWER's Highest Honor." *POWER Magazine*, August 1, 2014. Accessed April 13, 2015. http://www.powermag.com/ivanpah-solar-electric-generating-system-earns-powers-highest-honor/.

Solar Energy Industries Association. "Solar Industry Data." Accessed April 13, 2016. http://www.seia.org/research-resources/solar-industry-data.

Trieb, Franz, Christoph Schillings, Marlene O'Sullivan, Thomas Pregger, and Carsten Hoyer-Klick. "Global Potential of Concentrating Solar Power." SolarPaces Conference, Berlin, September 2009. http://www.solarthermalworld.org/sites/gstec /files/global%20potential%20csp.pdf.

Chapter 8. Wind

ABB. "Five Key Characteristics Make Wind Farms More Financially Viable." Washington, D.C.: ABB White Paper, October 25, 2013. https://library.e.abb.com /public/25a1d23c5ba5ac0785257c0f00586db5/ABB-799-WPO__5keys-characteristics-wind__FINAL__11-15-13.pdf.

EurActiv.com. "EU Wind Industry Faces Tough Challenge—and Politicians Should Not Make It Worse." Wind Europe, press release, February 4, 2013. Accessed April 13, 2015. http://pr.euractiv.com/pr/eu-wind-industry-faces-tough-challenge-and -politicians-should-not-make-it-worse-93078.

International Electrotechnical Commission. *Wind Turbines—Pt. 1: Design Requirements*. 3rd ed. Report no. IEC 61400-1:2005(E). Geneva: International Electrotechnical Commission, 2005.

Rowlatt, Justin. "Rare Earths: Neither Rare, nor Earths." *BBC News*, March 23, 2014. Accessed April 13, 2015. http://www.bbc.com/news/magazine-26687605.

Wiser, Ryan, and Mark Bolinger. *2013 Wind Technologies Market Report*. U.S. Department of Energy, Energy Efficiency and Renewable Energy, August 2014. Accessed

February 26, 2015. http://emp.lbl.gov/sites/all/files/2013__Wind__Technologies __Market__Report__Final3.pdf.

Chapter 9. Nuclear

Berthélemy, Michel, and Lina Escobar Rangel. "Nuclear Reactors' Construction Costs: The Role of Lead-Time, Standardization and Technological Progress." *Energy Policy* 82, no. 1 (2015): 118–130. doi:10.1016/j.enpol.2015.03.015.

Biello, David. "Spent Nuclear Fuel: A Trash Heap Deadly for 250,000 Years or a Renewable Energy Source?" *Scientific American*, January 28, 2009. Accessed April 24, 2015. http://www.scientificamerican.com/article /nuclear-waste-lethal-trash-or-renewable-energy-source/.

California Energy Commission. *2014 Integrated Energy Policy Report Update*. Report no. CEC-100-2014-001-F. January 2015. http://www.energy.ca.gov/2014publications/ CEC-100-2014-001/CEC-100-2014-001-F.pdf.

Chang, Y. I., C. Grandy, P. Lo Pinto, and M. Konomura. *Small Modular Fast Reactor Design Description*. Report no. ANL-SMFR-1. Argonne National Laboratory, Commissariat à l'Energie Atomique, and Japan Nuclear Cycle Development Institute, July 1, 2005. Accessed June 4, 2015. http://www.ne.anl.gov/eda/Small__Modular __Fast__Reactor__ANL__SMFR__1.pdf.

"Chinese Fast Reactor Completes Full-Power Test Run." *World Nuclear News*, December 19, 2014. Accessed April 15, 2015. http://www.world-nuclear-news.org /NN-Chinese-fast-reactor-completes-full-power-test-run-1912144.html.

Deign, Jason. "Middle East Nuclear Fuel Cycles: An Open Rather than Closed Case?" Energy Industry Report, August 15, 2012. Accessed April 13, 2015. http:// energyindustryreport.blogspot.com/2012/08/middle-east-nuclear-fuel-cycles -open__15.html.

Edelstein, Norman, and Lester Morss. *Radiochemistry and Nuclear Chemistry, vol. II: Chemistry of the Actinide Elements*. Encyclopedia of Life Support Systems. EOLSS Publishers Co., 2009.

International Energy Agency. "Taking a Fresh Look at the Future of Nuclear Power." January 29, 2015. Accessed April 13, 2015. http://www.iea.org/newsroomandevents /news/2015/january/taking-a-fresh-look-at-the-future-of-nuclear-power.html.

International Energy Agency. *Technology Roadmap: Nuclear Energy—2015 Edition*. Paris: Organisation for Economic Co-operation and Development/International Energy Agency/Nuclear Energy Agency, 2015. https://www.iea.org/media /freepublications/technologyroadmaps/TechnologyRoadmapNuclearEnergy.pdf.

ITER. "ITER—The Way to New Energy." Accessed April 13, 2015. https://www.iter. org/.

Japan Atomic Industrial Forum. "Spent Fuel and Radiotoxicity." Accessed September 30, 2015. http://www.jaif.or.jp/ja/wnu__si__intro/document/2009/m __salvatores__advanced__nfc.pdf.

Mollard, Pascale. "Star Power: Troubled ITER Nuclear Fusion Project Seeks New Path." Phys.org, May 22, 2015. Accessed April 13, 2016. http://phys.org/news/2015 -05-star-power-iter-nuclear-fusion.html.

Nuclear Energy Agency. *Technology Roadmap Update for Generation IV Nuclear Energy Systems*. Paris: Organisation for Economic Co-operation and Development Nuclear Energy Agency for the Generation IV International Forum, January 2014. Accessed February 26, 2015. https://www.gen-4.org/gif/upload/docs/application/pdf/2014 -03/gif-tru2014.pdf.

Pearce, Fred. "Are Fast-Breeder Reactors the Answer to Our Nuclear Waste Nightmare?" Guardian, July 30, 2012. Accessed May 12, 2015. http://www.theguardian .com/environment/2012/jul/30/fast-breeder-reactors-nuclear-waste-nightmare.

RT International. "Fast Reactor Starts Clean Nuclear Energy Era in Russia." June 27, 2014. Accessed April 13, 2015. https://www.rt.com/news/168768-russian-fast-breeder-reactor/.

Stanford, George S. "What Is the IFR?" Nuclear Engineering Division, Argonne National Laboratory, May 2013. Accessed January 1, 2015. http://www.ne.anl .gov/About/reactors/ifr/What%20Is%20the%20IFR.25.pdf.

Till, Charles E. "Plentiful Energy and the Integral Fast Reactor Story." *International Journal of Nuclear Governance, Economy and Ecology* 1, no. 2 (2006): 212–221. doi:10.1504/ijngee.2006.011242.

Till, Charles E., and Yoon I. Chang. "The Integral Fast Reactor." Advances in Nuclear Science and Technology 20 (1988): 127–154. doi:10.1007/978-1-4613-9925-4__3.

World Nuclear Association. "International Framework for Nuclear Energy Cooperation." August 2015. Accessed April 13, 2015. http://world-nuclear.org /information-library/current-and-future-generation/international-framework-for-nuclear-energy-coopera.aspx.

World Nuclear Association. "Molten Salt Reactors." April 2016. Accessed April 13, 2016. http://www.world-nuclear.org/information-library/current-and-future -generation/molten-salt-reactors.aspx.

World Nuclear Association. "Thorium." September 2015. Accessed April 13, 2015. http://www.world-nuclear.org/information-library/current-and-future -generation/thorium.aspx.

Chapter 10. Advanced Energy Storage Solutions

HYDROGEN

Air Products and Chemicals. *Breakthrough for Hydrogen Fuel Storage Is Like a "Liquid Battery."* Pub. no. 352-10-001-GLB. Allentown: Air Products and Chemicals, 2010. http://www.airproducts.com/~/media/Files/PDF/company/tech-energy-liquid -battery.pdf?la=en.

Energy.gov, Office of Energy Efficiency and Renewable Energy. "Hydrogen Storage." Accessed June 1, 2015. http://energy.gov/eere/fuelcells/hydrogen-storage.

h-tec. "Solar-Hydrogen Energy System." Lübeck: h-tec, 2003. http://www.h-tec.com/ fileadmin/content/edu/Lehrmaterialien/Transparencies.pdf.

Koroneos, C., A. Dompros, G. Roumbas, and N. Moussiopoulos. "Life Cycle Assessment of Hydrogen Fuel Production Processes." *International Journal of Hydrogen Energy* 29, no. 14 (2004): 1443–1450. doi:10.1016/j.ijhydene.2004.01.016.

National Hydropower Association. "Developing Hydro." Accessed May 1, 2015. http:// www.hydro.org/tech-and-policy/developing-hydro/.

Okada, Yoshimi, and Mitsunori Shimura. "Development of Large Scale H2 Storage and Transportation Technology with Liquid Organic Hydrogen Carrier (LOHC)." Chiyoda Corporation, February 5, 2013. https://www.jccp .or.jp/international/conference/docs/15rev-chiyoda-mr-shimura-chiyoda-h2 -sturage-and-transpor.pdf.

Sandia National Laboratories. "Sunshine to Petrol: Solar Recycling of Carbon Dioxide into Hydrocarbon Fuels." Report no. SAND 2009-5796P. U.S. Department of Energy, Sandia National Laboratories, 2009. http://energy.sandia.gov/wp-content/ gallery/uploads/S2P__SAND2009-5796P.pdf.

Taylan, Onur, and Halil Berberoglu. "Fuel Production Using Concentrated Solar Energy." In *Application of Solar Energy*, edited by Radu Rugescu. *InTech*, February 6, 2013. doi:10.5772/54057.

Teichmann, Daniel, Wolfgang Arlt, Peter Wasserscheid, and Raymond Freymann. "A Future Energy Supply Based on Liquid Organic Hydrogen Carriers (LOHC)." *Energy and Environmental Science* 4, no. 8 (2011): 2767–2773. doi:10.1039 /c1ee01454d.

U.S. Department of Energy. *A Nation's Vision of America's Transition to a Hydrogen Economy—To 2030 and Beyond*. February 2002. https://www.hydrogen.energy.gov /pdfs/vision__doc.pdf.

U.S. Department of Energy. *2015 Annual Progress Report: DOE Hydrogen and Fuel Cells Program*. Washington, D.C.: Department of Energy, December 2015. https://www. hydrogen.energy.gov/annual__progress15.html.

BATTERY

Ambri. "Storing Electricity for Our Future." Conference presentation, National Governors Association, July 27, 2015. http://www.nga.org/files/live/sites/NGA/files /pdf/2015/1507LearningLabAmbriOverview__Giudice.pdf.

Battery University. "BU-101: When Was the Battery Invented?" Accessed April 12, 2015. http://batteryuniversity.com/learn/article/when__was__the__battery__invented.

Bowen, Catherine Drinker. *The Most Dangerous Man in America: Scenes from the Life of Benjamin Franklin*. Boston: Little, Brown, 1974.

Casey, Tina. "Flow Battery vs. Tesla Battery Smackdown Looming." *CleanTechnica*, June 21, 2015. Accessed April 12, 2015. http://cleantechnica.com/2015/06/21 /flow-battery-vs-tesla-battery-smackdown-looming/.

Energy Storage Association. "Redox Flow Batteries." Accessed April 12, 2015. http:// energystorage.org/energy-storage/technologies/redox-flow-batteries.

Fusionteq. "Battery 101—The Basics." Accessed April 12, 2015. http://www.fusionteq. com/html/battery__101__-__the__basics.html.

Ginsberg, Judah. "The Columbia Dry Cell Battery: A National Historic Chemical Landmark." Washington, D.C.: American Chemical Society, September 27, 2005. https://www.acs.org/content/dam/acsorg/education/whatischemistry/landmarks /drycellbattery/columbia-dry-cell-battery-historical-resource.pdf.

Hou, Jianhua, Chuanbao Cao, Faryal Idrees, and Xilan Ma. "Hierarchical Porous Nitrogen-Doped Carbon Nanosheets Derived from Silk for Ultrahigh-Capacity Battery Anodes and Supercapacitors." *ACS Nano* 9, no. 3 (2015): 2556–2564. doi:10.1021/nn506394r.

MIT Electric Vehicle Team. "A Guide to Understanding Battery Specifications." December 2008. http://web.mit.edu/evt/summary__battery__specifications.pdf.

"Round Trip Efficiency." *Energymag*, February 8, 2014. Accessed August 31, 2015. https://energymag.net/round-trip-efficiency/.

"Silk May Be the New 'Green' Ultra-High-Capacity Material for Batteries." Kurzweil Accelerating Intelligence, March 11, 2015. Accessed April 12, 2015. http://www.kurzweilai.net/silk-may-be-the-new-green-ultra-high-capacity-material-for-batteries.

Wiley, Ted, Eric Weber, Mike Eshoo, Jay Whitacre, Elizabeth Pond, Aaron Marks, and Jonathan Matusky. "Enabling Solar Generation and Reducing Diesel Consumption at the Redwood Gate Ranch Microgrid with Aqueous Hybrid Ion (AHI) Batteries." Paper presented at Battcon 2014 Stationary Battery Conference and Trade Show, Boca Raton, May 2014. http://www.battcon.com/PapersFinal2014/3%20Wiley%20Paper%202014%20Final.pdf.

Chapter 11. Smart Grids

California Independent System Operator. "Renewables Watch." Accessed April 14, 2016. http://www.caiso.com/green/renewableswatch.html.

ComEd: An Exelon Company. "ComEd's Hourly Pricing Program." Accessed April 14, 2015. https://hourlypricing.comed.com/.

International Energy Agency. *Technology Roadmap: Smart Grids*. Paris: Organisation for Economic Co-operation and Development/International Energy Agency, April 2011. https://www.iea.org/publications/freepublications/publication/smartgrids__roadmap.pdf.

Smalley, Joshua. "A Smart Approach: Integrating Distributed Energy Resources." Windpower Engineering and Development, June 3, 2015. Accessed July 14, 2015. http://www.windpowerengineering.com/policy/environmental/a-smart-approach-integrating-distributed-energy-resources/.

U.S. Bureau of Labor Statistics. "Occupational Employment and Wages, May 2015: 43-5041 Meter Readers, Utilities." Accessed April 14, 2015. http://www.bls.gov/oes/current/oes435041.htm.

U.S. Department of Energy. *The Smart Grid: An Introduction*. Washington, D.C.: U.S. Department of Energy. http://energy.gov/sites/prod/files/oeprod/DocumentsandMedia/DOE__SG__Book__Single__Pages(1).pdf.

Chapter 12. Transportation

Air Products. "SmartFuel® Hydrogen Energy." Accessed April 14, 2015. http://www.airproducts.com/industries/Energy/Hydrogen-Energy.aspx.

"American Railways: High-Speed Railroading." Economist, July 22, 2010. http://www.economist.com/node/16636101.

CA.gov. "California High-Speed Rail Authority." Accessed April 14, 2015. http://www.hsr.ca.gov/.

ChartsBin. "Worldwide Total Motor Vehicles (per 1,000 People)." Accessed April 14, 2015. http://chartsbin.com/view/1114.

C2ES Center for Climate and Energy Solutions. "Freight Transportation." Accessed April 14, 2015. http://www.c2es.org/technology/factsheet/FreightTransportation.

C2ES Center for Climate and Energy Solutions. "Transportation Overview." Accessed April 14, 2015. http://www.c2es.org/energy/use/transportation.

Electric Auto Association. "EV History." Accessed April 14, 2015. http://www.elec-tricauto.org/?page=evhistory.

"Fuel Cells to Power Regional Trainsets." *Railway Gazette*, September 24, 2014. Accessed April 14, 2015. http://www.railwaygazette.com/news/technology/single-view/view/fuel-cells-to-power-regional-trainsets.html.

Honda. "The 2008 FCX Clarity." Accessed April 14, 2015. http://automobiles.honda.com/fcx-clarity/.

Hyundai. "The Amazing 2016 Tucson Fuel Cell." Accessed April 14, 2015. https://www.hyundaiusa.com/tucsonfuelcell/index.aspx.

International Air Transport Association. "IATA Sustainable Alternative Aviation Fuels Strategy." 2016. https://www.iata.org/whatwedo/environment/Documents/sustainable-alternative-aviation-fuels-strategy.pdf.

Lo, Chris. "Hydrail and LNG: The Future of Railway Propulsion?" Railway-technology.com, May 13, 2013. Accessed April 14, 2015. http://www.railway-technology.com/features/featurehydrail-lng-future-railway-propulsion-fuel/.

Make Biofuel. "Bioethanol Production." Accessed April 14, 2015. http://www.makebiofuel.co.uk/bioethanol-production.

Melaina, Marc W. "Retail Infrastructure Costs Comparison for Hydrogen and Electricity for Light-Duty Vehicles." SAE Technical Paper 2014-01-1969, April 1, 2014. doi:10.4271/2014-01-1969.

Nuvera. "PowerTap Hydrogen Generation Appliance." Accessed April 14, 2016. http://www.nuvera.com/products-services/powertap-hydrogen-generation-appliance/powertap-hydrogen-station-specfications-page.

O'Dell, John. "8 Things You Need to Know about Hydrogen Fuel-Cell Cars." Edmunds.com, May 8, 2015. Accessed June 14, 2015. http://www.edmunds.com/fuel-economy/8-things-you-need-to-know-about-hydrogen-fuel-cell-cars.html.

Renewable Fuels Association. "E85." Accessed April 14, 2015. http://www.ethanolrfa.org/resources/blends/e85/.

Royal Academy of Engineering. *Future Ship Powering Options: Exploring Alternative Methods of Ship Propulsion.* London: Royal Academy of Engineering, July 2013. http://www.raeng.org.uk/publications/reports/future-ship-powering-options.

Shahan, Zachary. "100% Electric Car Sales Up 58% in US in 2014." *CleanTechnica*, January 7, 2015. Accessed April 14, 2015. http://cleantechnica.com/2015/01/07/100-electric-cars-58-us-2014/.

SpaceX. "Hyperloop." August 12, 2013. Accessed January 15, 2016. http://www.spacex.com/hyperloopalpha.

Tesla. "Model S." Accessed April 14, 2015. https://www.teslamotors.com/models.

Toyota. "Toyota Mirai—The Turning Point." Accessed April 14, 2015. https://ssl.toyota.com/mirai/fuel.html.

U.S. Department of Energy, Energy Efficiency and Renewable Energy. "Alternative Fuels Data Center." Accessed March 1, 2015. http://www.afdc.energy.gov.

U.S. Department of Energy, Energy Information Administration. "How Many Alternative Fuel and Hybrid Vehicles Are There in the United States?" October 21, 2015. Accessed January 14, 2016. https://www.eia.gov/tools/faqs/faq.cfm?id=93.

U.S. Department of Transportation, Federal Highway Administration and Bureau of Transportation Statistics. *Freight Facts and Figures 2013*. Report no. FHWA-HOP-14-004. Washington, D.C.: U.S. Department of Transportation, January 2014. http://ops.fhwa.dot.gov/Freight/freight__analysis/nat__freight__stats /docs/13factsfigures/pdfs/fff2013__highres.pdf.

U.S. Department of Transportation, Federal Railroad Administration. "National Rail Plan: Moving Forward." September 2010. Accessed February 26, 2015. https:// www.fra.dot.gov/Elib/Document/1336/.

U.S. Energy Information Administration. "Biofuels: Ethanol and Biodiesel Explained." Accessed April 14, 2015. https://www.eia.gov/energyexplained/index .cfm?page=biofuel__home#tab1.

U.S. Environmental Protection Agency. "E15 Fuel Registration." September 28, 2015. Accessed November 14, 2015. https://www.epa.gov/fuels-registration -reporting-and-compliance-help/e15-fuel-registration.

U.S. Environmental Protection Agency. *Fast Facts: U.S. Transportation Sector Greenhouse Gas Emissions 1990–2011*. Report no. EPA-420-F-13-033a. September 2013. Accessed February 26, 2015. http://nepis.epa.gov/Exe/ZyPDF.cgi?Dockey= P100GYH6.pdf.

Chapter 13. Heat Generation

Apricus. "Evacuated Tube Solar Collectors." Accessed April 14, 2016. http://www .apricus.com/html/solar__collector.htm#.Vw-vw2P7WFI.

Florides, G. A., S. A. Kalogirou, S. A. Tassou, and L. C. Wrobel. "Design and Construction of a LiBr–Water Absorption Machine." *Energy Conversion and Management* 44, no. 15 (2003): 2483–2508. doi:10.1016/s0196-8904(03)00006-2.

H2 Tools. "Hydrogen Compared with Other Fuels." Accessed April 14, 2015. https:// h2tools.org/bestpractices/h2properties.

International Energy Agency. *Technology Roadmap: Solar Heating and Cooling*. Paris: Organisation for Economic Co-operation and Development/International Energy Agency, 2012. http://www.oecd-ilibrary.org /docserver/download/6112241e.pdf?expires=1469136184&id=id&accname= guest&checksum=4348267C32B6C6E487C7A436D35568FB.

Ivancic, Aleksandar, Daniel Mugnier, Gerhard Stryi-Hipp, and Werner Weiss. *Solar Heating and Cooling Technology Roadmap*. Brussels: RHC Renewable Heating and Cooling European Technology Platform, June 2014. http://www.rhc-platform .org/fileadmin/user__upload/Structure/Solar__Thermal/Download/Solar __Thermal__Roadmap.pdf.

Navarro-Rivero, Pilar, and Björn Ehrismann. *Durability Issues, Maintenance and Costs of Solar Cooling Systems*. V2.0. Report no. 5.3.2. QAiST Quality Assurance in Solar Heating and Cooling Technology, May 21, 2012. http://www.estif.org /solarkeymarknew/images/downloads/QAiST/qaist%20d5.3%20tr5.3.2%20 durability%20issues%20maintenance%20and%20costs%20of%20solar%20 cooling%20systems.pdf.

Ramgopal, M. "Vapour Absorption Refrigeration System." Lecture 17, Department of Mechanical Engineering, Indian Institute of Technology, Kharagpur.

Solar Energy Industries Association. *Solar Heating and Cooling Roadmap.* Washington, D.C.: Solar Energy Industries Association, October 2, 2013. http://www.seia.org/us-solar-heating-cooling-shc-alliance/solar-heating-cooling-shc-roadmap.

U.S. Department of Energy, Energy Efficiency and Renewable Energy. "Hydrogen Properties." http://hydrogen.pnl.gov/hydrogen-data/hydrogen-properties.

Part III. A Scan of Country Energy Plans: The Globe We Share

Chapter 16. The United States

DeMarban, Alex, and Annie Zak. "ConocoPhillips Cracks Open Giant Petroleum Reserve, with Good Results." *Alaska Dispatch News*, February 20, 2016. Accessed April 14, 2016. http://www.adn.com/article/20160220/conocophillips-cracks-open-giant-petroleum-reserve-good-results.

Holt, Mark, and Carol Glover. *Energy Policy Act of 2005: Summary and Analysis of Enacted Provisions.* Report no. CRS-RL33302. Washington, D.C.: Congressional Research Service, March 8, 2006.

Joling, Dan. "Lawsuit Challenges Alaska Petroleum Reserve Access." March 1, 2013. https://www.yahoo.com/news/lawsuit-challenges-alaska-petroleum-access-153931960.html?ref=gs.

Jones, Jeffrey M. "Record-High 42% of Americans Identify as Independents." Gallup, January 8, 2014. Accessed April 14, 2016. http://www.gallup.com/poll/166763/record-high-americans-identify-independents.aspx.

Mackinder, Evan. "Pro-Environment Groups Outmatched, Outspent in Battle over Climate Change Legislation." OpenSecrets RSS, August 23, 2010. Accessed April 24, 2015. http://www.opensecrets.org/news/2010/08/pro-environment-groups-were-outmatc/.

Motel, Seth. "Polls Show Most Americans Believe in Climate Change, but Give It Low Priority." Pew Research Center RSS, September 23, 2014. Accessed April 14, 2015. http://www.pewresearch.org/fact-tank/2014/09/23/most-americans-believe-in-climate-change-but-give-it-low-priority/.

Stockholm International Peace Research Institute. "SIPRI Military Expenditure Database." Accessed April 24, 2015. http://www.sipri.org/research/armaments/milex/milex__database.

U.S. Department of Energy. "Energy Policy Act of 2005." PsN:PUBL058, August 8, 2005.

U.S. Department of Energy. Strategic Plan 2014–2018. Report no. DOE/CF-0067. Washington, D.C.: U.S. Department of Energy, April 2014. http://www.energy.gov/sites/prod/files/2014/04/f14/2014__dept__energy__strategic__plan.pdf.

U.S. Environmental Protection Agency. *Regulatory Impact Analysis for the Proposed Carbon Pollution Guidelines for Existing Power Plants and Emission Standards for Modified and Reconstructed Power Plants.* Report no. EPA-452/R-14-002. Research Triangle Park, N.C.: U.S. Environmental Protection Agency, June 2014. https://www.epa.gov/sites/production/files/2014-06/documents/20140602ria-clean-power-plan.pdf.

Wooley, John, and Gerhard Peters. "Political Party Platforms of Parties Receiving Electoral Votes 1840–2016." American Presidency Project. Accessed April 14, 2015. http://www.presidency.ucsb.edu/platforms.php.

Chapter 17. Canada

British Columbia. *The BC Energy Plan: A Vision for Clean Energy Leadership*. Report no. 250.952.0241. Victoria: Ministry of Energy, Mines and Petroleum Resources, 2009. Accessed February 27, 2015. http://www2.gov.bc.ca/assets/gov/farming-natural-resources-and-industry/electricity-alternative-energy/bc__energy__plan__2007.pdf.

"British Columbia's Carbon Tax: The Evidence Mounts." *Economist*, July 31, 2014. Accessed April 15, 2015. http://www.economist.com/blogs/americasview/2014/07/british-columbias-carbon-tax.

Canada's Premiers, Council of the Federation. *Canadian Energy Strategy: Progress Report to the Council of the Federation*. Ottawa: Council of the Federation, July 2015. http://www.canadaspremiers.ca/phocadownload/publications/canadian__energy__strategy__eng__fnl.pdf.

Canada's Premiers, Council of the Federation. *A Shared Vision for Energy in Canada*. Ottawa: Council of the Federation, August 2007. http://canadaspremiers.ca/phocadownload/publications/energystrategy__en.pdf.

Energy Policy Institute of Canada. *A Canadian Energy Strategy Framework*. August 2012. http://weg.ge/wp-content/uploads/2013/05/Canada-Energy-Strategy-Framework-2012.pdf.

Plumer, Brad. "How Would a Carbon Tax Work? Let's Ask British Columbia." *Washington Post*, Wonkblog, September 19, 2012. Accessed April 15, 2015. https://www.washingtonpost.com/news/wonk/wp/2012/09/19/how-would-a-carbon-tax-work-lets-ask-british-columbia/.

SaskPower CCS. "Boundary Dam Carbon Capture Project." February 20, 2015. Accessed April 15, 2015. http://saskpowerccs.com/ccs-projects/boundary-dam-carbon-capture-project/.

U.S. Department of Energy. *Canada—Energy Brief*. Washington, D.C.: U.S. Department of Energy, Energy Information Administration, 2015.

World Nuclear Association. "Nuclear Power in Canada." February 2016. Accessed April 15, 2016. http://www.world-nuclear.org/information-library/country-profiles/countries-a-f/canada-nuclear-power.aspx.

Chapter 18. Brazil

Brazilian Government. "Electricity in the 2024 Brazilian Energy Plan." Report no. PDE 2024. Office of Strategic Energy Studies/Ministry of Mines and Energy, December 2, 2015.

Government Offices of Sweden. *Energy Policy in Brazil: Perspectives for the Medium and Long Term*. Östersund, Sweden: Swedish Agency for Growth Policy

Analysis, 2013. https://www.tillvaxtanalys.se/download/18.201965214d8715af
d113b87/1432548740127/Energisystem%2Bbortom%2B2020%2BBrasilien.pdf.

Tulloch, James. "Biofuelled Brazil Takes Its Foot Off the Gas." Allianz, May 4, 2011.
Accessed April 15, 2015. https://www.allianz.com/en/about__us/open-knowledge
/topics/mobility/articles/110504-biofuelled-brazil-takes-its-foot-off-the-gas.html/.

U.S. Department of Energy. *Brazil—Energy Brief*. Washington, D.C.: U.S. Department of Energy, Energy Information Administration, 2015.

World Nuclear Association. "Nuclear Power in Brazil." October 2015. Accessed April 15, 2016. http://www.world-nuclear.org/information-library/country-profiles/countries-a-f/brazil.aspx.

Chapter 19. Germany, Norway, Spain, and Italy

Buchan, David. *The Energiewende—Germany's Gamble*. Report no. SP 26. Oxford: Oxford Institute for Energy Studies, June 2012. https://www.oxfordenergy.org/wpcms/wp-content/uploads/2012/06/SP-261.pdf.

Federal Ministry of Economics and Technology. *Germany's New Energy Policy: Heading towards 2050 with Secure, Affordable and Environmentally Sound Energy*. Berlin: Federal Ministry of Economics and Technology, April 2012. http://www.bmwi.de/English/Redaktion/Pdf/germanys-new-energy-policy.

Fraunhofer Institute for Solar Energy Systems. "New World Record for Solar Cell Efficiency at 46%." Press release 26/14, December 1, 2014. Accessed April 15, 2015. https://www.ise.fraunhofer.de/en/press-and-media/press-releases/press-releases-2014/new-world-record-for-solar-cell-efficiency-at-46-percent.

International Energy Agency. *Energy Policies of IEA Countries: Germany 2013*. Paris: Organisation for Economic Co-operation and Development/International Energy Agency, 2013. doi:10.1787/9789264190764-en.

Koranyi, Balazs. "Insight—End of Oil Boom Threatens Norway's Welfare Model." *Reuters UK*, May 8, 2014. Accessed June 8, 2015. http://uk.reuters.com/article/uk-norway-economy-insight-idUKKBN0DO07520140508.

Kost, Christoph, Johannes N. Mayer, Jessica Thomsen, Niklas Hartmann, Charlotte Senkpiel, Simon Philipps, Sebastian Nold, Simon Lude, Noha Saad, and Thomas Schlegl. *Levelized Cost of Electricity Renewable Energy Technologies*. Freiburg: Fraunhofer Institute for Solar Energy Systems, November 2013. Accessed June 8, 2015. https://www.ise.fraunhofer.de/en/publications/veroeffentlichungen-pdf-dateien-en/studien-und-konzeptpapiere/study-levelized-cost-of-electricity-renewable-energies.pdf.

Linde Group. "Green Light for Green Hydrogen at Energiepark Mainz." July 2, 2015. Accessed November 15, 2015. http://www.the-linde-group.com/en/news__and__media/press__releases/20150701__news.html.

Marx, Eric. "Clean Energy: After Years of Theorizing, the Hydrogen Economy Is Emerging from Excess Wind Power in Germany." E&E Publishing, LLC, September 18, 2015. Accessed April 15, 2016. http://www.eenews.net/stories/1060024908.

"Public Social Spending." *Economist*, December 2, 2010. Accessed April 15, 2015. http://www.economist.com/node/17632977.

U.S. Department of Energy. *Germany—Energy Brief*. Washington, D.C.: U.S. Department of Energy, Energy Information Administration, 2015.

U.S. Department of Energy. *Italy—Energy Brief*. Washington, D.C.: U.S. Department of Energy, Energy Information Administration, 2015.

U.S. Department of Energy. *Norway—Energy Brief*. Washington, D.C.: U.S. Department of Energy, Energy Information Administration, 2015.

U.S. Department of Energy. *Spain—Energy Brief*. Washington, D.C.: U.S. Department of Energy, Energy Information Administration, 2015.

Chapter 20. Russia

Goodrich, Lauren, and Marc Lanthemann. "The Past, Present and Future of Russian Energy Strategy." Stratfor, February 12, 2013. Accessed April 15, 2015. https://www.stratfor.com/weekly/past-present-and-future-russian-energy-strategy.

Government of the Russian Federation. *Energy Strategy of Russia for the Period Up to 2030*. Moscow: Ministry of Energy of the Russian Federation, 2010. http://www.energystrategy.ru/projects/docs/ES-2030__(Eng).pdf.

International Energy Agency. *Russia 2014*. Paris: Organisation for Economic Co-operation and Development/International Energy Agency, 2014. https://www.iea.org/publications/freepublications/publication/Russia__2014.pdf.

Reed, Stanley, and Sebnem Arsu. "Russia Presses Ahead with Plan for Gas Pipeline to Turkey." New York Times, January 21, 2015. Accessed April 15, 2015. http://www.nytimes.com/2015/01/22/business/international/russia-presses-ahead-with-plan-for-gas-pipeline-to-turkey.html?__r=0.

U.S. Department of Energy. *Russia—Energy Brief*. Washington, D.C.: U.S. Department of Energy, Energy Information Administration, 2015.

Chapter 21. Saudi Arabia

Alshahrani, Saad ben Ali. "Saudi Arabia Reassesses Energy Subsidies." *Al-Monitor*, December 11, 2013. Accessed April 15, 2015. http://www.al-monitor.com/pulse/business/2013/12/saudi-arabia-energy-subsidies-assessment.html.

Clover, Ian. "Solar Power Key for Saudi Future, Says Energy Chief." *pv Magazine*, October 28, 2014. Accessed April 15, 2015. http://www.pv-magazine.com/news/details/beitrag/solar-power-key-for-saudi-future--says-energy-chief__100016969/#axzz45ud6O7Ry.

"Saudi Arabia Set to Become Electricity Exporter." *Saudi Gazette*, May 27, 2014. Accessed April 15, 2015. http://english.alarabiya.net/en/business/energy/2014/05/27/Kingdom-set-to-become-electricity-exporter.html.

"Saudi Arabia's Nuclear, Renewable Energy Plans Pushed Back." Reuters, January 19, 2015. Accessed April 15, 2015. http://www.reuters.com/article/saudi-nuclear-energy-idUSL6N0UY2LS20150119.

Saudi Government. "Saudi Arabia's Renewable Energy Strategy and Solar Energy Deployment Roadmap." Riyadh: King Abdullah City for Atomic and Renewable Energy, April 7, 2013. https://www.irena.org/DocumentDownloads/masdar/Abdulrahman%20Al%20Ghabban%20Presentation.pdf.

U.S. Department of Energy. *Saudi Arabia—Energy Brief*. Washington, D.C.: U.S. Department of Energy, Energy Information Administration, 2015.

World Nuclear Association. "Nuclear Power in Saudi Arabia." January 2016. Accessed April 15, 2016. http://www.world-nuclear.org/information-library/country -profiles/countries-o-s/saudi-arabia.aspx.

Chapter 22. Japan

Japan's Ministry of Economy, Trade and Industry. Strategic Energy Plan. April 11, 2014. http://www.enecho.meti.go.jp/en/category/others/basic__plan/pdf/4th __strategic__energy__plan.pdf.

Tokyo Gas. "Development of the New Model of a Residential Fuel Cell, 'ENE-FARM.'" Accessed April 15, 2015. http://www.tokyo-gas.co.jp/techno/english/menu3/2 __index__detail.html.

U.S. Department of Energy. *Japan—Energy Brief*. Washington, D.C.: U.S. Department of Energy, Energy Information Administration, 2015.

Chapter 23. China

"Chinese Fast Reactor Completes Full-Power Test Run." *World Nuclear News*, December 19, 2014. Accessed April 15, 2015. http://www.world-nuclear-news .org/NN-Chinese-fast-reactor-completes-full-power-test-run-1912144.html.

Chinese Government. China's Energy Policy 2012. Beijing: Information Office of the State Council, People's Republic of China, October 2012. http://www.iea.org /media/pams/china/ChinaEnergyWhitePaper2012.pdf.

KPMG China. "China's 12th Five-Year Plan: Sustainability." April 2011. https://www .kpmg.com/CN/en/IssuesAndInsights/ArticlesPublications/Documents/China -12th-Five-Year-Plan-Sustainability-201104-v2.pdf.

Lewis, Joanna. "Energy and Climate Goals of China's 12th Five-Year Plan." C2ES Center for Climate and Energy Solutions, March 2011. http://www.c2es.org/docUploads /energy-climate-goals-china-twelfth-five-year-plan.pdf.

U.S. Department of Energy. *China—Energy Brief*. Washington, D.C.: U.S. Department of Energy, Energy Information Administration, 2015.

World Bank and the Development Research Center of the State Council, People's Republic of China. *China 2030: Building a Modern, Harmonious, and Creative Society*. Washington, D.C.: International Bank for Reconstruction and Development/World Bank and the Development Research Center of the State Council, People's Republic of China, March 23, 2013. http://www-wds.worldbank.org/external /default/WDSContentServer/WDSP/IB/2013/03/27/000350881__201303271631 05/Rendered/PDF/762990PUB0china0Box374372B00PUBLIC0.pdf.

World Nuclear Association. "China's Nuclear Fuel Cycle." March 2016. Accessed April 15, 2016. http://www.world-nuclear.org/information-library/country-profiles /countries-a-f/china-nuclear-fuel-cycle.aspx.

Chapter 24. India

Ahn, Sun-Joo, and Dagmer Graczyk. *Understanding Energy Challenges in India*. Paris: Organisation for Economic Co-operation and Development/International Energy Agency, 2012. https://www.iea.org/publications/freepublications /publication/India__study__FINAL__WEB.pdf.

Bajoria, Jayshree, and Esther Pan. "The U.S.–India Nuclear Deal." Council on Foreign Relations, November 5, 2010. Accessed April 15, 2015. http://www.cfr.org /india/us-india-nuclear-deal/p9663.

Pidd, Helen. "India Blackouts Leave 700 Million without Power." *Guardian*, July 31, 2012. Accessed April 15, 2015. http://www.theguardian.com/world/2012/jul/31 /india-blackout-electricity-power-cuts.

Prime Minister's Office. "PM's Inaugural Address at the International Conference on Peaceful Uses of Nuclear Energy." Press Information Bureau, Government of India, September 29, 2009. Accessed April 15, 2015. http://pib.nic.in/newsite /erelease.aspx?relid=52858.

U.S. Department of Energy. *India—Energy Brief*. Washington, D.C.: U.S. Department of Energy, Energy Information Administration, 2014.

Chapter 25. South Korea

Bojanczyk, Kamil. "The Price Is Right: Advanced Electricity Pricing in South Korea." Greentech Media, June 20, 2012. Accessed June 7, 2015. http://www.greentechmedia .com/articles/read/the-price-is-right-advanced-electricity-pricing-in-south-korea.

Cheon, Son Jong. "Smart Grid in Korea." Korea Smart Grid Institute, November 26, 2013. http://www.iea.org/media/training/bangkoknov13/session__7c__ksgi __korea__smart__grids.pdf.

Cheong, Seung I. "South Korea: A Paradigm Shift in Energy Policy." *Living Energy* 8 (2013): 76–81. http://www.energy.siemens.com/hq/pool/hq/energy-topics /publications/living-energy/pdf/issue-08/essay-south-korea-SeungIlCheong -Living-Energy-8.pdf.

International Energy Agency. *Energy Policies of IEA Countries: The Republic of Korea*. Paris: Organisation for Economic Co-operation and Development/International Energy Agency, 2012. https://www.iea.org/publications/freepublications /publication/Korea2012__free.pdf.

Patel, Sonal. "Sihwa Lake Tidal Power Plant, Gyeonggi Province, South Korea." *POWER Magazine*, December 1, 2015. Accessed April 15, 2016. http://www .powermag.com/sihwa-lake-tidal-power-plant-gyeonggi-province-south-korea/.

U.S. Department of Energy. *South Korea—Energy Brief*. Washington, D.C.: U.S. Department of Energy, Energy Information Administration, 2014.

World Nuclear Association. "Nuclear Power in South Korea." January 2016. Accessed April 15, 2016. http://www.world-nuclear.org/information-library/country -profiles/countries-o-s/south-korea.aspx.

Part IV. History, Inertia, Interventions, and Adaptation: Energy Policy Conundrums

Chapter 27. Our Response to Limited Resources

Davies, Emma. "Endangered Elements: Critical Thinking." *Chemistry World* 8, no. 1 (January 2011): 50–54. http://www.rsc.org/images/Endangered%20 Elements%20-%20Critical%20Thinking__tcm18-196054.pdf.

European Commission. "Critical Raw Materials." Accessed April 16, 2015. http://ec .europa.eu/growth/sectors/raw-materials/specific-interest/critical/index__en.htm.

Heimlich, Russell. "Public Sees a Future Full of Promise and Peril." Pew Research Center for the People and the Press RSS, June 22, 2010. Accessed April 16, 2015. http://www.people-press.org/2010/06/22/public-sees-a-future-full-of-promise-and-peril/.

Helium Scarcity. "He2 4.0026." Accessed April 16, 2015. http://www.heliumscarcity.com/.

Howell, Elizabeth. "How Long Have Humans Been on Earth?" Universe Today, January 19, 2015. Accessed April 16, 2015. http://www.universetoday.com/38125/how-long-have-humans-been-on-earth/.

Kramer, Andrew E. "Russia Cuts Off Gas Deliveries to Ukraine." *New York Times*, January 1, 2009. Accessed February 27, 2015. http://www.nytimes.com/2009/01/02/world/europe/02gazprom.html.

MacFarquhar, Neil. "Gazprom Cuts Russia's Natural Gas Supply to Ukraine." *New York Times*, June 16, 2014. Accessed April 16, 2015. http://www.nytimes.com/2014/06/17/world/europe/russia-gazprom-increases-pressure-on-ukraine-in-gas-dispute.html.

Makhijani, Shakuntala, and Alexander Ochs. "Renewable Energy's Natural Resource Impacts." In *State of the World 2013: Is Sustainability Still Possible?* 84–98. Washington, D.C.: Worldwatch Institute/Island Press, 2013.

Ponting, Clive. "The Lessons of Easter Island." Eco-Action. Accessed April 16, 2015. http://www.eco-action.org/dt/eisland.html.

Pope Francis. *Praise Be to You = Laudato Si': On Care for Our Common Home.* Vatican City: Libreria Editrice Vaticana, 2015.

Siegler, Kirk. "Owens Valley Salty as Los Angeles Water Battle Flows into Court." NPR, March 11, 2013. Accessed April 16, 2015. http://www.npr.org/2013/03/11/173463688/owens-valley-salty-as-los-angeles-water-battle-flows-into-court.

U.S. Department of State, Office of the Historian. "Milestones: 1969–1976. Oil Embargo, 1973–1974." Accessed April 16, 2015. https://history.state.gov/milestones/1969-1976/oil-embargo.

Chapter 28. Global Energy Price and Cost Points

Auffhammer, Maximilian. Review of *Hidden Costs of Energy: Unpriced Consequences of Energy Production and Use*, by the National Research Council. *Environmental Health Perspectives* 119, no. 3 (March 2011): a138. doi:10.1289/ehp.119-a138.

ChartsBin. "Worldwide Retail Prices of Gasoline (US Cents per Litre)." Accessed April 16, 2015. http://chartsbin.com/view/1115.

GlobalPetrolPrices.com. "Gasoline Prices, US Gallon." Accessed April 16, 2015. http://www.globalpetrolprices.com/gasoline_prices/.

"Lazard's Leveled Cost of Energy Analysis—Version 8.0." Lazard, September 2014. Accessed March 9, 2015. https://www.lazard.com/media/1777/levelized_cost_of_energy_-_version_80.pdf.

Mahapatra, Lisa. "Gas Prices at the Pump: Europeans Pay Almost Twice as Much as US Residents." *International Business Times*, June 25, 2013. Accessed April 16, 2015. http://www.ibtimes.com/gas-prices-pump-europeans-pay-almost-twice-much-us-residents-1322727.

McMahon, Tim. "Inflation Adjusted Gasoline Prices." InflationData.com. Accessed April 16, 2016. http://inflationdata.com/articles/inflation-adjusted-prices/inflation-adjusted-gasoline-prices/.

U.S. Energy Information Administration. "Levelized Cost and Levelized Avoided Cost of New Generation Resources in the Annual Energy Outlook 2015." June 2015. https://www.eia.gov/forecasts/aeo/pdf/electricity__generation.pdf.

Whitley, Shelagh. *Time to Change the Game: Fossil Fuel Subsidies and Climate*. London: Overseas Development Institute, November 2013. Accessed November 1, 2014. https://www.odi.org/sites/odi.org.uk/files/odi-assets/publications-opinion-files/8668.pdf.

Chapter 29. Energy Policy Tool Kit

"Carbon Copy." *Economist*, December 14, 2013. Accessed September 22, 2015. http://www.economist.com/news/business/21591601-some-firms-are-preparing-carbon-price-would-make-big-difference-carbon-copy.

Clark, Patterson. "Unexpected Loose Gas from Fracking." *Washington Post*, April 14, 2014. Accessed April 16, 2016. https://www.washingtonpost.com/apps/g/page/national/unexpected-loose-gas-from-fracking/950/.

Clements, Benedict, David Coady, Stefania Fabrizio, Sanjeev Gupta, Trevor Alleyne, and Carlo Sdralevich. *Energy Subsidy Reform: Lessons and Implications*. Washington, D.C.: International Monetary Fund, 2013.

C2ES Center for Climate and Energy Solutions. "California Cap-and-Trade Program Summary." January 2014. http://www.c2es.org/docUploads/calif-cap-trade-01-14.pdf.

Cusick, Daniel. "Business: Ranks of Companies Setting an Internal Price on CO2 Emissions Almost Triple in a Year." E&E Publishing, LLC, September 21, 2015. Accessed April 16, 2016. http://www.eenews.net/stories/1060025003.

Ecofys. "World GHG Emission Flow Chart 2010." May 28, 2013. Accessed June 16, 2016. http://www.ecofys.com/files/files/asn-ecofys-2013-world-ghg-emissions-flow-chart-2010.pdf.

Ellerman, A. Denny, and Paul L. Joskow. *The European Union's Emissions Trading System in Perspective*. Arlington, Va.: Pew Center on Global Climate Change, May 2008. http://www.c2es.org/docUploads/EU-ETS-In-Perspective-Report.pdf.

"Energy Subsidies: Scrap Them." Economist, June 14, 2014. http://www.economist.com/news/leaders/21604170-there-are-moves-around-world-get-rid-energy-subsidies-heres-best-way-going.

Hotten, Russell. "Volkswagen: The Scandal Explained." BBC News, December 10, 2015. Accessed April 16, 2016. http://www.bbc.com/news/business-34324772.

International Energy Agency. "Energy Subsidies." January 1, 2013. Accessed April 16, 2015. http://www.worldenergyoutlook.org/resources/energysubsidies/.

International Renewable Energy Agency. "Renewable Energy Employs 7.7 Million People Worldwide, Says New IRENA Report." Press release, May 19, 2015. Accessed June 4, 2015. http://www.irena.org/News/Description.aspx?NType=A&mnu=cat&PriMenuID=16&CatID=84&News__ID=407.

Keep Tap Water Safe. "List of Bans Worldwide." June 2, 2015. Accessed August 16, 2015. https://keeptapwatersafe.org/global-bans-on-fracking/.

McCarthy, Niall. "The Renewable Energy Industry Employs Nearly 8 Million People Worldwide [Infographic]." *Forbes*, May 20, 2015. Accessed June 16, 2015. http://www.forbes.com/sites/niallmccarthy/2015/05/20/the-renewable-energy-industry-employs-nearly-8-million-people-worldwide-infographic/#278158ea40d9.

Pyper, Julia. "Could Fuel Cells Solve the Emissions Problem for Coal Plants?" Greentech Media, September 8, 2015. Accessed April 16, 2016. http://www.greentechmedia.com /articles/read/Could-Fuel-Cells-Solve-the-Emissions-Problem-for-Coal-Plants.

"Renewable Energy Tendering Schemes." Energypedia. December 9, 2014. Accessed April 16, 2015. https://energypedia.info/wiki/Renewable__Energy__Tendering__Schemes.

Romm, Joe. "It's Not Too Late to Stop Climate Change, and It'll Be Super-Cheap." ThinkProgress RSS, January 29, 2015. Accessed April 16, 2016. http://thinkprogress .org/climate/2015/01/29/3616382/solving-climate-change-cheap/.

Samaras, Constantine, Jay Apt, Inês L. Azevedo, Lester B. Lave, M. Granger Morgan, and Edward S. Rubin. "Cap and Trade Is Not Enough: Improving U.S. Climate Policy." Department of Engineering and Public Policy, Carnegie Mellon University, March 2009. http://www.cmu.edu/gdi/docs/cap-and-trade.pdf.

Sijm, Jos, Hector Pollitt, Wanter Wetzels, Unnada Chewpreecha, and P. Koutstaal. *Splitting the EU ETS: Strengthening the Scheme by Differentiating Its Sectoral Carbon Prices*. Report no. ECN-E-13-008. Petten, the Netherlands: ECN, 2013.

Taylor, Julie. "Feed-in Tariffs (FIT): Frequently Asked Questions for State Utility Commissions." National Association of Regulatory Utility Commissioners, June 2010. http://pubs.naruc.org/pub/536E2E36-2354-D714-51AC-C9C1DADE08F6.

Tweed, Katherine. "Coal Loses Nearly 50,000 Jobs, Wind and Solar Add 79,000." Greentech Media, April 3, 2015. Accessed April 16, 2015. http://www.greentechmedia.com /articles/read/coal-loses-nearly-50000-jobs-wind-and-solar-add-79000.

U.N. Framework Convention on Climate Change. "Transformative Action on Methane toward COP 21: Join the CCAC Oil and Gas Methane Partnership." August 19, 2015. Accessed April 16, 2016. http://newsroom.unfccc.int/unfccc-newsroom /transformational-initiative-ccac-oil-gas-methane-partnership/.

U.S. Energy Information Administration. "Feed-in Tariff: A Policy Tool Encouraging Deployment of Renewable Electricity Technologies." May 30, 2013. Accessed April 16, 2015. https://www.eia.gov/todayinenergy/detail.cfm?id=11471#.

U.S. Environmental Protection Agency. "Carbon Dioxide Capture and Sequestration." Accessed April 16, 2015. https://www3.epa.gov/climatechange/ccs/.

U.S. Environmental Protection Agency. "Carbon Pollution Emission Guidelines for Existing Stationary Sources: Electric Utility Generating Units." *Federal Register* 80, no. 205 (October 23, 2015): 64661–65120. https://www.gpo.gov/fdsys/pkg /FR-2015-10-23/pdf/2015-22842.pdf.

U.S. Environmental Protection Agency. "Global Greenhouse Gas Emissions Data." Accessed April 16, 2015. https://www3.epa.gov/climatechange/ghgemissions /global.html.

U.S. Environmental Protection Agency. "Overview of Greenhouse Gases: Methane Emissions." Accessed April 16, 2016. https://www3.epa.gov/climatechange /ghgemissions/gases/ch4.html.

U.S. Environmental Protection Agency Office of Transportation and Air Quality. "EPA Sets Tier 3 Motor Vehicle Emission and Fuel Standards." Report no. EPA-420-F-14-009, March 2014. https://www3.epa.gov/otaq/documents /tier3/420f14009.pdf.

World Bank. "Zero Routine Flaring by 2030." May 22, 2015. Accessed September 16, 2015. http://www.worldbank.org/en/programs/zero-routine-flaring-by-2030.

Index

Index

airplanes, 143–44

alternate energy sources, 18–19; discoveries relevant to, 162–64; recycling critical materials for, 77, 87; technology readiness for, 160–61. *See also* biofuels; geothermal power; hydrogen; hydropower; nuclear power; renewable energy; solar photovoltaic (solar PV); solar power; tidal power; wind power

ammonia, 115

Aristotle, 33

Asimov, Isaac, 60, 122

Australia, 13, 49, 85

batteries: characteristic needed for new energy market, 119; chemical, 116; flow, 118–19; lithium ion, 117–18

Becquerel, Edmond, 58, 162

Becquerel, Henri, 58

Beijing Institute of Technology, 118

Bhabha, Homi, 240

biofuels, 137–39; in Brazil, 205–6, 256; as interim solution for light-duty vehicles, 138; mandates for, 291; use on airplanes of, 143; will go the way of petroleum, 322

BP *Outlook 2030*. See *Outlook 2030* (BP)

Brazil: biofuel production in, 138, 205–6, 256; is energy-conflicted, 208; energy intensity of, 257; petroleum in, 205, 206, 207, 208; predictions about energy use of, 206–7; primary energy mix of, 178; renewable energy capacity of, 177

British Columbia, 201–2, 203, 298–300

Bush, George W., 337, 338

Bush, George H. W., 336

California: greenhouse gas cap and trade program of, 293, 294, 295; as leader in promoting use of renewables in transportation, 140–41, 142

Canada: contradictory approach to energy of, 204; electricity pricing in, 280; energy intensity of, 257; energy plan of, 201; high use of noncarbon energy sources of, 199–200; petroleum in, 199, 201; potential wind energy of, 46; predictions about energy use of, 200; primary energy mix of, 178; renewable energy capacity of, 177. *See also* British Columbia

carbon allowance systems, 292–94

carbon capture and sequestration (CCS), 179–81, 194; at Boundary Dam, Canada, 202, 318; high cost of, 181; lack of pressures to invest in, 318

carbon emissions: efforts to reduce don't address goal of transitioning away from fossil fuels, 132–33, 180; used by firms to internally value projects, 311. *See also* CO_2; greenhouse gases (GHG)

carbon tax: in British Columbia, 203, 298–300; progressive, 202–3, 320; worldwide, is ideal policy tool, 320

Carré, Ferdinand, 153

Cervantes, Miguel de, 89

Chernobyl accident, 221

China: climate change concerns subordinate to economic development of, 235–36; dependence on fossil fuels of, 179, 233–35, 236–37; electricity energy mix of, 236–37; energy intensity of, 257; energy mix determined by economics in, 235, 237, 238; as largest supplier of rare earth elements, 274; as leader in thermal solar, 152; manufacturing base at risk from rising energy costs in, 174; petroleum in, 235, 238; plans to reduce energy intensity of, 236; potential wind energy in, 46; predictions about fuel mix of, 235–36; primary energy mix of, 178; renewable capacity rankings of, 177; tensions created with neighbors from interest in fossil fuels in South China Sea, 235, 269–70; use of solar PV and thermal of, 85, 237; values nuclear power, 64, 238–39; as world's largest energy consumer, 233

chlorofluorocarbons, 306

Clean Air Act, 187, 188–89

Clean Power Plan (CPP), 187–92; assumes continued use of fossil fuels, 189–90; CO_2 emission reduction goals of, 188–89; is well-intentioned but wrong, 298

climate change: concerns about are inversely related to fossil fuel reserves, 32; disadvantages of just focusing on, 168–69; recognized as terminal sentence by Paris agreement unless collective action is taken, 303; regulation wrong tool to combat, 190–92; replacement of fossil fuels most powerful long-term solution to, 192; trumped by economics in U.S., 194

climate policy: needs to have longer planning horizon, 184; shouldn't be limited to the energy sector, 167. *See also* Clean Power Plan

Clinton, Bill, 336

Clinton, Hilary, 340

CO_2: atmospheric, 297; emission-based charges, 292; emissions by country of, 305; problem with measuring at smokestack, 191. *See also* greenhouse gases (GHG)

coal, xv; is being replaced by natural gas in power sector, 2, 10, 189, 288; CO_2 emissions of, 26–27; combustion of, 24; consumption of, 26; efficiency of energy conversion of, 24, 61, 148; global sources of, 26; heat values of, 25; jobs lost in, 315–16; likely to be first fossil fuel to see demand shrink, 15; most used to generate electricity, 26; projected growth rates of, 15. *See also* carbon capture and sequestration (CCS)

combined heat and power systems (CHP): centralized, 151; in Denmark, 124; micro-, 148–49; solar, 151–52

Curie, Marie, 58

decentralization: benefits of, 319–20; of combined heat and power systems, 148; of electrical grids, 123–24, 129–30, 255; of energy sources in Denmark, 124–25; through collection from temporals, 74, 319

Delucchi, Mark A., 58

Democratic Party, energy platforms of, 335–36, 338–39, 340

Denmark, 123–24, 280

Dickinson, Emily, 314

Dole, Robert, 337

Doyle, Arthur Conan, 308

dysprosium, 273

Easter Island, 264–65

economics: determines energy sources used, 184, 185, 238; guides behavior, 9; as major obstacle to transition from fossil fuels, 323; is most potent policy tool for change in energy sourcing, 288; placed above environment in U.S., 194, 202; plays stronger hand than security in energy decisions, 185

Edison, Thomas, 122

Einstein, Albert, 154, 163, 286

electrical grids: direct vs. alternating current in, 122; investment in not keeping up with demand for, 127–28; loss of energy in, 319; micro-, 128–29, 319; oversupply leads to curtailment of renewables, 94, 159; reliability of, 108; smart, 124–28,

245–46, 248, 255; use of renewables points to decentralization of, 123–24

electricity: expected increases in consumption of, 158; fivefold range in price of, 279–80; Franklin's experiments with, 116; levelized cost of, 280–84; as mainstay of how the world delivers energy, 253; most thermal energy wasted in producing, 147–48; net energy metering of, 290–91; one billion people lack, 3; portion generated by fossil fuels hasn't changed since 1980, 18; portion generated by nonfossil sources in study group, 258–59; supply of currently demand-driven, 74; transitory-phase architecture of, 75–77. *See also* coal; geothermal power; hydrogen; hydropower; nuclear power; solar photovoltaic (solar PV); solar power; wind power

emission trading systems, 293–94

energy: appetite for increasing faster than transition to alternates, 18; definition of, xii–xiv; efficiency, 60–62; every choice of comes with negative impacts, 63–64; global potential of, 67–68; levelized cost of (LCOE), 280–84; nearly as fundamental as food and water, 263, 316; primary vs. secondary, 60; there are no free passes in harnessing, 79. *See also* alternate energy sources; power sector; renewable energy

energy architecture: post-fossil fuels, 110; transitionary-phase of, 75–76

energy decisions: economics plays stronger hand than security in, 185; value systems and, 63–64, 157

energy employment, 315–16

Energy Information Administration (EIA), 14, 15, 27, 186, 237, 251, 257, 259, 280

energy infrastructure: is like building the Internet, 317–18; taxes to support changes in, 299, 302; will take decades to adapt to change in energy market, 2, 160; wind and solar power at disadvantage with current, 76

energy intensity, 218–19, 221, 236–38, 245, 257–58

energy literacy, 201

energy mix: not much has changed since 1980 in, 18; potential for carbon tax to change, 215, 299; single formula not practical for, 157; of study group, 178–79; of world, 251–53. *See also under* individual countries

energy planning: distorted influence of status quo incumbents on, 196; greatest challenge is overcoming shortsightedness in, 315; needs to span multiple generations, 272; need to consider local values and choice in, 157–58; time horizon often too short for, 286, 309; unintended consequences of, 293; wide variation between countries of, 308. *See also* energy policies

energy policies: biofuel mandates, 291; is closely related to whether country has fossil fuel reserves, 169–70; feed-it tariffs, 289; investment incentives, 291; major ones available to leadership, 289; net energy metering, 290–91; renewable portfolio standards, 192–93, 248, 289; shouldn't be narrowly defined by greenhouse gas emissions, 167–68; tendering, 290; two classes of, 287–89. *See also* energy planning

Energy Policy Act of 2005, 197, 300

energy pricing: cost lines of renewables and fossil fuels are beginning to cross, 281–83, 284; cost of fossil fuels is incorrectly measured, 282, 288; of electricity, residential view, 284, 285; of electricity, utility view, 279–80; of gasoline, 278

energy production: decentralization of gives energy security, 124, 319–20; externalities in, 294, 296–97

energy security: decentralization of energy production gives, 124,

319–20; fossil fuels create false sense of, 18; global security helped by domestic, 316; renewables have potential to provide most countries with, 173, 196–97, 311; in U.S. economics plays stronger hand than, 185

energy sourcing: seven signals of changes in, 4–10; world needs to agree on timing for decarbonization of, 109–10

energy storage: categories of, 113; is crucial with temporals, 49, 76, 111, 113, 317–18; failure to focus on is indication of short-term planning, 107; as key to breaking hold of fossil fuels, 112, 120; large investment in, 119–20; necessary even with changes in consumer behavior, 109; round-trip efficiencies for current, 111–12; underinvestment in blocks large-scale adoption of temporals, 317–18; will be required in post-fossil fuel period, 111

energy subsidies, 148–49, 224, 241, 300–302

energy transition: architecture of, 75; developed countries have an obligation to lead, 320; extending is expensive way, 120; fossil fuels necessary for smooth, viii, 17; in Germany, 209–14, 254, 256; increasing per capita consumption will make it difficult, 18; lack of long planning horizon for, 309; likely to be disruptive worldwide, 251–53, 263, 272, 312; major obstacle is economics not technology, 323; in Norway, 214–15; requires infrastructure and improved storage, 160, 282–83, 291–92; seven signals of, 4–11; smart grids critical to, 124, 128–29; will be hardest for heat generation and transportation, 76; will follow a ground-up rather than a top-down pattern, 253; will occur slower in fossil fuel-wealthy countries, 174;

. See also alternate energy sources; decentralization; energy policies; fossil fuels; renewable energy

enhanced oil recovery (EOR), 5, 8

environment: economics, not technology, prevent adequate protection of, 318. See also climate change; climate policy

Environmental Protection Agency (EPA). See Clean Power Plan

ethanol. See biofuels

European Technology Platform on Renewable Heating and Cooking, 155–56

European Union: critical raw materials watch list of, 94, 270–71; Emission Trading System of, 293, 295

ExxonMobil Outlook for Energy. See Outlook for Energy (ExxonMobil)

feed-in-tariff (FIT), 289, 290

fission, 34–35, 37–38, 40–41

fossil fuels: alternate energy sources will limit expansion of, 17; as center of world's energy supply because they can be transported and used on demand, 107; CO_2 emissions of, 27; are concentrated in a few countries, 32, 273; cost of using incorrectly measured, 288; create false sense of security, 18; dampened creative energy, 57; depletion of, viii, 3, 14–15, 16; depletion of places lid on the harm we can self-inflict, 297; difficulties in estimating what is economically recoverable, 3; easy extraction of undermining sense of urgency, 18; economics as major obstacle to transition from, 323; environmental regulation rather than climate change policy is best way to replace, 319; energy storage as key to breaking hold of, 112–13, 120; are far less efficient than temporals with storage, 111, 120; fast transition to stop using is most appealing, 312; four forces driving decline in,

309, 310; increasing demand for, 14, 18; are increasingly unattractive as energy sources, 1–2, 9–10, 209; increasing marginal cost of, 6–7, 310; major exporters of will need to transition their economies, 311; most powerful solution to climate change is to replace, 192; possibility of mitigating effects on climate without abandoning, 179–82; production of will get more expensive and contentious, 6–7, 310; ranking of countries by dependence on, 178; reducing carbon emissions doesn't address transitioning away from, 132–33, 180; reserves of, 12–13; reserves of are inversely related to concerns about climate change, 32; true costs of in U.S. distorted by regulatory imbalance, 282; use of compared with smoking, 260; we are lingering too long in era of, 258–60; will be necessary for smooth transition to renewables, viii, 17. *See also* coal; gasoline; natural gas; petroleum

Fossil Fuel Wealth Factor (FFWF): calculation of, 173; data for "Dually Challenged" countries, 328–331; data for "Fortunate" countries, 332; data for "Underconsumer" countries, 333; data for "Vulnerable" countries, 326–27

France, 65, 85, 101

Francis, Pope, 263

Franklin, Benjamin, 116

fuels. *See* coal; fossil fuels; hydrogen; natural gas

Fukushima disaster, 97, 227–28, 229, 230

fusion, 39–41, 104–6, 160, 161, 195

gasoline, 28, 136, 277–79, 299, 310

GE Hitachi Nuclear Energy, 100

geo-engineering, 181–82

geo-exchange, 152–53

geo-solar heat pumps, 152–53

geothermal power, 51–54, 151, 177; in Japan, 230; potential of, 66–67, 68; potential vs. actual utilization of, 69; provinces of, 53; in Spain, 216

Germany: car ownership and road density in, 133; chose to discontinue use of nuclear power, 64, 209–10, 221; electricity pricing in, 280; energy transition in, 209, 254, 256; has embarked on a path that we will all have to take, 256; energy intensity of, 257–58; goal of reducing greenhouse gas emissions of, 298; hydrogen power in, 42, 135, 212–13; predictions about fuel mix of, 210; primary energy mix of, 278; renewable energy capacity of, 177; solar power in, 49, 85, 214; use of feed-in tariff by, 289

Gore, Al, 337

Great Depression, 268

greenhouse gases (GHG): geo-engineering to reduce, 181–82; regulation doesn't really solve problem of, 132, 168, 187–88; U.S. commitment to reducing, 186–87, 298; U.S. contribution to, 131, 298. *See also* carbon capture and sequestration (CCS); CO_2; Clean Power Plan; methane

Grove, Sir William Robert, 114, 162

Hawaii, 192, 193

Hawa Mahal, India, 78–79

health and environmental policies, 292–306. *See also* carbon allowance systems; carbon tax; Clean Power Plan; energy subsidies

heating, 146–47; district, 151, 215. *See also* combined heat and power systems (CHP); hydrogen; natural gas; *under* solar power

helium-3, 40

hydrogen: can use natural gas infrastructure, 230; compared with natural gas, 149, 150–51; electrolysis of, 113–14; ENE-FARM appliance, 231, 254; fuel cell vehicles, 134–35,

136–37; as means to deliver and store energy, 253–54; place of production of, 149; to power micro-CHP systems, 148–49; to power trains, 142; safety issues of, 150; is sustainable and perfect complement to renewables, 322; transportation issues of, 114–15; will take decades to build full-cycle system using, 115–16

hydraulic fracturing (fracking), 6, 10, 13

hydropower, 65–66, 68; in Brazil, 206, 207; in Canada, 199, 200; efficiency of, 61; as first major renewable energy source, 50–51; potential vs. actual capacity of, 69;

India: disproportionately high role of coal in, 178, 241, 242, 244; is energy-challenged, 244; energy insecurity of, 240–41; energy intensity of, 257; losses in power sector in, 241; nuclear plants in, 243; petroleum imports to, 242, 244; power is largest and fastest growing sector, 242–43; predictions about fuel mix of, 242; primary energy mix of, 178; renewable energy capacity of, 177; renewable power in, 243

Indonesia, 301

insolation, 47, 48, 50

Intergovernmental Panel on Climate Change, 298, 299

International Energy Agency, 301, 302

International Energy Outlook (EIA), 251–52, 297–98, 320

International Renewable Energy Agency, 315

International Thermonuclear Experimental Reactor, 195

investment incentives, 92–93, 287–88, 291

Iroquois, 161–62

Italy, 85, 177, 178, 216–17, 257–58

ITER project, 104–5

Ivanpah, California, 80–81

Jacobson, Mark Z., 68

Japan: concerns about self-sufficiency of, 229; effect of Fukushima disaster on, 227–28, 229, 230; electricity pricing in, 280; energy intensity of, 257; feed-in tariffs in, 85; hydrogen fuel plans of, 230–31, 232; micro-CHP systems in, 148; offshore wind farm plans of, 45–46, 258; plans to build renewable capacity in, 229–32; plan to build a "Hydrogen Society" in, 254; predictions about fuel mix of, 228; primary energy mix of, 178; renewable energy capacity of, 177; solar PV prices in, 85

Kafka, Franz, 23, 157

Kerry, John, 338

Lawrence Livermore National Laboratory, 105

Lazard Ltd., 281–82

Leonardo da Vinci, 218

Lincoln, Abraham, 183

lithium, 274

lithium-bromide refrigeration cycle, 154–55

McCain, John, 339

methane, 30, 181, 191–92

Mexico, 280

natural gas, 30–31; CO_2 emissions of less than coal, 27, 189; compared with hydrogen, 24, 149, 150–51; contribution to greenhouse gases of may be underestimated, 190–92; demand for, 14–15, 31; depletion of exploitable supplies of, 14–15, 16; flaring, 191; methane in, 30, 181, 191–92; is replacing coal in power sector, 2, 10, 189, 288; reserves of, 3, 13–14, 31, 184; use by sector in U.S., 186; vehicles using, 139; will be last fossil fuel standing, 10

net energy metering, 290–91

Newton, Sir Isaac, 54

Nicholson, William, 113, 162

Norway, 311; electricity pricing in, 280; energy intensity of, 257; fossil fuels in, 214–15; gasoline prices in, 278; primary energy mix of, 178

nuclear power: can contribute to the retirement of fossil fuels, 176; China places high value on, 64, 238–39; countries using, 37, 103; doubts about, 33, 40–41; easy to dismiss when potentials for renewables are unrealistically high, 64; future of shouldn't be condemned by the mistakes of the past, 317; Germany has chosen to discontinue, 64, 209–10; hazards of, 38–39; in holding pattern in U.S., 158, 195; physics of, 33–34; projected growth of, 174–76; solving fusion will be one of humankind's most significant developments in energy, 105; status in study group countries of, 254–55; timeline of, 163–64; U.S. investment in, 195

nuclear power plants: hydrogen production from, 195, 221; issue of waste from, 35–36, 39, 96–97, 98–99, 103–4, 194; loss of thermal energy from, 148; spent fuel content of waste from, 35–36

nuclear reactors: current and planned, 103; Generation IV (Fast Reactor), 98, 99–101, 103, 195, 221, 238, 317; modular fast (nuclear battery), 102; thermal-neutron, 35, 98, 103; thorium, 102; waste from the thermal-neutron reactor is fuel for the fast reactor, 99, 101, 104, 317

Obama, Barak, 186–87, 338–39
ocean thermal energy, 56
Oil Embargo of 1973–74, 138, 267
Organization of the Petroleum Exporting Countries (OPEC), 267
Outlook for Energy: A View to 2040 (ExxonMobil), 5, 14
Outlook 2030 (BP), 3
Owens Valley, California, 265–67

Paris Climate Accord, 303–6
personal vehicles, 73; California a leader in supporting alternative, 140; contribution to greenhose gases of, 131, 132; cost per mile for three fuels of, 136; electric-powered, 128, 130, 132, 135–36; electric vs. hydrogen, 136–37; flex-fuel, 138, 139, 208; hydrogen fuel cell, 134–35, 213, 231; lack of fuel infrastructure for alternative, 135, 138–39; as most common use of petroleum, 28, 29; number of alternative by class in U.S., 138–39; ownership of and road density, 133; vehicle mass emission regulations for, 296–97

petroleum, 27–30; is already losing it's appeal, 2, 189; enhanced recovery of, 5, 8; exporters of face challenges in adapting to post-fossil fuel era, 2; marginal cost of is increasing, 6–7, 310; oversupply of, 7, 184; personal transportation fueled by, 131; politics and, 267–70; reserves of, 12–13, 29–30; U.S. strategic reserves of, 149–50. *See also* fossil fuels; gasoline; *under* individual countries

platinum, 274
plutonium, 98, 100–101, 103
power: definition of, xiv–xv; inequity in access to, 3; on-demand, 111, 112, 125; scale of, xv–xvi. *See also* biofuels; geothermal power; hydrogen; hydropower; nuclear power; solar power; tidal power; wind power

power grids. *See* electrical grids
power sector: decentralization of through collection from temporals, 74, 319; regulations are local, not global, 297; renewables will first find application in, 186; will change with integration of large portions of temporals, 108–9. *See also* electricity; electrical grids; nuclear power plants

Primary Energy Consumption Factor (PECF): calculation of, 173; data for "Dually Challenged" countries, 328–

331; data for "Fortunate" countries, 332; data for "Underconsumer" countries, 333; data for "Vulnerable" countries, 326–27

Pringle, Rev. Arthur, 130

Quebec, carbon market, 295

radioactivity, 35, 36
railways, 141–42
rare earth elements: changing prices of, 272; as essential constituents of alternate energy sources, 273–74; need for recycling of, 274
raw materials: top critical in European Union, 271; response to limited supplies of, 270, 271–72. *See also* rare earth elements
recycling: of critical materials necessary for alternate energy to be sustainable, 77, 87; of rare earth elements and other raw materials, 272, 274, 275
regulation: of greenhouse gases doesn't really solve problem, 132, 168, 187–88; not the best way to transition away from fossil fuels, 187–88, 190, 298; as preferred tool for air quality and other environmental concerns, 132, 292, 319; of vehicle mass emissions, 296–97; as wrong tool to combat climate change, 190–92

renewable energy: adoption of slowed by lack of storage and infrastructure, 139; aligns with responsibility for sustainability, 9; capability to scale, 319; capacity in 2014 of, 158–59; capacity ranking of countries for, 177; cost lines for beginning to cross fossil fuels, 281–83, 284; current vs. potential use of, 68–69; current vs. predicted portion of electricity generated with in 2050, 127; decentralization possible with most types of, 123–24, 319; definition of, 42; effect of assigning unrealistic potentials to, 63–64; extraction "costs" of, 62–63, 321; is far more equitably distributed than fossil fuels, 174; is far older than petroleum as energy source, 314–15; fluctuating vs. steady sources of, 49; has been holding ground since 1980, 250; heat generation and transportation most difficult with, 76–77; hourly variation in output of different sources of, 126; large tracts of land often required for, 321; minimal use in transportation sector of, 131; potential to provide most countries with energy security of, 173, 196–97, 311; storage key in, 69, 76, 126, 292; technology readiness for replacing fossil fuel of, 160–61; timeline of, 163; tipping point for adoption approaching, 283–85; value systems may limit use of, 64, 68; will not be available for "eternity," 275. *See also* biofuels; geothermal power; hydrogen; hydropower; solar photovoltaic (solar PV); solar power; tidal power; wind power
renewable portfolio standards (RPS), 192–93, 248, 289
Renewables Global Status Report (REN21): *2014,* 156; *2015,* 8, 66, 83, 159, 177, 260
Republican Party, energy platforms of, 335, 336–37, 338, 339–40, 341
Romney, Mitt, 339–40
Russia: effect of politics on export of natural gas by, 269; energy intensity of, 218–19, 257; nuclear power in, 101, 220–22; predictions about fuel mix of, 220; relies on natural gas for most of its primary energy, 179; petroleum in, 218, 219; potential wind energy in, 46; primary energy mix of, 178; will not lead the way in renewable deployment, 220

Saudi Arabia: is completely dependent on fossil fuels, 258; domestic subsidies in, 223–24; energy intensity of, 219, 257; petroleum in, 223–24, 225; plans to diversify energy sources in, 8, 225–26, 256–57; predictions about fuel mix of, 256–57; primary energy mix of, 178

security. *See* energy security

"Shale Boom," 184, 185

Shell Company, 17

Siemens, 210, 213

solar cooling, 156

solar heating, 151–52

solar photovoltaic (solar PV): airplane, 144; brings energy sourcing closer to consumer, 87–88; comparison of categories of, 82; capacity rankings by country of, 177; cost of, 90, 282; in developed spaces, 86–87; energy loss in, 83, 84; pricing of, 84–86; thin film, 83–84; use of raw materials in, 274–75

solar power: compared with fuels, 74; concentrated (CSP), 79–82, 177; consistency of, 47; for cooling, 153–55; early use of, 47, 78–79; efficiency of, 61; for heating, 151–52, 155–56; land area needed for, 48, 80–81; potential of, 47–48, 65, 68; potential vs. actual use of, 69; world areas best for, 48, 49

South Korea: advanced pricing mechanisms for electricity of, 279–80; electricity pricing in, 280; energy intensity of, 245, 257; export of nuclear reactors by, 246; minor use of renewable energy by, 245; nuclear power in, 247; petroleum refining in, 246; predictions about fuel mix of, 246–47; primary energy mix of, 178; renewal portfolio standards of, 248; use of smart grids planned in, 245–46, 248–49

Spain, 81, 177, 178, 179, 215–16, 257

Stirling engine, 80

storage. *See* batteries; energy storage

sulfur, 29

Szilard, Leo, 154

Taiwan, 235

technology: commercialization and change in, 158–59; economics, not technology, will determine tomorrow's energy sources, 184; often used to support bad habits, 63; readiness for alternate energy sources, 160–61; is subject to value systems, 157; will not allow us to avoid changes in our behavior, 273. *See also* batteries; carbon capture and sequestration (CCS); combined heat and power systems (CHP); electrical grids; nuclear reactors; solar photovoltaic (solar PV); storage

temporals. *See* solar power; wind power; *under* energy storage

Tesla, Nicola, 122, 163

thorium, 37, 102

tidal power, 54–56

Toffler, Alvin, 73

Tolkien, J. R. R., 260

transportation sector: airplanes, 143–44; decentralization of, 319; freight, 141; railways, 141–42; transition to renewables will be hard for, 76–77; waterways, 143. *See also* personal vehicles

Trump, Donald, 341

unintended consequences, 293

United Kingdom, 85, 278

United States: commitment to reducing greenhouse gases by, 186–87, 298; domestic fossil fuel supplies as a weight holding back change, 187; economics trumps climate change in, 194; electricity pricing in, 280; energy intensity of, 257; energy mix by source and sector in, 185–86; energy plan fossil-fuel centric, 197; national security of strengthened by becoming energy self-sufficient, 284–85; nuclear power in holding pattern in, 158, 195; petroleum in, 184; political differences regarding energy in, 183, 195–96, 335–41; potential wind energy in,

46; predictions about energy mix in, 185–86; primary energy mix of, 178; public opinion doesn't consider global warming top priority but does recognize that fossil fuels will run out, 168–69; residential vs. commercial energy usage in, 146–47; responsible for 22% of greenhouse gas emissions, 131, 298; role of energy in national security underestimated in, 196; security as top energy priority in, 183; solar PV prices in, 85; has whipsaw energy plan, 184. *See also* California; Clean Power Plan; personal vehicles
U.S. Department of Energy, 193–94
U.S. Energy Information Administration, 14, 15, 251–52
U.S. Geological Survey, 13–14, 53
U.N Framework Convention on Climate Change, 192, 303
uranium, 34: annual use of, 37; concentration in spent fuel of, 36, 99; energy release from fission of, 38; reserves of, 37

value systems: energy decisions and, 63, 129, 157; may limit use of renewables,

64, 68; technology preferences subject to, 157
vapor absorption refrigeration systems (VARS), 153–55
vehicle emission and fuel regulations, 296–97
Venezuela, 278–79, 301
Volkswagen, 131–32
Volta, Alessandro, 116, 162
Voltaire, 250

water power. *See* hydropower
Whitman, Walt, 78
wind power, 89–91; actual vs. potential of, 69; efficiency of, 61; estimated potential of, 65, 68; geographic potential of, 43–45; investment incentives for, 92–93; negative impacts of, 94; offshore, 93, 258; potential by country of, 45–46; power purchase agreements (PPA) and, 92–93. *See also* wind turbines
wind turbines, 89–92, 94
World Nuclear Association, 103

Xu Dazhe, 238